Acta Physica Austriaca
Supplementum XXV

Proceedings of the
XXII. Internationale Universitätswochen für Kernphysik 1983
der Karl-Franzens-Universität Graz
at Schladming (Steiermark, Austria)
February 23rd–March 5th, 1983

Sponsored by
Bundesministerium für Wissenschaft und Forschung
Steiermärkische Landesregierung
International Centre for Theoretical Physics, Trieste
Sektion Industrie der Kammer der
Gewerblichen Wirtschaft für Steiermark

1983

Springer-Verlag
Wien New York

Recent Developments in High-Energy Physics

Edited by
H. Mitter and C. B. Lang, Graz

With 125 Figures

1983

Springer-Verlag
Wien New York

Organizing Committee

Chairman

Prof. Dr. H. Mitter
Institut für Theoretische Physik
der Universität Graz

Committee Members

C. B. Lang
L. Mathelitsch
W. Plessas
H. Zankel

Secretary

Mrs. E. Neuhold
Miss E. Tandl

ISSN 0065-1559
ISBN-13:978-3-7091-7653-5   e-ISBN-13:978-3-7091-7651-1
DOI: 10.1007/ 978-3-7091-7651-1

CONTENTS

FOREWORD

This volume contains the written versions of the lectures
held at the "22 Internationale Universitätswochen für Kernphysik" in Schladming, Austria, in February 1983. The generous
support of the Austrian Federal Ministry of Science and
Research, the Styrian Government and other sponsors once
again made it possible for expert lecturers to be invited.
In choosing the topics, the aim was to achieve a balance
between the theoretical and phenomenological contributions;
on the theoretical side, discussions centred on the impact
of different approaches to quantum field theory on the elementary particle scenario, on the other, on the recent results in high energy physics which have provided fresh motivations for new kinds of experiments as well as having had
a profound influence on cosmology. Limited space has made it
impossible to include manuscripts of the many interesting
seminars presented. The lecture notes were reexamined by
the authors after the school and are now published in their
final form. It is a pleasure to thank all the lecturers for
their efforts, which made it possible to speed up publication.
Thanks are also due to Mrs. Neuhold for the careful typing
of the notes.

H. Mitter

C.B. Lang

Acta Physica Austriaca, Suppl. XXV, 3–70 (1983)
© by Springer-Verlag 1983

THE EARLY UNIVERSE - FACTS AND FICTION[+]

by

G. BÖRNER

Max-Planck-Institut für Physik und Astrophysik
Institut für Astrophysik
Karl-Schwarzschild-Str. 1
8046 Garching b. München, FRG

1. INTRODUCTION

Cosmology is a part of the natural sciences which has remained in the philosophical realm for a very long time. Only comparatively recently have observations taken part in shaping our picture of the universe. The modern belief of a "hot big bang" cosmology is based on a few fundamental results: i) the discovery that the nebulae are systems external to our own galaxy (Hubble 1926) together with the realization that our galaxy is a common type of spiral system in a rather homogeneous distribution of galaxies made it very unlikely that we occupy a preferred position in the universe. ii) In 1929 Hubble found evidence that the distant galaxies are receding from us with a velocity proportional to their distance. This linear relation between recession velocity and distance - appropriately named "Hubble's Law" - must be considered as one of the outstanding discoveries of modern phy-

---

[+]Lectures given at the XXII. Internationale Universitätswochen für Kernphysik,Schladming,Austria,February 23-March 5,1983.

sics. We shall discuss its present observational status in detail (sec. 3). If everything is flying apart now, it must have been closer together earlier, and therefore it seems quite natural to deduce that the universe was denser in the past.

iii) The measurement of an isotropic cosmic radiation field with a Planck spectrum at $3^{o}K$ (Penzias and Wilson 1965) lends strong support to the idea of a hot dense phase in the history of the universe. The natural interpretation of this "$3^{o}K$ black-body background" as a relict of the "hot big bang" origin of the universe has gained wide acceptance now. There are other data which support the "big bang" picture, such as the number counts of radio sources, the helium abundance, the age of old stars, and the age of the elements (see sec.3).

iv) Another gratifying aspect is the fact that the simplest cosmological models(Friedmann[+]) obtained from Einstein's General Theory of Relativity can accomodate the features of a general expansion, and a hot dense origin.

Thus a "standard cosmological model",i.e. an isotropic, homogeneous Friedmann-model expanding from a singularity is used now.

It is the aim of these lectures to present an account of this standard model, and the relevant observations. Therefor in section 2 I describe some properties of the FRW cosmological models, and in section 3 I discuss the observations and how they can be accomodated by these models. Since the aim is to separate facts and fiction, the description will be critical. The question to what extent the observations directly determine the cosmology model will be addressed in (3.5). We shall see that cosmology is still far from being an observational science.

In section 4 the physics of the early universe is considered, and in section 5 some problems or rather unknown

---

[+] Friedmann, Lemaître, Gamov

features of the standard model and possible ways to the resolution by modern particle theories are pointed out.

2. THE COSMOLOGICAL MODELS

2.1 The Friedmann-Robertson-Walker Space-Times (FRW)

The motion of all bodies in the universe is controlled
by the gravitational force, and Einstein's theory of General
Relativity (GR) is taken as the fundamental theory of gravitation throughout these lectures.[+] There is not really much
choice - so far this theory has passed all experimental tests
gloriously, while competitors, as e.g. the Jordan-Brans-Dicke
theory have been reduced to insignificance by recent observations (for an extensive review on the experimental tests of
GR see Will 1980).

A space-time is a 4-dim. differentiable topological
manifold M with a metric g_{ab}; g_{ab} has the signature (+---).
The metric tensor determines distances, time intervals,
and the causal order of events. Einstein's field equations
relate g_{ab} to the energy-momentum tensor T_{ab}:

$$R_{ab} - \frac{1}{2} R\, g_{ab} + \Lambda\, g_{ab} = 8\pi G\, T_{ab} \quad , \tag{1}$$

(c=1, unless it appears explicitly in a formula) ,

(R_{ab} is the contracted curvature tensor, $R = R_a{}^a$).
Λ is an arbitrary constant, often called the "cosmological
constant" because it was introduced by Einstein to obtain a
static cosmological model as a solution of (1).

The solution to (1) will be reasonable only if a physically sensible matter content is described by it. Therefore
T_{ab} is taken as the energy-momentum tensor of some specified

---

[+] (This is our first working hypothesis: WH1)

form of matter, and then a solution of (1) gives a corresponding space-time. In particular one may choose $T_{ab} = 0$ (empty
space),

$$T_{ab} = \frac{1}{4\pi}(F_{ac} \ F_{bd} \ g^{cd} - \frac{1}{4} \ g_{ab} \ F_{ij} \ F_{kl} \ g^{ik} \ g_{il}) \quad \text{(electromagnetic}$$
$$\text{field)},$$

$$T_{ab} = (\rho + p)V_a \ V_b - p \ g_{ab} \qquad \text{(Perfect fluid)} ,$$

or

$$T_{ab}(x) = \int p^a \ p^b \ f(x,p)d\pi,$$

where $f(x,p)$ is the distribution function of some particle
system (the kinetic theory within GR has been developed by
several authors, see e.g. Ehlers 1971).

Most known exact solutions of (1) describe spaces of
high symmetry. The FRW-spaces are such highly symmetric space-
times. They have exact spherical symmetry about every point.
This implies (Walker 1944; Hawking and Ellis 1973) that the
spacetime is spatially homogeneous and admits a six-parameter
group of isometries whose surface of transitivity are space-
like 3-surfaces of constant curvature. Minkowski space, de
Sitter space are special examples of such FRW space-times.

One can choose coordinates so that the line element
reads

$$ds^2 = dt^2 - R^2(t) \ d\sigma^2 \qquad (2)$$

where $d\sigma^2$ is the line-element of a 3-space of constant curvature, independent of time. The line-element can be written as

$$d\sigma^2 = d\chi^2 + f^2(\chi) \ (d\theta^2 + \sin^2\theta \ d\phi^2) \qquad (3)$$

where $f(\chi)$ depends on the sign of the constant curvature. By

a suitable rescaling of the factor R(t) the curvature can be
normalized to k = +1; -1, or 0.

$$f(\chi) = \begin{cases} \sin \chi & \text{for } k= +1 \\ \chi & \text{for } k= 0 \\ \sinh \chi & \text{for } k= -1 \end{cases} \qquad (4)$$

The coordinate χ runs from 0 to ∞ if k = 0 or k = -1. If
k = +1 it runs from 0 to π. When k = 0 or k = -1 the 3-spaces
are diffeomorphic to R^3 and so are infinits ("open"), but when
k = +1 they are diffeomorphic to a 3-sphere S^3 and so are
compact ("closed", "finite").

This metric requires via (1) that the energy-momentum
tensor is of the perfect-fluid type

$$T_{ab} = \rho V_a V_b - p(g_{ab} - V_a V_b) \qquad (5)$$

with a mean motion V^a. The density and pressure are functions
of t only, and the flow lines are the curves (χ,θ,ϕ) = constant.
The coordinates chosen in (3) are "comoving coordinates".

The scale factor R(t) changes like the separation of
neighbouring flow lines of the perfect fluid, i.e. it is a
measure of the distance of 2 comoving "galaxies". For two
points with comoving coordinates (χ,θ,ϕ) and (χ_0,θ,ϕ) the
coordinate distance at a time t is:

$$d = R(t) (\chi - \chi_0) .$$

The Hubble expansion requires an increase of the relative
distance at the present epoch: $\dot{R}(t_0) > 0$.

Light propagates along the null-geodesics of space-times.
It can most easily be studied by changing to a new time
coordinate :

$$dt = R(t)d\eta . \qquad (6)$$

Then

$$ds^2 = R^2(\eta)\{d\eta^2 - d\chi^2 - f^2(\chi)d\Omega^2\}$$

"Radial" propagation of signals, i.e. $\theta = \phi = $ const. gives $d\eta = \pm d\chi$ along the lightcone, or for rays towards the observer

$$\chi = \eta_o - \eta \tag{7}$$

We take $\chi_o = 0$ to be our present position, η_o the present epoch. Two signals emitted at times η_A and $\eta_A + d\eta_A$ by a source following the mean motion at χ_A will be received at times η_o and $\eta_o + d\eta_o$ at $\chi_o = 0$. Because of (7) $d\eta_A = d\eta_o$ (Fig. 1). Expressed in the proper time t this condition reads

$$\frac{dt_o}{R(t_o)} = \frac{dt_A}{R(t_A)}$$

or expressed in terms of a frequency $\omega \propto \frac{1}{t}$:

$$\omega_o R_o = \omega_A R(t_A) \quad .$$

There is a shift of the emitted frequency ω_A to the observed one ω_o:

$$\frac{\omega_A}{\omega_o} = \frac{R_o}{R(t_A)} \quad .$$

This corresponds to a shift of spectral lines

$$z = \frac{\lambda_o - \lambda_A}{\lambda_A} = \frac{\omega_A}{\omega_o} - 1 = \frac{R_o}{R(t_A)} - 1 \quad .$$

We know from Hubble's observations that the shift is a redshift:

$$1 + z = \frac{R_o}{R(t)} > 1 \quad . \tag{8}$$

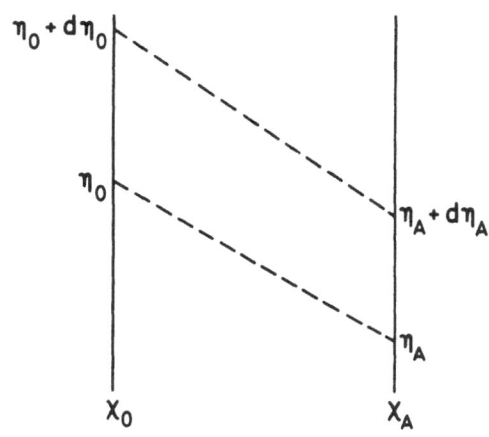

Fig.1: Signals emitted at times η_A and $\eta_A + d\eta_A$ at χ_A are received at times η_o and $\eta_o + d\eta_o$ at χ_o. $d\eta_o = d\eta_A$.

This redshift of the spectral lines of comoving objects cannot be separated into a part due to gravitational effects and a part due to motion. Within Einstein's theory these are the same thing.

For nearby galaxies we can, however, see that (8) corresponds to the Doppler formula:

Expand $R(t) = R_o + (t-t_o) \dot{R}_o$, and

$$Z = \frac{R_o}{R_o + \Delta t \dot{R}_o} - 1 \approx -\Delta t \frac{\dot{R}_o}{R_o} \quad .$$

Hence $v = Z = dH_o$ \hfill (9)

where $d = -\Delta t$ (c=1) and

$$H_o \equiv \frac{\dot{R}_o}{R_o} \quad , \text{ the "Hubble constant".}$$

Eq. (9) is "Hubble's Law", the linear relation between redshift and distance.

There is a universal time-coordinate in these cosmological models. This cosmic time t corresponds to the proper
time of observers following the mean motion V_a.

When we use these models to accomodate cosmological observations, we assume that they are in some sense good approximations to the real universe. In other words, we use as a
working hypothesis:

WH2: At any time and at any point in the universe there exists
a mean motion such that all cosmological properties are isotropic with respect to the reference frame following that
mean motion (cf. Ehlers 1976). The presently known measurements do not contradict this hypothesis, the isotropy of the
$3^{\circ}K$ background in fact gives strong support to it. Whether
or not WH2 is actually tested by the observations will be discussed in section 3.5.

Inserting (2), (3), (4) into the field equations (1)
leads to 3 equations:

$$\ddot{R} = -\frac{4\pi}{3}(\rho + 3p)GR + \frac{1}{3}\Lambda R \ , \tag{10}$$

$$\dot{R}^2 - \frac{8\pi G}{3}\rho R^2 - \frac{1}{3}\Lambda R^2 = -k \ , \tag{11}$$

$$(\rho R^3)^{\cdot} + p(R^3)^{\cdot} = 0 \ , \tag{12}$$

$$(\dot{R}: = \frac{dR}{dt}) \ .$$

Whenever $\dot{R} \neq 0$ any two of eqs. (10), (11), (12) imply the
third. These are the Friedmann equations (Friedmann 1922), and
they determine the dynamics of the cosmological model if an
equation of state, e.g. p = f(ρ) is given.

At the present epoch t = t_o the pressure is very small.
Let us take p = 0, then from (12)

$$\frac{4\pi}{3}\rho = \frac{M}{R^3} \tag{13}$$

where M is a constant. (11) then can be written as

$$\dot{R}^2 - \frac{2M}{R} - \frac{1}{3}\Lambda R^2 = -k \equiv E/M \quad . \tag{14}$$

This is an energy conservation equation for a comoving volume of matter; The constant E is the sum of kinetic and potential energy, the term ΛR^2 is a kind of oscillator energy (for $\Lambda > 0$). Equation (13) shows the conservation of mass, when $p = 0$. At the present epoch $t = t_0$ ($p = 0$) the equations (10) and (11)

$$\ddot{R}_0 = -\frac{4\pi}{3}\rho_0 G R_0 + \frac{1}{3}\Lambda R_0 \quad ,$$

$$\dot{R}_0^2 - \frac{8\pi}{3} G \rho_0 R_0^2 - \frac{1}{3}\Lambda R_0^2 = -k \quad ,$$

can be rewritten in terms of the "Hubble constant" $H_0 := \dfrac{\dot{R}_0}{R_0}$ and the "deceleration parameter" $q_0 = \left.\dfrac{\ddot{R}R}{\dot{R}^2}\right|_0$ as

$$\frac{1}{3}\Lambda = \frac{4\pi}{3}G \rho_0 - q_0 H_0^2 \quad ,$$

$$\frac{k}{R_0^2} = H_0^2 (2q_0 - 1) + \Lambda \quad . \tag{15}$$

For $\Lambda = 0$, we see that

$$2q_0 = \frac{8\pi G}{3H_0^2} \rho_0 = \frac{\rho_0}{\rho_c} \quad , \tag{16}$$

$\rho_c = \dfrac{3H_0^2}{8\pi G}$ plays the role of a critical density, and eq. (15) shows that the sign of k is determined by the ratio $\dfrac{\rho_0}{\rho_c}$.

12

$$k = +1 \qquad \text{for} \quad \rho_o > \rho_c \quad ,$$
$$k = 0 \qquad \text{for} \quad \rho_o = \rho_c \quad ,$$
$$k = -1 \qquad \text{for} \quad \rho_o < \rho_c \quad . \tag{17}$$

Since the present main density ρ_o is in principle a measurable quantity, it is possible to determine whether our universe is an open or closed FRW spacetime (see section 3). Indeed for $\Lambda = 0$, H_o and q_o determine the spacetime and the present age completely: Inserting (15) and the relation $\rho = \rho_o (R/R_o)^{-3}$ into (11) we find

$$\left(\frac{\dot{R}}{R_o}\right)^2 = H_o^2 \{ 1 - 2q_o + 2q_o \left(\frac{R_o}{R}\right) \} \quad ,$$

substituting $x = R/R_o$,

$$t - t_M = \frac{1}{H_o} \int_{R_M/R_o}^{R/R_o} [1 - 2q_o + \frac{2q_o}{x}]^{-1/2} \, dx \quad . \tag{18}$$

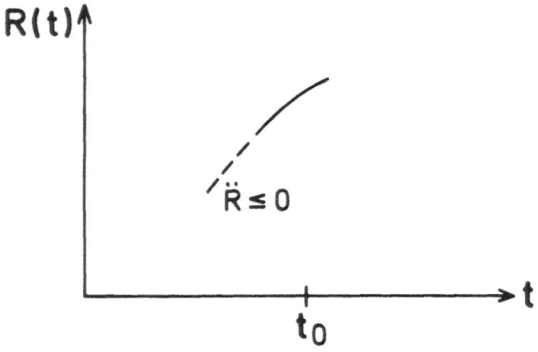

$\ddot{R} \leq 0$

Fig.2: R(t) is a concave function of t. Therefore R(t)↦0 for a finite t, if $\ddot{R} \leq 0$.

where t_M is a time in the past for which the approximation $p = 0$ still holds. $t_M \ll t_o$ and $R_M \ll R_o$ lead to the approximate relation (for times close to the present epoch t_o):

$$t = \frac{1}{H_o} \int_0^{R/R_o} [1-2q_o + \frac{2q_o}{x}]^{-1/2} \, dx \; . \tag{19}$$

The approximation $p \ll \rho$ is reasonable now and for some time into the past. But the universe contains the 3°K background radiation field, and its energy density increases as $\rho_\gamma \, R^4 =$ const. (see 2.3). Thus at a sufficiently early epoch the equation of state $p = 0$ has to be replaced by $p = 1/3 \, \rho_\gamma$.

2.2 The Initial Singularity

Without solving the Friedmann equations it can be deduced that the present expansion $\dot{R}_o > 0$, together with the requirement $\rho + 3p - \frac{\Lambda}{4\pi G} \geq 0$ lead to a concave graph of $R(t)$ vs. t with $\ddot{R} \leq 0$. If the condition $\rho + 3p - \frac{\Lambda}{4\pi G} \geq 0$ holds for all times, then $\ddot{R} \leq 0$ always, and $R(t)$ must necessarily be zero at some finite time in the past (fig. 2).

Let us call this point $t = 0$, then $R(0) = 0$, and our present epoch t_o is the cosmic time elapsed since the initial singularity $t = 0$. This point is a true singularity of the space-time: The energy density $\rho \to \infty$ for $R \to 0$, and therefore one can construct an invariant quantity from the components of the curvature tensor which will also diverge as $R \to 0$.

The point $R = 0$ has to be cut out of the space-time, and it is not clear at present how close to $R = 0$ we should accept the validity of the classical theory of GR.

This initial singularity is a remarkable feature of the FRW-cosmologies. It means that the universe we know came into being some finite time ago.

The condition $\rho + 3p - \frac{\Lambda}{4\pi G} \geq 0$ is a sufficient condition for the existence of a singularity (Hawking and Penrose 1970, Ellis and Hawking 1973). $\rho > 0$ and $p \geq 0$ is certainly satisfied

by the kinds of matter known to us. The inequality $\rho + 3p > 0$
is certaily satisfied up to nuclear matter density. We shall
see in (2.3) that the singularity can be avoided only by a
special choice of Λ or $p < 0$, and then $\ddot{R}$ is always positive.
The maximum density in these models would not have been very
much larger than ρ_0. This would make it difficult to incor-
porate the background radiation and the cosmic abundance of
helium. Observations today do not unambiguously give
$\rho + 3p - \frac{\Lambda}{4\pi G} \geq 0$; but they are sufficient to conclude that
this inequality was satisfied at some past epoch. This implies
a singularity, provided $\rho + 3p$ was never smaller before that
epoch.

Thus the description of the universe by a FRW model leads
to the conclusion that there was an early epoch of high density, and high temperature (because of a high density of
radiation), when all the structures we see now - galaxies,
stars, etc. - did not yet exist.

What was there before the initial singularity? This
question cannot be answered in this context, because time
and space came into existence with the singular event in our
past.

The structure of the space-time can conveniently be
studied by using the form of the line element

$$ds^2 = R^2(\eta)\{ d\eta^2 - d\chi^2 - f^2(\chi)d\Omega^2\}$$

for $k = +1$, $f(\chi) = \sin\chi$, and this is conformal to the Einstein static space.

This spacetime can be visualised - suppress θ and ϕ, i.e.
each point represents a 2-sphere - as the cylinder $x^2 + y^2 = 1$
in 3-dim. Mikowski-space

$$ds^2 = dt^2 - dx^2 - dz^2 .$$

The conformal structure of the FRW space-times then corresponds to a part of the Einstein static model, determined by

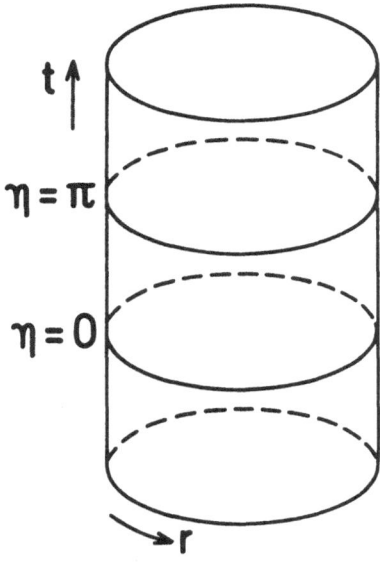

Fig.3a: The Einstein cylinder,
$\eta = 0$ and $\eta = \pi$ are the boundaries
of a k = +1 FRW spacetime.

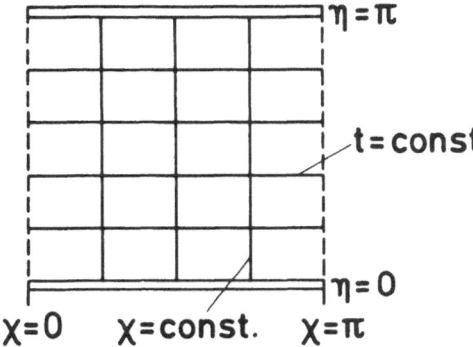

Fig.3b: Penrose diagram of a k =+1
FRW spacetime.

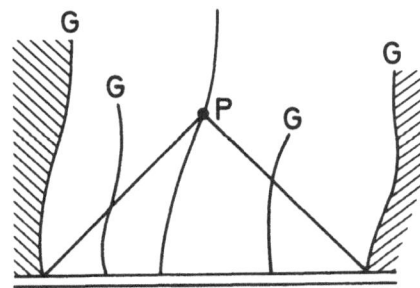

Fig.3c: The past lightcone of an
observer at P contains only a part
of the objects in the universe.

the values taken by η.

For k = +1, p = Λ = 0 one has 0 < η < π (see 2.3), i.e.
the initial singularity η = 0, is spacelike, as well as the
singularity in the future η = π. The Penrose diagram for
this spacetime is Fig. 3b. There are particle horizons in this
spacetime (Fig. 3c). An observer at P can only see objects
inside his past light cone. The shaded area is the particle
horizon - it contains all objects that are yet invisible to
the observer in P. Since the future singularity η = π is also
spacelike there are events which an observer O will never see
(future event-horizon), and events which he can never reach
by signals (past event horizon) (Fig. 3d).

For k = 0, k = -1 and p = Λ = 0 the Penrose diagram
is shown in Fig. 4 (all figures from Hawking and Ellis 1973).

Future infinity is a null surface, and there are no
future event horizons for the comoving observers in these
spaces.

Light signals emitted at $t(\eta_o - \chi)$ and received at time
$t_o(\eta_o)$ at the point $\chi = 0$ come from points located on the
sphere with area

$$O_k = 4\pi \, R^2(\eta_o - \chi) \begin{cases} \sin^2\chi \\ \chi^2 \\ \sin^2 h\chi \end{cases} \tag{20}$$

O_k increases from 0 (at $\chi = 0$), reaches a maximum, and decreases to disappear at $\chi = \eta_o$ (where $R(\eta_o - \chi) = 0$). Fig.5
shows this reconverging past lightcone.

2.3 Explicit Solutions

Let us finish the discussion of the FRW cosmological
models with a short representation of explicit solutions.

At present the universe contains matter and radiation
(3^oK background) without appreciable interaction with equations
of state p = 0 (for matter) and $p = \frac{1}{3}\rho_\gamma c^2$ (for radiation).

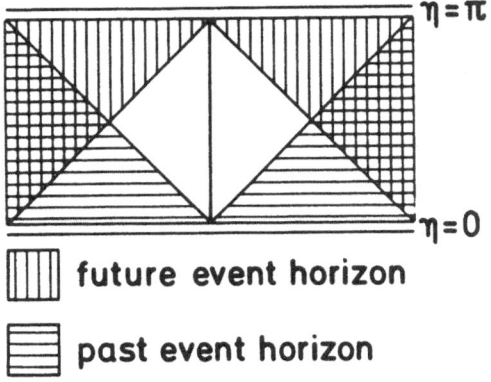

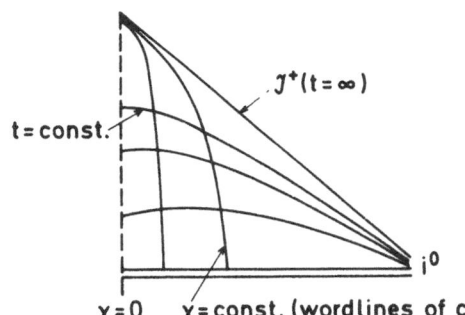

future event horizon

past event horizon

Fig.3d: ||| future event
horizon and ≡ past event
horizon in a k = +1 model

Fig.4: Penrose diagram of
k = 0,-1 models in RW coordinates. Future infinity is a
null surface.

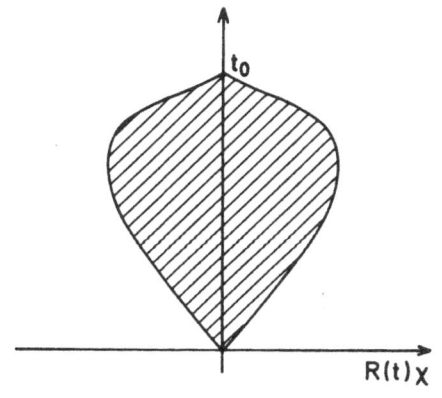

Fig.5:The past light-cone at
t_0 reconverges for $t \to 0$.

Therefore $\rho_m R^3$ = const. and $\rho_\gamma R^4$ = const., and

$$\rho = \rho_{mo} (\frac{R_o}{R})^3 + \rho_{\gamma o} (\frac{R_o}{R})^4 \quad . \tag{21}$$

Insert this into the Friedmann equation, and obtain for $\Lambda = 0$, $k = +1$:

$$(\dot{R})^2 = \frac{8\pi G}{3c^2} \rho_{mo} R_o^3 \frac{1}{R} + \frac{8\pi G}{3c^2} \rho_{\gamma o} R_o^4 \frac{1}{R^2} - 1 \quad . \tag{22}$$

The solution is given in Ref. 32, p. 741.
Set $\frac{8\pi G}{3c^2} \rho_{mo} R_o^3 = R_{max}$ ("Schwarzschild radius $\frac{2GM}{c^2}$")

$$\frac{8\pi G}{3c^2} \rho_{\gamma o} R_o^4 = R_*^2$$

$$\rightarrow \quad \dot{R}^2 - \frac{R_{max}}{R} - \frac{R_*^2}{R^2} = -1 \quad .$$

The solution is

$$R = (\frac{R_{max}}{2}) - [(\frac{R_{max}}{2})^2 + R_*^2]^{1/2} \cos(\eta + \delta) \quad ,$$

$$t = \frac{R_{max}}{2} \eta - [(\frac{R_{max}}{2})^2 + R_*^2]^{1/2} \{\sin(\eta+\delta) - \sin\delta\} \quad , \tag{23}$$

where δ= arc $\tan[2R_* / R_{max}]$, $0 < \eta < 2\pi$

For $k = -1$ let $(1-\cos \eta) \rightarrow (\cosh \eta - 1)$,

$$(\eta - \sin \eta) \rightarrow (\sinh \eta - \eta) \quad ,$$

$$0 < \eta < \infty \quad .$$

For $k = 0$ we find

$$\frac{2(R_{max}R - 2R_*^2)}{3R_{max}^2} (R_{max}R + R_*^2)^{1/2} = t - \frac{4}{3} \frac{R_*^3}{R_{max}^2} \quad .$$

This simplifies considerably for the extreme cases of no radiation (matter dominated) $\rho_\gamma = 0$:

$$\frac{2}{3} (R)^{3/2} = (R_{max})^{1/2} t \tag{24}$$

and no matter (radiation dominated) $\rho_m = 0$:

$$t = \frac{R^2}{2R_*} . \tag{25}$$

The relation (25) holds also for the $k = +1$ and $k = -1$ models in the radiation-dominated phase.

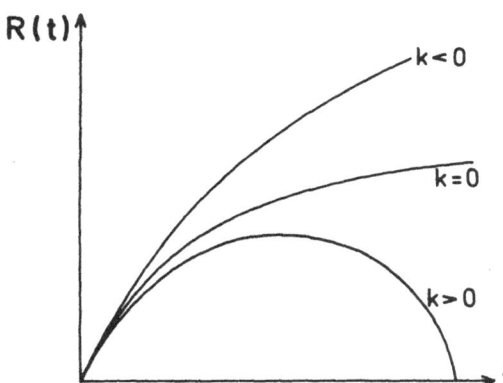

Fig.6: R vs. t for $k = +1, -1, 0$ models.

In Fig.6 we have plotted R(t) for the 3 different cases. For $k = +1$ the scale factor goes to zero again, i.e. this universe falls into a singularity in the future. $k = 1$, $k = 0$ models expand forever.

Finally the solutions with $\Lambda \neq 0$ shall be briefly discussed. The early phases will also be radiation dominated - if $R \to 0$ - , and relation (25) applies also for $\Lambda \neq 0$. Therefore only a matter-dominated phase will be considered:

$$(\dot{R})^2 = \frac{R_{max}}{R} + \frac{1}{3} \Lambda R^2 - k .$$

Introduce dimensionless quantities by setting

$$x: = \frac{R}{R_{max}} \quad , \quad \tau := \frac{t}{R_{max}} \quad , \quad \lambda := \frac{1}{3} \Lambda R_{max}^2 \quad .$$

Then

$$(\frac{dx}{d\tau})^2 = \frac{1}{x} + \lambda x^2 - k \quad . \tag{26}$$

This equation can be discussed like the one-dimensional Newtonian motion in a potential $-\frac{1}{x} -\lambda x^2$ with energy $-k$. For $\lambda < 0$ the potential is monotonic, and all possible solutions evolve from an initial singularity into a final singularity (Fig.7a). For $\lambda > 0$ the potential has a maximum at $x_m = (2\lambda)^{-1/3}$. The potential is < 0 for all x, i.e. k = 0, -1 solutions correspond to expansion from a singularity.
For k = +1 there is a critical value $\lambda_c = 4/27$ such that $\frac{dx}{d\tau}=0$ at x_m.

For $\lambda > 4/27$, and k= +1, we still find expansion forever, starting from an initial singularity (Fig.7b). When λ= 4/27, k = +1 there is the possibility of a static solution, the "Einstein Universe", with a constant radius $x_E = (\frac{8}{27})^{-1/3}= \frac{3}{2}$. From eq.(10) $\ddot{x}$ = 0 for $\Lambda = 4\pi G\rho$, i.e.

$$R_E = \frac{1}{\sqrt{\Lambda}} \quad .$$

This static universe is unstable insofar as a small change in λ leads to expansion: There is a solution starting from x = 0, and "creeping" towards x_E. A third solution starts expanding from the finite radius x_E ("Eddington-Lemaitre" (Fig.7c).

If λ is only slightly larger than 4/27, an intermediate case between Einstein and Eddington-Lemaitre arises: The cosmological model starts from x = 0, and then stays for a long time (depending on $\lambda = \lambda_c +\epsilon$) near x_E, with $\dot{x} \approx 0$, before it expands as in the E.-L. case (Fig.7d). The last possibility

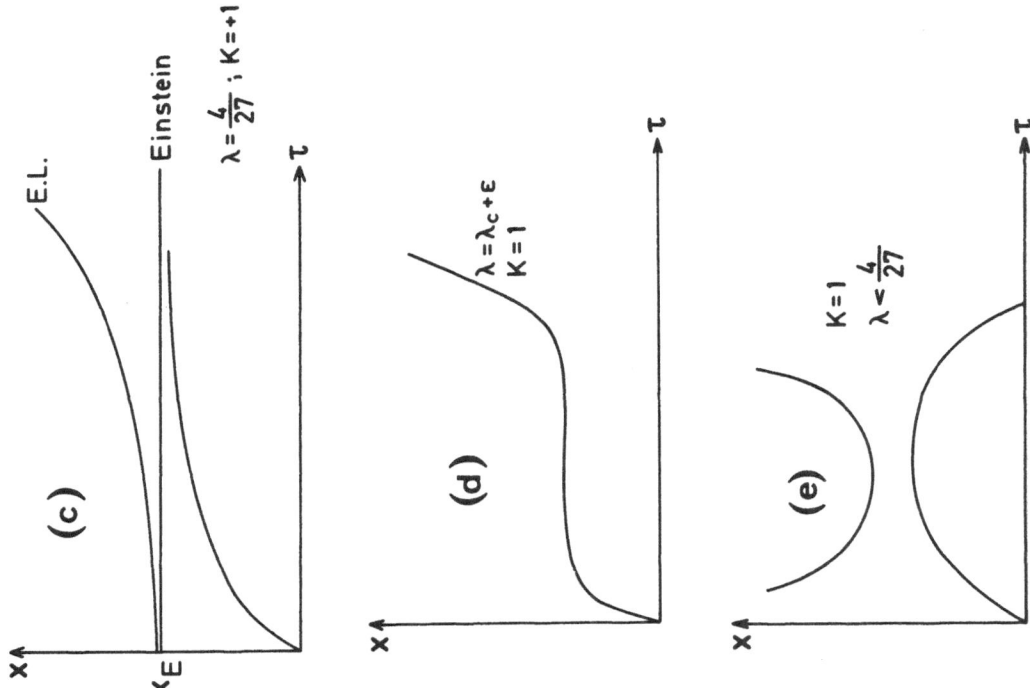

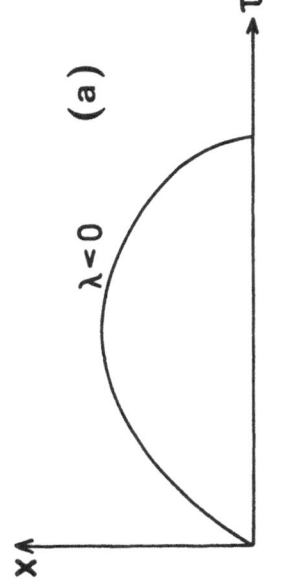

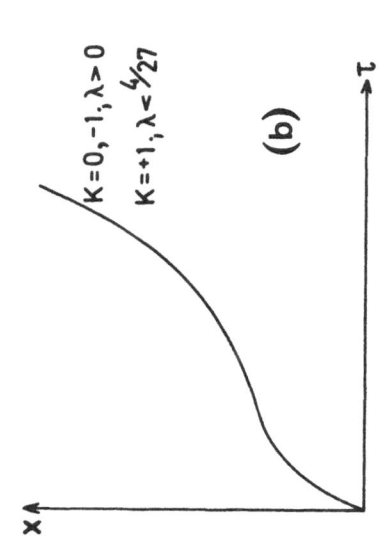

<u>Fig.7a to 7e:</u> Various cases of x vs.
τ for λ ≠ 0 FRW models.

k = +1, λ < 4/27 leads to a model expanding from and recollapsing to x = 0, and to a model contracting from infinity to a minimal x and expanding again (Fig.7e).

It seems that the models that do not pass through a phase with very small scale factor cannot easily explain the $3^{\circ}K$ background radiation, and can thus be excluded.

3. FACTS

In this section I want to introduce the relevant cosmological observations. We shall see how they fit into the "standard model", and to what extent they determine the parameters. We have seen in the preceding section that the cosmological model (if it is a FRW space-time) and the present epoch can be determined by the measurement of 2 dimensionless numbers: q_o, $\Omega \equiv \dfrac{\rho_o}{\rho_c}$, and a scale constant, H_o.

$$(\Lambda = 0 \quad \leftrightarrow \quad 2q_o = \Omega) \quad .$$

3.1 Age Determinations

Ages of meteors, lunar rocks, and minerals on earth are obtained through measurements of isotopic ratios of radioctively decaying elements. The "age" measured corresponds to dating an event of isolation of a system of parent and daughter elements which are afterwards following the exponential laws of radioactive decay.

The α decay pairs ^{238}U - ^{206}Pb (half life $\tau = 6.5 \times 10^9$y),

$$^{235}U - ^{207}Pb \quad (\tau = 10^9 y) \ ,$$

$$^{232}Th - ^{208}Pb (\tau = 2 \times 10^{10} y)$$

are often used in the measurement of rock ages. Their half-lives are comparable to the intervals of interest, so that isotopic ratios will be changed noticeably. In general the isotopic abundance by mass of a daughter element ^iD is the sum of an initially present part ^iD_o, and a part ^iD_r produced (at the point investigated) by the radioactive decay of a parent isotope ^kP:

$$^iD = {}^iD_r + {}^iD_o = {}^iD_o + {}^kP(e^{\lambda t} - 1) \ .$$

To eliminate the unknown ^iD_o, one refers all abundances to a <u>stable</u> isotope ^jD, and has

$$(\frac{^iD}{^jD}) - (\frac{^iD}{^jD})_o = \frac{^kP}{^jD} (e^{\lambda t} - 1) \ . \tag{27}$$

This is a straight line in a plot of $\frac{^iD}{^jD}$ vs. $\frac{^kP}{^jD}$ with angle $\tan \alpha = (e^{\lambda t} - 1)$ (see Fig.8).

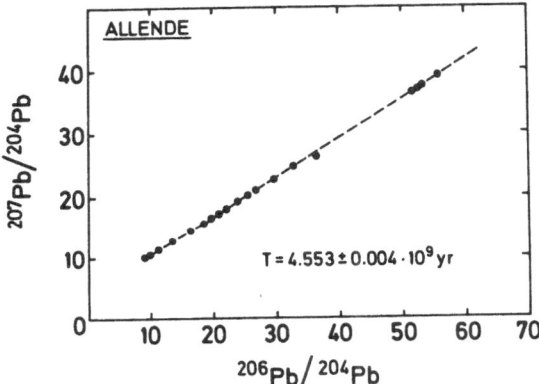

<u>Fig.8:</u> A ^{207}Pb $-$ ^{206}Pb isochron obtained for various inclusions of the Allende meteorite. All measured points lie on a straight line, speaking for a common origin about 4.553×10^4 years ago.

To determine such an isochrone from experimental data
only (with the assumptions of constant $(^iD/^jD_b)$), one needs
at least two measured points, i.e. $^iD/^jD$ for 2 different
values of $^iD/^jD$. One must be sure, that these 2 points have
the same history. Very often different minerals from the same
rock or meteorite can be used to determine isochrones. For
the U, Th decay chains the stable isotope used is ^{204}Pb.

The oldest lunar rocks were formed $(4.5 - 4.6) \times 10^9$
years ago, meteorites have ages of 4.57×10^9 years (Fig.8).
[T. Kirsten 1978]

These rock measurements also give the abundance at
the time when the solar system was formed. The comparison
of such isotopic abundances with abundances derived from models of nucleosynthesis (Fowler and Hoyle 1980) gives ages
for the elements. The basic idea is that the heavy elements
were cooked in the hot interiors of a first generation of
very massive stars or in explosive supernova events. Models
of galactic evolution are needed to describe the relation
between the mass of higher elements converted into stars,
and the elements reejected by supernovae, winds, and novae
(e.g. Tinsley 1980). Thus there is a large amount of theoretical input before the nucleosynthesis production ratios
and the abundance ratios at the time of formation of the solar system can be related.

Recently (Thielemann et al. 1983) the production ratios
of the actinide chronometer pairs ^{232}Th/^{238}U, ^{235}U/^{238}U,
^{244}Pu/^{238}U have been recalculated. These elements are produced - according to current belief - in processes of rapid
neutron capture (r-process) which occur e.g. in supernova
explosions. These calculations require a knowledge of
β-strength functions, and a consideration of β-delayed fission
and neutron emission. Thielemann et al.(1983) have - for the
first time - incorporated these effects. The ratios of their
work are compared below to the early estimates by Fowler
and Hoyle (1960):

$^{232}Th/^{238}U$	$^{235}U/^{238}U$	$^{244}Pu/^{238}U$	
1.39	1.24	0.12	Th (1983)
1.65	1.65	-	FH (1960)

The new nucleosynthesis calculations - widely publicized in newspapers - involve theoretical extrapolations into a regime away from the experimentally known. Thus these numbers are theory-dependent, and will possibly change with new experiment data. Errors are difficult of estimate.

Assuming a constant effective nucleosynthesis production rate, the duration of nucleosynthesis Δ is found to be

$$\Delta = (12 - 18) \times 10^9 \text{ y.} \tag{28}$$

The age of the galaxy follows as

$$t_G = \Delta + 4.6 \times 10^9 \text{ y} = (17-23) \times 10^9 \text{ y .} \tag{29}$$

The error estimates given (Thielemann et al.1983) are certainly not very reliable. A lower limit might perhaps be estimated - with possible reservations about nuclear physics uncertainties - by assuming that nucleosynthesis of all elements occurred in a single event. Then Δ reduces to a time $\tau_{SE} = 6 \times 10^9$ y, and

$$t_G = 10 \times 10^9 \text{ y .} \tag{30}$$

The present epoch, i.e. the age of the universe, must be larger than t_G. We do not know what to add to t_G - perhaps 10^9 y as is commonly done - because we do not know, when galaxy formation started. Be that as it may, the important result emerges that the matter around us is very old, some 10 to 20 million years, and therefore the universe must be old too.

This value is consistent with the ages of globular clusters. In these dense clusters of stars (in our galaxy and in others) all stars have been born at the same time. The low mass stars are still on the main sequence (hydrogen burning), while the very massive ones have moved away from the main sequence (started the He burning). One tries to determine the mass of the stars in the cluster which are just leaving the main sequence. Then the theory of stellar evolution can be employed to find the age of these stars - and of the cluster. The initial He abundance is a parameter to be specified. Recent work has been published by Sandage (1982) on the age of galactic globular clusters

$$t_{GC} = (17 \pm 2) \quad 10^9 \ y \tag{31}$$

and Nissen (1982) on the age of NGC 6397

$$t_N = (12 \div 20) \times 10^9 \ y. \tag{32}$$

Again error estimates have to be treated with caution, but the agreement between the nucleosynthesis age and the stellar ages is impressive.

3.2. The Hubble Constant H_o

The observations of distant galaxies have lead (Hubble, 1929) to the view of an expanding universe. How can this overall expansion be measured?

Three different pieces of knowledge help to interpret the incoming electromagnetic signals
i)   The redshift z can be measured as the redshift of spectral lines.
ii) The combination of Liouville's eq. and Einstein's GR (Ehlers 1971) allows to prove a simple relation for the specific intensity I_ω (erg/cm^{2}sec unit solid angle unit

frequency interval):

$$I_{\omega/\omega^3} = \text{const.}$$

When we compare intensities at the source $I_{\omega,e}$ and at
the observer $I_{\omega,o}$ this gives

$$I_{\omega,o} = \frac{\omega_o^3}{\omega_e^3} \, I_{\omega,e} = \frac{I_{\omega,e}}{(1+z)^3} \, . \tag{33}$$

iii) The flux from a point source is observed along a bundle
of null geodesics with a small solid angle $d\Omega_S$ at the
source and cross-sectional area dS_o at the observer.
Therefore an area distance for a point source can be defined by (Ellis 1971)

$$dS_o = r_p^2 \, d\Omega_S \, . \tag{34}$$

The solid angle $d\Omega_S$ cannot be measured, however, and
therefore r_p is not a measurable quantity either. An analogous
area distance can be defined which is in principle observable by considering an extended object of cross-sectional
area dS_S, subtending the solid angle $d\Omega_o$ at the observer.
Then by

$$dS_S = r_A^2 \, d \, \Omega_o \, . \tag{35}$$

an area distance r_A is defined. r_A can be measured, in principle, if the solid angle subtended by some object is measured,
where the cross-sectional area can be found from astrophysical
consideration.

When there are no anisotropics, $r_A \delta = D$, where D is the
linear extent, δ the angular diameter of some object. In a
FRW-cosmology

$$D = \int R(t) f(\gamma) d\theta = R(t) f(\gamma) \delta \, , \quad \text{and}$$

therefore

$$r_A = R(t)f(\gamma) \quad .\tag{36}$$

These two different distance definitions between a given galaxy and the observer r_P and r_A are essentially equivalent (Etherington 1933; Penrose 1966; Ellis 1971, p.153):

$$r_P^2 = r_A^2 (1+z)^2 \qquad \text{" Reciprocity theorem".}\tag{37}$$

This result is a consequence of the geodesic deviation equation in a Lorentz manifold. For $z = 0$, equal surface elements dS_O and dS_S subtend equal solid angles $d\Omega_S$ and $d\Omega_O$, irrespective of the curvature of space-time. The factor $(1+z)^2$ is the special relativistic correction to solid-angle measurements.

Let us use r_A to relate luminosity and flux observed: From (33):

$$\frac{F}{d\Omega_O} = \int I_{\omega,o}\,d\omega_o = \int \frac{I_{\omega,e}\,d\omega_o}{(1+z)^3}$$

$$= \frac{1}{(1+z)^4}\,\frac{L}{dS_S 4\pi}\tag{38}$$

where L is the total luminosity of the source.

$$\text{Then } F = \frac{L}{(1+z)^4 r_a^2 4\pi} \quad .\tag{39}$$

Now if $\Lambda = 0$ and $f(\chi)$ can be expressed in terms of z, q_o, H_o (e.g. Mattig 1958), and r_A then as

$$r_A = R(t)f(\chi) = \frac{1}{H_o q_o^2(1+z)^2}\{q_o z+(q_o-1)[(1+2q_o z)^{1/2}-1]\},\tag{40}$$

for $q_o \neq 0$

$$r_A = \frac{1}{2H_o} \left(1 - \frac{1}{(1+z)^2}\right) \ .$$

Expanding for small z gives the linear relation

$$(1+z)^2 r_A \equiv d = z/H_o \ . \qquad (41)$$

We see, how by measuring F, L, z we obtain r_A and then H_o.
In practice the situation is a bit more involved:
First the flux F is translated into a magnitude - a fossile
scale used by astronomers to classify the apparent brightness
of objects. A total flux $F[\text{erg cm}^{-2} \text{ sec}^{-1}]$ received corresponds to a bolometric magnitude

$$m = -2.5 \log_{10}\left(\frac{F}{2.52 \ 10^{-5}\text{erg cm}^{-2}\text{s}^{-1}}\right) = -2.5 \log_{10} F + N \ .$$
$$(42)$$

Insert (39) and (40) to obtain

$$m = -2.5 \log_{10} L + 5 \log_{10}(r_A(1+z)^2) + N =$$

$$= -2.5 \log_{10} L + 5 \log_{10} \frac{1}{H_o q_o^2} \{q_o z + (q_o - 1)[(1+2q_o z)^{1/2} - 1]\} + N =$$

$$= -2.5 \log_{10} L + 5 \log_{10} \frac{1}{H_o q_o^2} \{q_o z + (q_o - 1)[+q_o z - .. -,]\} + N =$$

$$= -2.5 \log_{10} L + 5 \log_{10}\left(\frac{z}{H_o} + \frac{1}{2}(q_o - 1)\frac{z^2}{H_o} + ..\right) + N \ . \qquad (43)$$

Incorporating the normalization N into the definition of an
"absolute magnitude" M to correspond to the intrinsic luminosity L one can rewrite (43) as

$$m - M = 25 - 5 \log H_o[\text{Km s}^{-1} \text{ Mpc}^{-1}] + 5 \log cz + 1.086(1-q_o)z \ ,$$
$$(44)$$

H_o is given in somewhat strange units $[Km\ s^{-1}Mpc^{-1}]$ (although H_o^{-1} has the dimensions of a time) 1 Mpc = 3.086 × 10^{24} cm (a typical cosmological distance unit). By looking at some objects with the same M but different m, we can determine the straight line given by (44) of m vs. log z, and so find H_o. Usually the galaxies will be distributed over some range of luminosities L (or magnitudes M) with a distribution function $\phi(M)$ (the "luminosity function"), and with a distribution function f(u) with respect to the local irregular radial velocity u. Then look at an average redshift for a fixed m (this is the usual procedure, e.g. Tammann and Sandage 1974)

$$<cz(m)> = \frac{\int_0^\infty r^2 dr \int_{-\infty}^\infty \phi(M)dM \int_{-\infty}^\infty f(u)du [H_o r_A + u]\delta(M+5\log r/r_o-m)\Theta(m_c-m)}{\int_0^\infty r^2 dr \int_{-\infty}^\infty \phi(M)dM \int f(u)du \delta(M+5\log(r/r_o)-m)\Theta(m_c-m)}$$

where we have incorporated the relation

m = M + 5 log (r/r_o), $(r_o$ = 10 pc)

and the linear "Hubble Law"

cz = H_o r.

m < m_c, because usually the galaxy catalogues contain objects only down to a limiting magnitude m_c. We want to look at a range of redshifts where the peculiar velocities can be neglected. Set x = $(r/r_o)10^{-m/5}$, and obtain

$$\log(<cz(m)>) = 0.2m + \log H_o + \log r_o \frac{\int_0^\infty x^3 dx \phi(-5\log x)}{\int_0^\infty x^2 dx \phi(-5\log x)} \tag{45}$$

To find H_o, one has to obtain the value of the integral over the luminosity function $\phi(M)$. Knowledge of this luminosity distribution is derived from close-by objects, where distances can be determined or estimated (about the cosmic distance scale see Tammann and Sandage 1974, de Vancouleurs 1977,

Weinberg 1972). Another possibility is to single out objects with a well-defined intrinsic luminosity (dangerous: see Sandage and Tammann 1982).

An important observational difficulty is the finite bandwidth of the receivers. In strongly redshifted sources one is looking at a completely different part of the spectrum than in sources with a small redshift. The correction of this effect (the so-called "K" correction) can be done, if the intrinsic spectral shape at the shorter wavelengths is known, either through theoretical guesses or through UV observations.

Faint extended objects (large m) almost merge into the background, and it is difficult to estimate where their edge is ("Aperture" effect).

Finally the local value of the Hubble constant, as e.g. measured by Hubble himself (he had obtained $H_O = 500$ km/sec Mpc), is expected to show systematic perturbations due to the large density contrast of the Virgo cluster. Therefore, precise distance measurements - or equivalently luminosity estimates - have to be found for galaxies beyond the Virgo cluster ($v \gtrsim 3000$ km sec^{-1}, i.e. $Z > 0.01$). Various methods have been used to push a distance ladder into that region. Sandage and Tammann (1976) found a value of

$$H_O = (55 \pm 5) \text{ Km s}^{-1}\text{Mpc}^{-1} , \tag{46}$$

de Vaucouleurs (1977) gave

$$H_O = (100 \pm 10) \text{ Km s}^{-1}\text{Mpc}^{-1} . \tag{47}$$

The discrepancy of a factor 2 can be traced back to a distance estimate of the Virgo cluster differing by a factor 2 (i.e. $\Delta M \simeq 1.5$). Fig. 9 shows a typical Hubble diagram (Sandage and Tammann 1976). Recently (Sandage and Tammann 1982) the light-curves of type I Supernovae have been used to

32

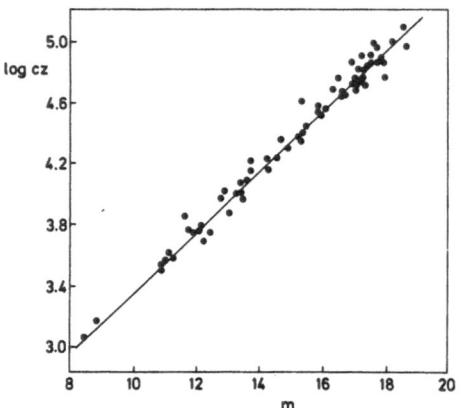

Fig.9: The Hubble diagram
for 66 first ranked cluster
galaxies, after Sandage &
Tammann (1976).

determine H_O (here the distance step to the Virgo cluster is avoided), and

$$H_O = (50 \pm 7) \, \text{Km s}^{-1}\text{Mpc}^{-1} \, , \tag{48}$$

H_O^{-1} is the present expansion time scale. For the $k = 0$, $\Lambda = 0$ FRW cosmologies, the age of the universe $t_O = \frac{2}{3}H_O^{-1}$, and taking $H_O = h \times 100$,

$$t_O = \frac{2}{3h} \times 10^{10} \, \text{years, i.e.}$$

$$7 \times 10^9 y < t_O < 1.3 \times 10^{10} y \, . \tag{49}$$

One could discover already a slight discrepancy between this range of t_O and the values given in (3.1). I would like to stress, however, how much more important is the remarkable agreement between these different ages. (An universe with $\Lambda \neq 0$ could give $H_O^{-1} < t_O$)

Distances have often been mentioned above. Just to impress on you that distance is a derived concept, without invariant meaning, and also to confuse you a bit more, let me show you several possible ways to define distance in relativistic cosmologies. (The following presentation is due to J. Ehlers (personal communication)).

An Einstein-de Sitter model in the matter-dominated era has

$$(k = 0,\ \Lambda = 0,\ p = 0):\ \frac{R(t)}{R_o} = (\frac{t}{t_o})^{2/3};\ H_o = \frac{2}{3t_o};$$

$$\rho_o = \rho_c = \frac{3}{8\pi G}\ H_o^2\ .$$

If t_e is the time of emission, z the redshift of the emitted photon, then the metric distances of the source from "us" at

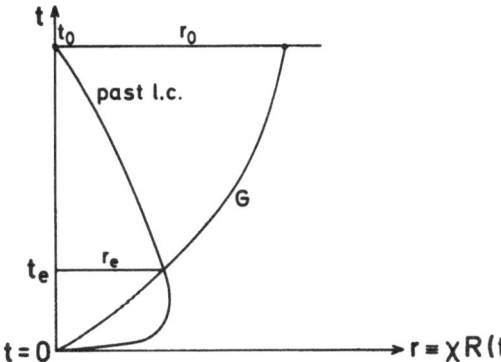

Fig.10: Schematic diagram of our past lightcone and the world line of a quasar G.

t_o and t_e are:

$$t_e = \frac{t_o}{(1+z)^{3/2}} \; ; \qquad r_o = (1+z)r_e \; , \quad \text{and}$$

$$r_o = r_H(1 - \frac{1}{\sqrt{1+z}}) \; ; \qquad r_e = \frac{r_H}{1+z}(1 - \frac{1}{\sqrt{1+z}}) \; ;$$

where $r_H = 3ct_o = \frac{2c}{H_o}$ is the radius of our particle horizon at t_o ("now"). The past null cone of $(t_o, r = 0)$ is given by $r_e = r_H((\frac{t_e}{t_o})^{2/3} - \frac{t_e}{t_o})$. The light travel time of the signal is $\Delta t = t_o - t_e = \frac{2}{3H_o}(1 - \frac{1}{(1+z)^{3/2}})$ (see Fig. 10). In addition we can define a Doppler velocity (special relativity)

$$\frac{Vr}{c} = z \frac{1+z/2}{1+z(1+z/2)}$$

and a conventional distance

$$d = \frac{cz}{H_o} = \frac{1}{2}r_H z \; .$$

With $H_o = (50 \div 100)$ the horizon radius $r_H = (6000 \div 12\,000 \text{ mpc})$. Take for example the quasar OQ 172 with a redshift $Z = 3.5$:

$$Vr = 0.9c; \qquad \bar{V}_{Kin} := \frac{r_o - r_e}{\Delta t} = 1.4 \; c; \quad d = 1.7 \; r_H(!)$$

$$r_o = 0.53 \; r_H; \qquad r_e = 0.12 \; r_H; \qquad d = 15 \; r_e = 3.3 \; r_o(!)$$

$$t_e = 0.10 \; t_o \; ; \qquad \Delta t = 0.9 \; t_o \; .$$

3.3 The Mean Density ρ_o

The distribution of matter in our neighbourhood is very inhomogeneous: Galaxies are made up of stars, they are arranged in groups, in clusters, and in superclusters. The large-scale structures (cf. Peebles 1980, Zeldovich et al.

1982, Einasto et al.1980) have dimensions of 50 up to 100 Mpc, a flattened shape (e.g. a flat disk shape of 20 Mpc × 12 Mpc and 5 Mpc thickness was derived by Ford et al. (1982) for 1451 + 22). The spatial structure is obtained by assuming an undisturbed Hubble expansion for the members of the supercluster. Typical masses of the luminous matter in superclusters are 10^{15} - 10^{16} M_O. In some cases string-like structures are observed (Perseus supercluster; the local supercluster seems to contain a chain of galaxies) (Zel'dovich et al.1982). The neighbouring superclusters to our local supercluster are at a distance of $\sim$100 Mpc. In between the clusters and chains of clusters there are giant volumes almost empty of visible objects (Zel'dovich et al.1982, Kirschner et al. 1981 - "A one million cubic megaparsec hole in the constellation Bootes?" see N.Y. Times and other newspapers). The picture of the large-scale structure of the universe that emerges from these observations is a distribution of the luminous matter in adjoining cells. The interior of the cells (of 100 Mpc typical dimension) does not contain luminous matter, the walls consist of a thin layer of galaxies, sometimes arranged in strings of clusters. The strings come together in knots which contain many galaxies and usually a prominent cluster of galaxies.

There are many questions concerning the evolution of such superstructures. A basic problem is to find out, whether the superstructures appeared first, followed by galaxy formation, or whether galaxies did form and then by their interaction the superclusters were built. Numerical simulations of the second alternative show good agreement with the observations on scales of 50 Mpc.

All these complex structures do not really look like a uniform density distribution. Can these observations be fitted into a homogeneous model of the universe? In my opinion there is only evidence for structure from the observations of luminous matter, not for homogeneity. The data do not contradict, however, the assumption that the mean density, found by averaging over scales of $\sim$500 Mpc, can be used

as the homogeneous density ρ_o of a FRW universe. This belief
is somewhat supported by the apparently uniform distribution
of bright radio sources. The distribution of the 5000 radio
sources (flux $S \geq 2Jy$) of the 4C catalogue appears to be
random (Peebles 1980; Peebles and Seldner 1978). The mean
number of 4C sources in a $3^o \times 3^o$ cell is $<N> \approx 2$, and the rms
fluctuation $\frac{\delta N}{<N>} \approx 0.7$ (compared to 0.045 for random distribution). Since many of the sources are at distances of the
horizon dimension ($c/H_o \sim 3000$ to 6000 Mpc) there is a good
test of isotropy on large scales: If the northern hemisphere
is divided into two equal parts the number of 4C sources
(N_1, N_2) in each satisfies $\frac{|N_1 - N_2|}{(N_1 + N_2)} \lesssim 0.015$. In contrast, the
bright galaxies give ≈ 2 for this ratio.

The diffuse X-Ray background is to a large part the flux
from active galaxies and clusters. It measures directly the
column density of matter out to c/H_o. The isotropy of the
background sets a limit on fluctuations of $\frac{\delta\rho}{\rho} \sim 0.01$ to 0.1
on scales of 100 to 10^3 Mpc (a few ^{o} to 20°)(Fabian 1982).
Finally the $3^o K$ background (see 3.4) is isotropic to $\frac{\delta T}{T} \leq 10^{-3}$
on angular scales from minutes to 180°. The matter distri-
bution is not measured directly by this value, since matter
and radiation are only weakly coupled at present. But the
large scale motions of the matter distribution must have been
isotropic to an accuracy better than 1 in 10^3 (since $T \propto \frac{1}{(1+Z)}$)!
Within the context of a FRW cosmological model this esta-
blishes a mean density ρ_o as a reasonable concept.

These recent observations show that matter distribution
and motion are very accurately isotropic around us on scales
$\approx c/H_o$. This leaves, of course, open the possibility that the
universe is inhomogeneous and isotropic only around our
wordline. It seems a useful working hypothesis that our position is typical, and not unique (see, however, 3.5).

The observations involved in determining a mean mass-
density ρ_o, are discussed in Peebles (1971), and Peebles
(1982). The galaxies in a certain angular cross-section of
the sky are measured down to a limiting magnitude, and a spa-

tial distribution is constructed by using the redshifts as distance indicators via the Hubble law. The luminosity density L of the sample is then derived from a well-tested luminosity function $\phi(M)$ (compare 3.3). In addition a mass-to-luminosity ratio M/L for galaxies is used to obtain $\rho_0(G) = L$ M/L g/cm^3, the mean density corresponding to luminous matter. A typical M/L value for spiral galaxies is $\simeq 4$, and a typical value obtained from this analysis is (Peebles 1971)

$$\Omega \equiv \frac{\rho_0}{\rho_c} = 0.01, \text{ corresp. to } \rho_0(G) = 5 \times 10^{-32} (\frac{H_0}{50})^2 [gcm^{-3}] \quad (50)$$

Now in rich clusters of galaxies the application of the virial theorem - i.e. viewing the cluster as a gravitationally bound system - shows that the usual M/L values do not give enough mass (Rubin 1979). This "missing mass" (i.e. non-luminous matter) changes the M/L value by a factor of 10 (this is the value for the Coma cluster).

Furthermore measurements of the rotational velocities in several galaxies give an $\Omega(r)$ which does not decrease as $r^{-1/2}$ (Kepler) at the edge. This indicates that the luminous matter does not give a true picture of the gravitationally active mass distribution (Ostriker and Peebles 1973). Theoretical considerations of the stability of rotating flat disks (unstable if $E_{rot}/E_{Grev} \geq 0.27$) suggest a much larger M/α, i.e. a large fraction of unseen mass. Finally in rich clusters a higher background density is obtained, if the mass in galaxies is extrapolated according to $M(r) \alpha r$ - as suggested by the rotation curves - to the edge of the neighbouring galaxies. If this is done for all galaxies then

$$\Omega = \frac{\rho_0}{\rho_c} = 0.6 \pm 0.25 , \quad (51)$$

(Peebles 1982).

This is rather close to $\Omega = 1$, the critical value for a ($\Lambda = 0$) closed K = +1, FRW model. We must conclude that the observa-

tions point to $\Omega < 1$, but $\Omega = 1$ or even $\Omega \geq 1$ cannot be excluded, since any homogenous background density is not accoun-
ted for in these measurements. Therefore a decision whether
we live in a closed or open universe cannot be made yet.

There may be an enormous amount of non-luminous matter
in galaxies. What kind of matter? Black holes, neutrinos of
nonzero rest-mass, dark stars of small mass, rocks have all
been mentioned. We see that the value of the mean density is
extremely uncertain, and that part of this uncertainty is
due to a lack of knowledge of local physics.

3.4 The $3^{O}K$ Cosmic Black-Body Radiation

In 1964 Penzias and Wilson discovered an excess radio
background at $\lambda = 7.35$ cm corresponding to a black-body temperature of 2.5 to $4.5^{O}K$. Since its publication in 1965 this
discovery has been interpreted as a cosmological background
signal indicating the existence of a hot, dense early stage
of the universe, when radiation and ionized matter had been
in equilibrium. The history of this discovery is told in
Weinberg's book "The first 3 minutes".

In the last few years accurate measurements have verified
the high frequency tail of the Planck distribution of the
radiation (Woody and Richards 1979) (Fig. 11). The nobel prize
in physics has been awarded to Penzias and Wilson in 1978 -
a further acknowledgement of the importance of their measurement. The recent observations have established the high isotropy, the black-body spectrum, and the absence of polarization - a correspondence as close as one could hope to the
simple hypothesis that the radiation is the remnant of a
homogeneous and structureless primeval explosion.

There are two excellent recent reviews on the experimental and theoretical situation (Weiss 1980; Sunyaev and Zel'dovich 1980).

One important fact about a Planck radiation field in

a homogeneous FRW universe is that its spectrum keeps the
Planck shape during the expansion: the adiabatic expansion
law $T_\gamma (R^3)^{\kappa-1}$ = const. gives for $\kappa = 4/3$ (photons) $T_\gamma R$ = const.
Thus the Planck spectrum

$$\rho_\gamma d\nu = 8\pi h (\nu/c)^3 d\nu [\exp(\tfrac{h\nu}{kT}) - 1]^{-1} \tag{52}$$

stays a Planck spectrum, since $T_\gamma = T_{\gamma 0}(1+z)$, and $h\nu = h\nu_0(1+z)$.

The number density of photons is

$$n_\gamma = \int \tfrac{\rho\gamma}{h\nu} \, d\nu = \tfrac{3.7a}{k} T_\gamma^3 \ . \tag{53}$$

Therefore the ratio of photon to baryon number density is a
constant in FRW cosmologies:

$$\frac{n_\gamma}{n_B} = \frac{3.7aT_\gamma^3}{kn_R} = \frac{3.7aT_{\gamma 0}(1+z)^3}{kn_{Bo}(1+z)^3} = \frac{n_{\gamma 0}}{n_{Bo}} \ . \tag{54}$$

The entropy density of the radiation is

$$S = \frac{4aT_\gamma^3}{3k} \ , \tag{55}$$

and the entropy per baryon S/n_B is a constant too,

$$S/n_B = 0.36 \, n_{\gamma 0}/n_{Bo} \ .$$

In Fig. 11 the spectrum of the radiation is shown (Woody
and Richards 1979). The maximum intensity is at $\lambda = 2$mm, and
the temperature of the Planck spectrum is $T_{\gamma 0} = 2.8^\circ$K. Therefore the energy density is 0.25 eV/cm^3 i.e. $n_{\gamma 0} = 400/\text{cm}^3$.
It is dominating radiation energy density $\rho\gamma/c^2 = 3 \times 10^{-34}\text{g/cm}^3$.
Thus the number

$$\frac{S}{n_B} = 10^{9\pm1}$$

(the large uncertainty in n_{Bo} (see 3.3) is reflected here) is a constant, characteristic for our universe, if it is a FRW space-time. S/n_B has to be set down as an initial condition in any classical FRW cosmology. Modern particle theories applied to the early universe may find a way to remedy this undesirable situation of having to introduce a net baryon asymmetry at t = 0.

The measurements show deviations from a black-body spectrum - between $3\,\mathrm{cm}^{-1}$ to $7\,\mathrm{cm}^{-1}$ the intensity is above, between $9\,\mathrm{cm}^{-1}$ and $12\,\mathrm{cm}^{-1}$ below a blackbody curve of $2.8^{\circ}\mathrm{K}$. The reasons for these deviations are unknown at present.

Such a background radiation field has important consequences: Since the energy density of matter $\rho_M = \rho_{Mo}(1+z)^3$ and of radiation $\rho_\gamma = \rho_{\gamma o}(1+z)^4$, we have $\rho_\gamma/\rho_M = (1+z)\rho_{\gamma o}/\rho_{Mo}$. Matter is dominating now, but at some time in our past, radiation was dominating energy density. Before a redshift z_R between $z_R = 10^3$ and $z_R = 10^4$ (uncertainty of ρ_{Mo}!) the universe was dominated by hot radiation.

Before the time t_{dec} when the recombination of hydrogen was finished ($T \approx 4500^{\circ}\mathrm{K}$), i.e. $z_{dec} \approx 1500$, matter and radiation have been in thermodynamic equilibrium. It is a remarkable coincidence that these two times may coincide $z_R \approx z_{dec}$. The universe must have been completely different for $t < t_R$ from its present appearance: This radiation dominated phase of the early universe was a hot and dense gas of radiation and matter. Structures, such as galaxies and stars did not yet exist.

Although at z_{dec} the mean free path of photons is drastically increased, there is still a small interaction with matter via Thomson scattering. Therefore the photons that we observe come from a surface of their last scattering, which has a smaller redshift than $z_R \approx 1500$: The optical depth for Thomson scattering

$$\tau = \int_t^{t_o} \sigma_T n_e(t)\,dt, \quad \text{where } n_e(t) = n_e(t_o)\left(\frac{R_o}{R}\right)^3 ,$$

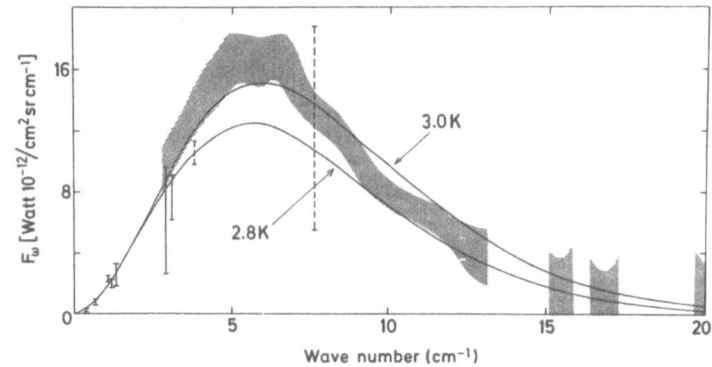

Fig.11: The spectrum of the 3° K background (after Woody & Richards 1979).

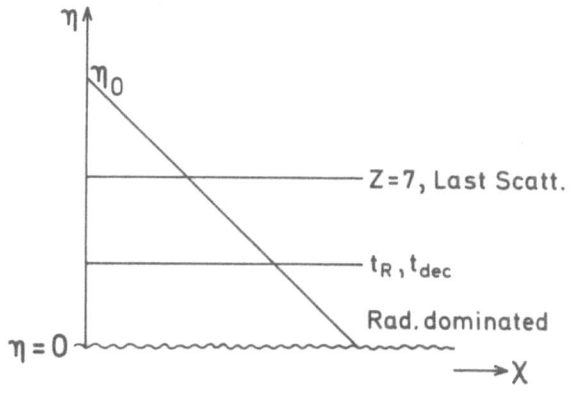

Fig.12: Schematic plot of surface of last scattering t=t$_S$, surface of decoupling t=t$_R$.

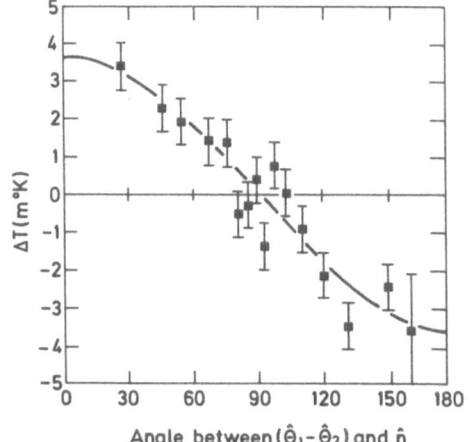

Fig.13: The dipole anisotropy $\alpha\frac{v}{c}\cos\theta$ due to the motion of the earth in the 3°K background restframe.

$$\tau = n_o R_o^3 \sigma_T \int\limits_t^{t_o} \frac{dt}{R^3(t)}$$

evaluation for $\Lambda = 0$, and a present density $\rho_o = \rho_c$ gives the result (Weinberg 1972) that $\tau = 1$ for $z = 7$, i.e. the last scattering surface was at $z = 7$ (see Fig.12).

This surface of last scattering is not necessarily sharply defined. Photons seen along one line of sight may have been last scattered in various uncorrelated regions. One expects nevertheless to see inhomogeneities in the matter distribution reflected in a variation of the temperature. (cf. Sunyaev and Zel'dovich 1980).

For adiabatic fluctuations, $S/n_B = $ const., i.e.

$$\frac{\delta n}{n} = 3 \frac{\delta T_\gamma}{T_\gamma} \ . \tag{56}$$

The angular scale associated with a typical galactic mass of $M = 10^{11} M_o$ at z_R is $\Theta = 30"$, and $\Theta \propto (M)^{1/2}$. The high isotropy of the 3^oK background proves that fluctuations in the homogeneous background around t_R must have been quite small: On large angular scales ($\geq 10^o$) there is an upper limit of $\frac{\delta T}{T} \leq 3 \times 10^{-4}$ (Weiss 1980), and a very recent measurement (Uson and Wilkinson 1982) puts a limit of $\frac{\delta T}{T} < 1.1 \times 10^{-4}$ on angular scales of 1.5'! These are unexpectedly strong limits.

In fact, a density contrast at t_R of $\frac{\delta \rho}{\rho} < 3 \times 10^{-4}$ can at most ($\rho_o = \rho_c$ universe) grow by a factor $(1+z_R)$ (Weinberg 1972), i.e. to $\frac{\delta \rho}{\rho} \leq 0.45$. But for our galaxy $\frac{\delta \rho}{\rho} >> 1$, and therefore these limits seem to destroy the possibility to let galaxies grow from small fluctuations.

Here is one reason why massive neutrinos have become so popular in cosmology (Schramm 1982): Fluctuations in the neutrino background can have larger amplitudes (not limited by $\frac{\delta T}{T}$ measurements) and may provide the gravitational potential wells for the formation of clusters of galaxies.

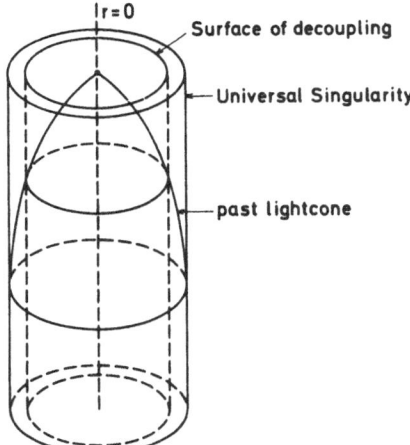

Fig.14: The static spherically symmetric cosmological model surrounded by a
time-like singularity.

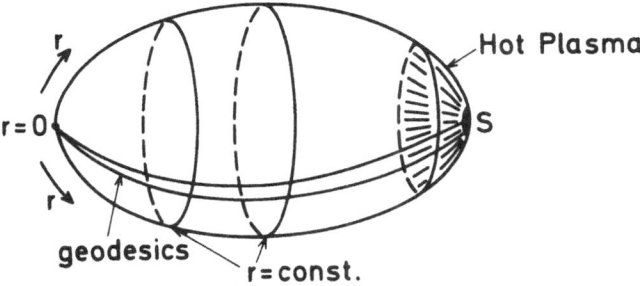

Fig.15: t = const. surface of the SSB. The time-like singularity S is surrounded by a hot plasma.

The 3°K background can be viewed as a perfectly isotropic medium, defining a perferred frame of reference. Any peculiar motion with respect to this frame can be measured. The expected anisotropy has a $\cos\theta$ behaviour:

$$T(\theta) = T_o(1-v^2/c^2)^{1/2}(1-v/c \cos\theta) = T_o(1+ v/c \cos\theta) . \qquad (57)$$

Observations at 33 MHz with a U-2 plane (Smoot et al. 1977; Muller 1978, 1980) showed indeed such a cosine behaviour of δT, as displayed in Fig. 13.

This is a measurement of the peculiar motion of the earth in the cosmic ether of the microwave background. The amplitude $\frac{\delta T}{T} \approx 4 \times 10^{-4}$ corresponds to

$$v_{\oplus} = 361 \times (\frac{3^{\circ}K}{T}) \text{ km/sec}$$

and adding the galactic rotation we find the proper velocity of our galaxy

$$v = (500 \pm 75) \text{ km/sec}$$

in the direction r.a. = 10.5 br., $\delta = -11$ deg.

The large magnitude of this velocity is disturbing, since the peculiar velocities of nearby galaxies - from optical observations - are small ~ 80 km/sec to 200 km/sec. Is the system of nearby galaxies not comoving as a whole with the expanding universe?

3.5 Can the Standard Model Be Verified Experimentally?

So far, we have considered a standard cosmological model, and have shown how the data can be accomodated within it. Now we want to ask to what extent the data actually determine the features of our universe (cf. Kristian & Sachs 1966). We still take GR as the correct theory of gravitation. It seems that the isotropy of the matter and radiation around us is fairly well established, while the homogeneity is not a well-tested assumption. It has been shown recently (Ellis 1980; Ellis et al.1978)how such a situation can be described by an inhomo-geneous, spherically symmetric spacetime:

$$ds^2 = -g^2(r)dt^2+dr^2+f^2(r)d\Omega^2. \tag{58}$$

The Killing vector $\frac{\partial}{\partial t}$, i.e. the space-time is static for observers with 4-velocity $u^a = g^{-1}\delta_0^a$. The model is spherically symmetric around $r = 0$ if f and g satisfy

$$f(0) = 0, \quad f'(0) = 1, \quad (f''/f)|_0 \text{ finite};$$

$$g(0) = 1, \quad g'(0) = 0 .$$

A "cosmic blackbody" radiation in this space-time has a tem-perature varying with r:

$$T(r) = T_0(1 + z) = T_0 g^{-1}(r) \tag{59}$$

where T_0 is the temperature along $r = 0$ ($T_0 = 3^0K$). Let g(r) decrease monotonically to zero on (0,R) with finite R. Then T(r) will grow as $r \rightarrow R$, and for some value r_d reach the de-coupling temperature $T_d = 4500$ K. The region beyond r_d would be occupied by a hot plasma ($r_d < r < R$) - the source of the black-body radiation.

The radial variation of the radiation and matter content of the space-time would closely correspond to the dependence on cosmic time t in the standard models. Successive values of r corresponding to decoupling, to nucleosynthesis, to pair production could be determined, as T(r) increases. Each varia-tion would here be ascribed to a spatial variation. The situa-tion here is that all observations - which are made on our past lightcone - are extended off the lightcone on time-like sur-faces (in contrast, in FRW cosmologies the extensions are made on space-like surfaces).

If R is finite, then $g(r) \to 0$ in a finite proper distance
from $r = 0$, and $T(r) \to \infty$, i.e. there is a singulartiy S at
$r = R$ (timelike). All past radial null-geodesics intersect S.
All space-like radial geodesics intersect it too. Therefore,
the singularity surrounds the central world line (Fig. 14).
The universe is spatially finite and bounded, each space-section
$(t = \text{const.})$ has diameter 2R. If $f(r) \to 0$ as $r \to R$ then the sin-
quality can be regarded as a point singularity on each space
section. The space-time is spherically symmetric about $r=0$,
and about S. In Fig. 15 a space-section $(t = \text{const.}, \Theta=\pi/2)$
is displayed.

Models of this sort have not been considered, because
the assumption of spatial homogeneity has been taken as a
cosmological principle ("Copernican principle"). This assumption has not been tested by observations, but it is introduced into any cosmological data analysis, because it is believed to be an unreasonable situation that we should be near
the centre of the universe. However, in such a spherical,
static model, conditions for life are most favourable near the
central line $r=0$, where the universe is cool. This "anthropic
principle" would give a reason why we should live near the
centre of our static universe.

The specific model presented above (Ellis et al. 1978)
can accomodate almost all the data, except for one crucial
point. When the field equations of GR are used on the line
element (58), together with a plausible equation of state,
then it is not possible to obtain a good fit to the observed
(m,z) relation. This difficulty of obtaining Hubble's law
in a static universe gives an idea of a proof that the universe
is expanding - a proof without resort to cosmological principles. Many more investigations of this kind should be carried
out, to explore various different possible ways of explaining
the measurements.

Ellis (1980) has summarized several important points
in this respect: All human astronomical observations have been
obtained in a very short time span, i.e. essentially at one

space-time point q. Information about the universe arrives
at q on time-like or null curves. These cosmological observations together with the field equations of GR and an equation
of state can "in principle" determine the metric tensor, the
matter distribution and velocities, on the null-cone through
q. If no incoming radiation interferes with this solution it
can be extended to a space-time region around the null-cone.

"In principle" it is possible to prove or disprove the
FRW nature of the universe by observations. In practice,
however, significant parts of the data cannot be obtained
directly. While isotropy around us is established fairly well
by the $3^{o}K$ background observations, the spatial homogeneity
is impossible to test without precise knowledge of the intrinsic properties and structure of all astrophysical objects
such as radiogalaxies, quasars, etc. At present we have things
the other way round: the assumption of a FRW cosmology is used
to derive limits on the physical parameters of radio sources.
Even with local physical laws we find a similar situation:
uncertainties in the microphysics reflect themselves in cosmological uncertainties, but e.g. the theory of element formation in FRW universes together with the observations of light
element abundances (see sec.4) have been used to give limits
on various local physical parameters such as the possible types
of heavy leptons, neutrino masses, variations in the coupling
constants. In other papers the abundances have been used to
provide constraints for FRW universes. But the same small
set of numbers cannot be used to elucidate both the local and
the cosmological physics. One data point is one data point,
squeezing more than one bit of information out of it requires
the artistic skill of modern cosmologists (e.g. Schramm 1982).

One basic feature is the increasing uncertainty with
increasing redshift, both because of source evolution, and
of scarcity of measurable information (decrease of flux received, decrease of angle subtended by the objects). Beyond
z=7 or so, electromagnetic signals can no longer reach us
(optical depth $\tau=1$).

The determination of the parameters of one particular FRW model is also made difficult by these uncertainties. H_o, q_o, Λ, and ρ_o are only known within rather wide limits today. In table 1 some characteristic numbers have been collected. The major problem here is the determination of ρ_o. Dynamical methods (see 3.3) give estimates of galactic and cluster masses but they do not determine the density of a uniform mass distribution which could have gone unobserved. Massive neutrinos are one possible candidate, and if they exist then 90% of the matter could be in neutrinos. Therefore a lack of knowledge of the fundamental matter component in the universe adds to the uncertainty in ρ_o.

Thus, using only observations we do not get very far. It is quite remarkable, on the other hand, that very far-reaching conclusions can be found, if the existence of the cosmological radiation background is supplemented by some reasonable hypotheses. I want to mention here two such results: The first is connected with the shape of the spectrum of the radiation (Hawking and Ellis 1973). It is assumed that the radiation has been at least partially thermalized by repeated scattering, as indicated by the black-body nature of the spectrum, and by the exact isotropy on small angular scales. Therefore the amount of matter on each past-directed null geodesic from us must be sufficient to cause a sufficiently high opacity. It can then be shown that this amount of matter is enough to make our past light-cone reconverge. The conditions of a theorem are then satisfied which proves that there should be a singularity somewhere in the universe.

The second result (Ehlers et al. 1968) involves the assumption of the Copernican principle that we do not occupy a privileged position in space-time. It is assumed that the 3°K background would appear equally isotropic to any observer moving on a time-like geodesic, and that the photons have been freely propagating for a long distance (supported by the observations. Then the universe has the Robertson-Walker metric.

4. THE EARLY UNIVERSE

4.1. Thermodynamic Equilibrium

During the radiation-dominated phase $\rho c^2 \propto R^{-4} \propto T^4$. Curvature term $\frac{k}{R^2}$ and cosmological constant Λ become negligible in eq. (11)

$$\dot{R}^2 = \frac{8\pi}{3c^2} G \rho R^2 \; . \tag{60}$$

Hence

$$\frac{32\pi}{3c^2} G\rho t^2 = 1 \; . \tag{61}$$

At early times all homogeneous and isotropic universes evolve like the k = 0 solution (Einstein-de Sitter universe).

An enormous simplification of the physics in the early universe is achieved by the assumption of a complete thermodynamic equilibrium state at some very early epoch (e.g. around $T \simeq 10^{11}$ °K). All the physical quantities of interest depend only on the temperature, and the chemical potentials. The previous history need not be known, except for the complication that the values of some chemical potentials may be determined by particle reactions earlier on.

It seems plausible that such a state of thermodynamic equilibrium exists. Consider a general reaction with cross section $\sigma(E)$ and write $\beta(T) = \langle \sigma v \rangle$ for the averaged product of the cross-section and the velocity. The reaction rate is $\Gamma(t) = \beta \; n \propto T^3 \beta(T)$, while $(n \propto R^{-3})$ the expansion rate $\frac{\dot{R}}{R} \propto t^{-1} \propto T^2$, and therefore

$$\Gamma(t)t \propto T\beta(T) \tag{62}$$

Since T is large for early times, expect $\Gamma t \gg 1$ at early times. As the universe expanded and cooled, collisions became less and eventually equilibrium could no longer be maintained for certain reactions ($\Gamma t \lesssim 1$). We shall see in the next section that modern theories of elementary particles complicate the picture by introducing an equilibrium state for $T \geq 10^{28}$ K, and reactions which drop out of equilibrium at this very early epoch already. The validity of the thermodynamic equilibrium description in this context has to be examined critically. In the spirit of complete honesty in which I try to talk to you about cosmology here, we should make a mental note. Another assumption is introduced (WH3): the universe has passed through a state of complete thermodynamic equilibrium.

In thermal equilibrium the number density of particles of type i, and their energy density at temperature T are (Jüttner 1928; Weinberg 1972; Steigman 1979)

$$n_i = (\frac{g_i}{2\pi^2}) \ (\frac{kT}{\hbar c})^3 \int\limits_0^\infty \{\exp(\frac{E_i - \mu_i}{kT}) \pm 1\}^{-1} z^2 dz \quad , \tag{63}$$

$$c^2 \rho_i = (\frac{g_i}{2\pi^2}) \ (\frac{kT}{\hbar c})^3 kT \int\limits_0^\infty \{\exp\ldots \pm 1\}^{-1} z^2 (\frac{E_i - \mu_i}{kT}) dz \tag{64}$$

g_i is the number of spin states;

$$z = \frac{pc}{kT} \ ; \qquad E_i = [(pc)^2 + (m_i c^2)^2]^{1/2} \quad ;$$

the +(-) sign applies to fermions (bosons).
Photons are bosons with $m_i = 0$, $g_i = 2$, $\mu_i = 0$,

$$n_\gamma = \frac{2\xi(3)}{\pi^2} \ (\frac{kT}{\hbar c})^3 = \frac{2.404}{\pi^2} \ (\frac{kT}{\hbar c})^3 = 31.8 \ T_{GeV}^3 [fm^{-3}] \quad , \tag{65}$$

$$\rho_\gamma = \frac{6\xi(4)}{\pi^2} \ (\frac{kT}{\hbar c})^3 (\frac{kT}{c^2}) = 2.7 (\frac{kT}{c^2}) \ n_\gamma \quad , \tag{66}$$

here ξ is Riemann's ξ-function.

For very high temperatures $kT \gg m_i c^2$, all particles i are relativistic. Then for bosons

$$n_B = (\frac{g_B}{2}) \; n_\gamma \; ; \qquad\qquad \rho_B = (\frac{g_B}{2}) \; \rho_\gamma \; ; \qquad\qquad (67)$$

and for fermions

$$n_F = \frac{3}{8} \; g_F \; n_\gamma \; ; \qquad\qquad \rho_F = (\frac{7}{16}) g_F \; \rho_\gamma \; . \qquad\qquad (68)$$

The chemical potentials μ_i are unknown parameters in this state. The convergence of the integrals in (63) and (64) requires $\mu_i \leq m_i c^2$ for bosons, For fermions there are no restrictions. The chemical potentials are nevertheless usually ignored (Weinberg 1972) - there are very few conserved quantities. The μ_i may be important only at $kT \approx m_i c^2$.

The total energy density in the radiation dominated phase is then

$$\rho = \rho_B + \rho_F \equiv (g(T)/2) \; \rho_\gamma \qquad\qquad (69)$$

where

$$g(T) := g_B + \frac{7}{8} \; g_F \; .$$

To determine $g(T)$ all particles in thermal equilibrium at temperature T are taken into account plus additional relativistic particles which have dropped out of equilibrium. (The temperature of these decoupled particles T_{dec} may be lower than the equilibrium temperature, and therefore their contribution to the density may be reduced by a factor $(\frac{T_{dec}}{T})^4$). We can write

$$t_{sec} = 2.42 \times 10^{-6} g^{-1/2} T_{GeV}^{-2} \qquad\qquad (70)$$

Take for example the equilibrium of photons, $e^\pm$ pairs, and 3 kinds of left-handed neutrinos $\nu_e, \; \nu_\mu, \; \nu_\tau$ at T = 1 MeV. $g_B = g_\gamma = 2$, and

$$g_F = g_e + g_{\nu_e} + g_{\nu_\mu} + g_{\nu_\tau} = 10, \text{ hence}$$

$g(1 \text{ MeV}) = 43/4$, and $t(1 \text{ MeV}) = 0.74$ sec.
The entropy in a comoving volume $V(\propto R^3)$ is approximately
conserved (nonadiabatic processes are usually not important)

$$\frac{S}{k} = \frac{4}{3}\left(\frac{\rho_I c^2}{kT}\right) V = \frac{2}{3} g_I(T)\left(\frac{\rho_\gamma c^2}{kT}\right) V \tag{71}$$

(I: particles in interaction).

A consequence of the conservation of the entropy S is
an additional heating of the gas of interacting relativistic
particles, whenever a particle species i drops out of thermal
equilibrium (Steigman 1974). Then the energy released by i-τ
annihilation produces additional photons, and other relativistic
particles. Entropy conservation gives

$$\frac{N_\gamma(T<m_i)}{N_\gamma(T>m_i)} = \frac{g_I(T>m_i)}{g_I(T<m_i)}$$

Above $T = 1$ MeV $g_I(T) \simeq g(T)$, but below $T = 1$ MeV the e, μ,
and τ neutrinos decouple. So, when e^+ pairs annihilate, the
neutrinos do not share in the energy released.

$$g_I(0.5 \text{ MeV} < T < 1 \text{ MeV}) = g_\gamma + 7/8 g_e = \frac{11}{2},$$

$$g_I(T < 0.5 \text{ MeV}) = g_\gamma = 2.$$

Hence

$$\frac{N_{\gamma 0}}{N_\gamma(T>0.5 \text{ MeV})} = \frac{11}{4}, \tag{72}$$

$N_{\gamma 0}$ is the number of photons in a comoving volume for temperatures below 0.5 MeV; the present number of photons is approximately $N_{\gamma 0}$ too. Then, at 0.5 MeV, the photon gas is heated by

$$\frac{T_\gamma^3(<0.5 \text{ MeV})}{T_\gamma^3(>0.5 \text{ MeV})} = \frac{11}{4} \tag{73}$$

The photon temperature increases by a factor $(\frac{11}{4})^{1/3}$. Below T = 1/2 MeV the temperature T_ν of the neutrino gas is equal to T_γ (>0.5 MeV), that is

$$\frac{T_{\nu o}}{T_{\gamma o}} = (\frac{4}{11})^{1/3} \simeq 0.6 \tag{74}$$

for the present epoch. The total number of effective degrees of freedom is (counting also the noninteracting relativistic particles)

$$g(T \geq 0.5 \text{ MeV}) = g_\gamma + \frac{7}{8}(g_e + 3g_\nu) = \frac{43}{4} \,,$$

$$g(T \leq 0.5 \text{ MeV}) = g_\gamma + \frac{7}{8} 3g_\nu (\frac{T_\nu}{T_\gamma})^4 = 3.36 \,.$$

For T > 1 MeV $g_I \simeq g$ is a slowly varying function of temperature. With increasing temperature more particles are in thermodynamic equilibrium, and g increases from 43/4 at T = 1 MeV to 423/4 at T = 100 GeV - if presently favoured designs for particle behaviour (e.g. Olive et al.1979) are assumed. At T = 1 MeV the time is 0.74 sec, and at T = 100 GeV we are at an epoch t = 0.74 × 10^{-10} sec.

Beliefers in GUT do not hesitate to approach - only in theory, of course - the initial singularity even closer: t ≤ 10^{-35} sec is their favourite time, and the place is very hot indeed: T = 10^{15} GeV! (more in section 5).

Such a hot and dense ($\rho \propto 10^{75}$ g/cm^3) early stage of the universe becomes a unique laboratory for high energy particle and nuclear physics. Unfortunately nobody is there to observe things, and we have to be content with a faint possibility to find a few vestiges of the original chaos. Monopoles, massive stable charged leptons, heavy hadrons or free quarks might be produced in the high temperature medium. They may survive to the present epoch, and may be detected (Cabrera 1982). A review by Steigmann (1979) covers many aspects of this scenario.

One important conclusion has been derived from the assumption of particle-antiparticle symmetry, namely that the ratio of surviving nucleons to photons would at present be (Steigman 1979)

$$(\frac{n_N + n_{\bar{N}}}{n_\gamma}) = 7 \times 10^{-19} . \tag{75}$$

This number is about 10 orders of magnitude smaller than the observed value. In classical cosmology with adiabatic expansion the ratio $\frac{n_\gamma}{n_B}$ is approximately a constant. The observed value $\frac{n_\gamma}{n_B} = 10^9$ is an initial condition of the specific model. The consideration of baryon-nonconserving interactions may change this, because it modifies any initially set ratio $\frac{n_\gamma}{n_B}$, and the theory should enable one to calculate this number-making theory and cosmological model subject to one more observational test (more in sec.5).

4.2 Nucleosynthesis and Abundance of He, D

The "hot big bang" model of the universe predicts a value for the cosmic abundance of helium by mass of

$$X(He^4) = 0.25 \text{ to } 0.30 . \tag{76}$$

This value is close to the observed number - a great success of the model. How does nucleosynthesis proceed in the early universe?

For $T > 1$ MeV the weak interactions

$$p + e^- \rightleftharpoons n + \nu_e ; \qquad p + \bar{\nu}_e \rightleftharpoons n + e^+$$

$$n \rightleftharpoons p + e^- + \bar{\nu}_e \tag{77}$$

have rates ($\approx 0.4 \text{ sec}^{-1} (\frac{T}{10^{10}})^5$) larger than the expansion rate ($\frac{1}{t}$ in eq.(70)), and therefore the neutron-to-proton ratio is kept at its thermal equilibrium value

$$\frac{n_n}{n_p} \stackrel{\sim}{\sim} \exp(-\Delta mc^2/KT) \;,$$

$$(\Delta mc^2 = 1.29 \text{ MeV}) \quad . \tag{78}$$

For $T \leq 1$ MeV the weak interactions are to slow to maintain equilibrium, and the neutrons start decaying.

$$X_n = \frac{n_n}{n_n+n_p} \propto (1+e^{\Delta mc^2/kT})^{-1} = \begin{cases} 0.5 & T \gg \Delta m \\ 0.2 & T = 1.07 \text{ MeV} \end{cases} \tag{79}$$

For $T \leq 1$ MeV β-decay decreases the fraction of neutrons from $X_n = 0.2$ (the "freeze-out" value).

Deuterium (D) is formed rapidly all the time, and also rapidly photodissociated by interaction with the hot radiation. Considerable amounts of the heavier elements can only be built up from 2-body reactions, and thus can only start when $T \leq 0.1$ MeV ($\leq 0.8 \times 10^9$ °K), i.e. when D is no longer photo-disintegrating. Then a number of nuclear reactions proceed very quickly up to the formation of He^4:

$$D + D \begin{cases} \longrightarrow He^3 + n \longrightarrow He^3 + p \\ \longrightarrow He^3 + p \end{cases}$$

$$He^3 + D \longrightarrow He^4 + p \; . \tag{79}$$

Heavier elements are not produced in significant amounts, because at the mass-number $A = 5$ there is no stable nucleus. Nucleosynthesis is completed, when all neutrons at $T = 0.1$ MeV ($X_n \simeq 1/7$) have been made into He^4 (a few minutes later: "The first 3 minutes"). The abundance of He^4 by mass is thus

$$X(He^4) = 2X_n \simeq 0.3 \quad . \tag{80}$$

This value depends crucially on the presence of a hot

radiation field preventing the formation of D before
T = 0.1 MeV. Thus the fact that at present most of the matter
is hydrogen, i.e. not all the matter is transformed into He^4
in the big bang, is another point in favour of the existence
of a cosmic background radiation.

Detailed calculations of D and He synthesis were first
carried out by Wagoner et al. (1967); (recent reviews by
Schramm and Wagoner 1977; Yang et al. 1979).

The resulting abundance of He^4 varies very little with
bayron density, and hence with present baryon density ρ_{Bo}
(since $\rho_B R^3$ = const.) for $2 \times 10^{-31} \le \rho_{Bo} \le 2 \times 10^{-29}$ [g cm^{-3}]
one finds a variation of $X(He^4)$ between 0.25 and 0.29 (see
Fig. 16). Deuterium is very sensitive to the density: in the
density interval 2×10^{-31} to 2×10^{-29} g/cm^3 the predicted
abundance of deuterium drops from 6×10^{-4} to less than 10^{-12}.

The observations of the cosmic abundance of He and
Deuterium are difficult - theories of the formation of spectral lines, as well as educated guesses about the initial
He content of stars are involved too (Yang et al. 1979;
Schmid-Burgk 1981). The observations give $0.22 \le X(He^4) \le 0.29$
with a tendency towards the smaller values. Models of galactic evolution estimate the amount of helium produced in stars
as $X(He^4)$ = (0.01 to 0.04), i.e. the primeval abundance as
deduced from the observations would be

$$0.18 \le X(He^4) \le 0.28 . \tag{81}$$

Deuterium measurements have been made by the Copernicus
satellite (Rogerson and York 1973) from interstellar absorp-
tionlines in the UV:

$$X(D) = (1.4 \pm 0.2) \times 10^{-5} . \tag{82}$$

If this is taken as the cosmic abundance, then the present
baryon density must be less than 6×10^{-31} g/cm^3 (Fig. 16),
i.e. $\Omega \le 0.12$. This would mean that we live in an open universe

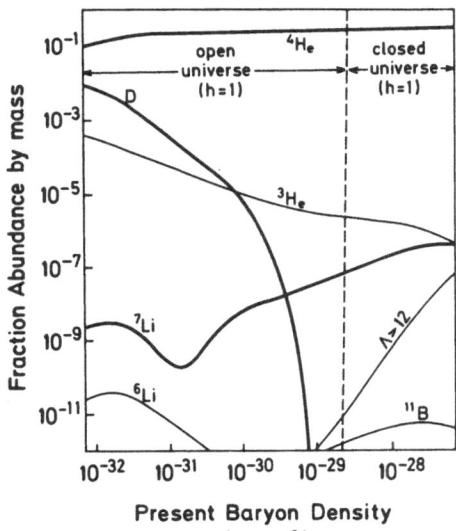

Fig.16: Abundances of P, Li, He and other light elements as functions of the present baryon density.

k = -1, if ρ_{Bo} was equal to the mean density. We have noticed, however, that unseen forms of matter may be present and invalidate this conclusion. It seems difficult to value the deuterium evidence correctly, because X(D) is such a small value. Therefore local production should not be excluded as a possibility.

Another interesting conclusion may be drawn from the He abundance (Schramm and Wagoner 1977; Steigman 1979). The weak reaction rate Γ for n↔p processes decreases with T (cf.eq.77), and when Γ < H(t), where H(t) is the expansion rate, the n/p ratio freezes out. Practically all neutrons present at freeze out time make helium. Thus the expansion rate H(t) around T ≈ 1 MeV determines the freeze out ratio, and the He abundance. Now H(t) ∝ρ , and all relativistic particles at T ≈ 1 MeV contribute to ρ. The helium abundance therefore depends sensitively on the number of types of relativistic particles present for T ≈ 1 MeV. If there are three 2-component neutrinos (e,μ,τ-neutrino) and $\rho_{Bo} \geq 2 \times 10^{-31}$ g/cm^3 then X(He) $\geq$ 0.25 (Yang et al.1979). Each new 2-component neutrino increases X(He):

$$\Delta X(He) \simeq 0.01(\Delta g_\nu/2) \quad . \tag{83}$$

The observations suggest $X(He^4) \leq 0.25$; this would mean only $e, \mu,$ and τ neutrinos exist. For example, right-handed neutrinos associated with the known left-handed neutrinos would lead to $\sum \Delta g_\nu = 6$, and $\Delta X = 0.03$. This is definitely in conflict with the data.

In Fig. 17 a schematic diagram is shown, picturing the history of our universe from $T \simeq 10^{12} \, {}^\circ K$ to the present epoch.

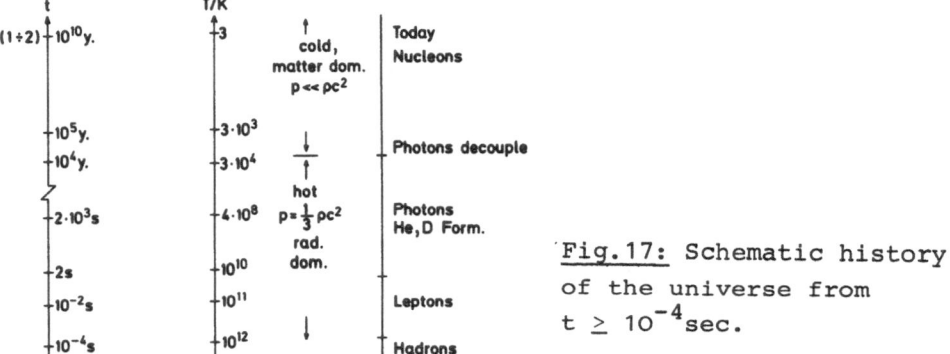

Fig.17: Schematic history of the universe from $t \geq 10^{-4}$ sec.

4.3 Massive Neutrinos

Light neutrinos with a non-zero rest mass $m_\nu c^2 \leq 1$ MeV have decoupled early, and

$$(\frac{N_\nu}{N_\gamma})_{T= 1MeV} = \frac{3}{8} g_\nu \quad . \tag{84}$$

Because of the reheating of the photon gas at e^+e^- annihilation, the present ratio is

$$(\frac{N_\nu}{N_\gamma})_0 = \frac{3}{22} g_\nu \quad , \tag{85}$$

and the mass density contributed by such neutrinos is

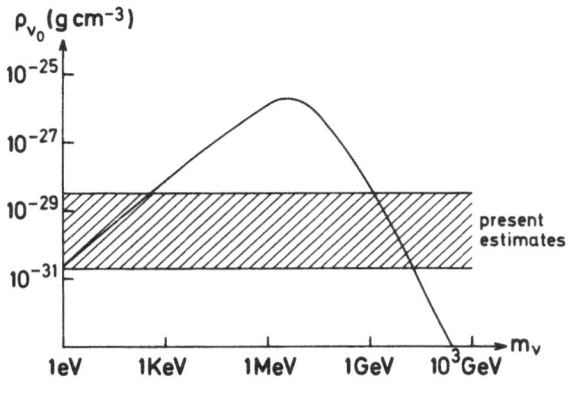

Present neu-
trino mass-density as
a function of neutrino
mass.

$$\rho_{\nu O} = 10^{-31} \, g_\nu m_\nu (eV) \quad [g \, cm^{-3}] \; . \tag{86}$$

If the restriction is made that $\rho_o \leq 2 \times 10^{-29} \, g \, cm^{-3}$ (the
closure density), then with $g_\nu = 4$ (for each massive $\nu\bar{\nu}$ type)
we obtain the constraint

$$m_{\nu_e} + m_{\nu_\mu} + m_{\nu_\tau} \leq 50 \; eV \quad . \tag{87}$$

In Fig. 18 the mass density in relic neutrinos is plotted as
a function of neutrino rest mass.

5. PARTICLE PHYSICS IN THE VERY EARLY UNIVERSE

Since this topic has been treated by Shafi (this volume)
I want to put in a few comments here from the cosmologist's
side. Several interesting new ideas have emerged from the
application of GUT to the very early, hot phase of a Friedmann
universe, but the subject is very exploratory in nature.

5.1 Baryon Synthesis

We have seen in sec. 3 that the observed ratio of pho-
ton to baryon density $\frac{n_\gamma}{n_B} \simeq 10^9$ is an initial condition in the
standard model of the universe. The implementation of a GUT

in the early phase may give such a one in 10^9 baryon-antibaryon asymmetry from a completely symmetric initial state.
In any such calculation there is complete thermodynamic equi-
librium at temperatures $T \geq 10^{28}$ $^{\circ}K$, i.e. at times $t \leq 10^{-35}$ sec.
The heavy gauge bosons of GUT, the X-particles (of mass
$m_x \simeq 10^{14}$ GeV) are still in thermodynamic equilibrium. Then
the universe expands and cools, the X-particles drop out
of equilibrium and decay. If the decay is CP violating, i.e.
if the decay channels of X and its antiparticles are different, a baryon-antibaryon asymmetry will be produced.
Calculations (Kolb 1981; Fry et al.1980; Kolb and Wolfram
1980) give the result

$$\frac{n_B}{n_\gamma} = \alpha \ \beta \ \varepsilon \qquad\qquad (87)$$

where α is $\approx 10^{-3}$ (describes the heating of the photon gas
by particle-antiparticle annihilation at
$t \simeq 10^{-35}$ sec).

and β is $\simeq 10^{-4}$ (due to the dynamical phase of expansion
and decays). ε is due to the CP-violating interaction. The
correct GUT of the universe and ε are essentially unknown.
Taking e.g. the SU(5)-GUT and an estimate of ε from K_o decays, one obtains (Kolb & Wolfram 1980)

$$\frac{n_B}{n_\gamma} \simeq 10^{-12} \text{ to } 10^{-16} \quad , \qquad\qquad (88)$$

(compare this to (75)!).

The calculations, so far, do not give the correct answer, but
they explain, that $\frac{n_B}{n_\gamma}$ is a small, but nonzero number. Another
interesting result is (Fry et al.1980) that any initial baryon
asymmetry is strongly damped by the thermal equilibrium phase.
If the baryon number is equal to the lepton number, $(n_B - n_L) = 0$,
as required by certain gauge theories, then the damping is
large enough to restore complete baryon-antibaryon symmetry.
We see that this result together with the discrepancies of

the calculated $\frac{n_B}{n_\gamma}$ with the observed value, can serve to eliminate certain models of GUT.

5.2 The "Inflationary" Universe

There are two features of the classical "standard model" which are connected with the initial conditions in a simi- larly undesirable way as $\frac{n_B}{n_\gamma}$:

i) Flatness

Our universe is very old, and the curvature term $\frac{k}{R^2}$ in the Friedmann equations is now of the same order as the "energy" terms (as is Λ):

$$(\frac{\dot{R}}{R})^2 = \frac{8\pi}{3} G\rho - \frac{k}{R^2} + \frac{1}{3} \Lambda$$

At earlier epochs therefore, $\frac{k}{R^2}$ must have been negligibly small. If we had $\frac{k}{R^2} \simeq \rho G$ at the "Planck time" $t_{Pl} = (\frac{G\hbar}{c^5})^{1/2}$ $\simeq 10^{-43}$ sec then after a few times t_{Pl}, the universe would recollapse.

Setting as before

$$\rho_c = (\frac{\dot{R}}{R})^2 \frac{3}{8\pi G} , \qquad \Omega \equiv \frac{\rho}{\rho_c} ,$$

we can write

$$\frac{\Omega-1}{\Omega} = \frac{k}{R^2} \frac{3}{8\pi G\rho} . \qquad (89)$$

Since $\rho \propto R^{-4}$ it follows $\frac{\Omega-1}{\Omega} \propto R^2$, hence

$$\Omega_i = (1 + \frac{R_i^2}{R_o^2} (\frac{\Omega_o-1}{\Omega_o})) . \qquad (90)$$

Now Ω_o is between 0.1 and 10; take R_o as the scale factor at the time of decoupling, then

$$\Omega_i = (1 + (\frac{T_R}{T_i})^2 \alpha) \qquad \text{with} \quad -9 \le \alpha \le 0.9 \qquad (91)$$

At $T = 1$ MeV   therefore $\Omega = 1 + O(10^{-15})$.

At $T = 10^{14}$ GeV $\Omega = 1 + O(10^{-49})$.

Our old universe must have been set up with very finely tuned initial data.

ii) Horizon "problem"

At t_R matter and radiation have decoupled, and have evolved independently to the present time. The isotropy of the 3°K background at present implies a similar isotropy at t_R. But we see the same temperature from regions that have not yet been in causal contact with each other.

The radial coordinate of the horizon at time t is:

$$\chi_H(t) = \int_0^t \frac{ds}{R(s)} .$$

The radial coordinate of the background is

$$\chi(t_o, t_R) = \chi_H(t_o) - \chi_H(t_R)$$

and two regions separated by 180° have the coordinate distance

$$\chi(t_{R_1}, t_{R_2}) = \chi(t_o, t_R) - \chi(t_R, t_o) = 2(\chi_H(t_o) - \chi_H(t_R)) .$$

This distance contains

$$N = 2 \frac{\chi_H(t_o) - \chi_H(t_R)}{\chi_H(t_R)}$$

horizon lengths at t_R. For $t > t_R$ we have $R \propto t^{2/3}$, (k = 0 model) $\chi_H(t) \propto t^{1/3} \propto R^{1/2}$, and

$$N = 2[(\frac{R_o}{R_R})^{1/2} - 1] = 2[(\frac{T_R}{T_o})^{1/2} - 1] \approx 75 \tag{92}$$

How is it possible that two regions which were separated by 75 horizon lengths acquire an identical temperature at t_R?

Both these "problems" are questions to the "naturalness"
of the initial data. Shouldn't it be possible to arrive at
the present situation from a wider range of initial data ?

These questions are addressed by the model of the
"inflationary" universe (Guth 1981; Linde 1982; Albrecht et
al 1982). The basic idea is a modification of the adiabatic
expansion, to produce entropy by a phase transition at a
very early epoch within the frame work of gauge theories
("GUT").This first order phase transition is the change from
a symmetric state at high temperatures $T > T_c$ to a less
symmetric Higgs phase for $T < T_c$. For $T < T_c$ the symmetric state
becomes metastable (in Fig.19 the corresponding effective
potentials are drawn). The vacuum energy density ρ_ν of the
symmetric state is renormalized to a finite value, and

$$\langle T_{\mu\nu} \rangle_{kT} = \rho_\nu g_{\mu\nu} \tag{93}$$

is assumed. Before and after the phase transition $T_{\mu\nu}$ is
supported to be of perfect fluid type with $p = \frac{1}{3} \rho$, $\rho \propto T^4$.
During the phase transition the vacuum energy density $\rho_\nu \simeq T_c^4$
starts to dominate (for $T < T_c$), $T_{\mu\nu} \simeq \rho_\nu g_{\mu\nu}$ and the expansion
changes into a de Sitter phase

$$(\frac{\dot{R}}{R})^2 = \frac{8\pi}{3} G \rho_\nu \simeq \frac{T_c^4}{M_p^2} \equiv \chi^2 . \tag{94}$$

(Guth 1981; Linde 1982), i.e.

$$R(t) = R(t_c) \exp \chi (t - t_c) ,$$

$$(M_p = \frac{1}{2}(\frac{\hbar c}{G})^{1/2} = 10^{19} \text{ GeV is the Planck mass}) . \tag{95}$$

If it is correct to take such a constant ρ_ν, then the universe
grows exponentially during this phase, when the symmetric
state is metastable. At some temperature $T_b << T_c$ the less
symmetric Higgs phase appears like bubbles in the symmetric
phase and expands (Fig.20 shows the rapid expansion $V \simeq c$ of

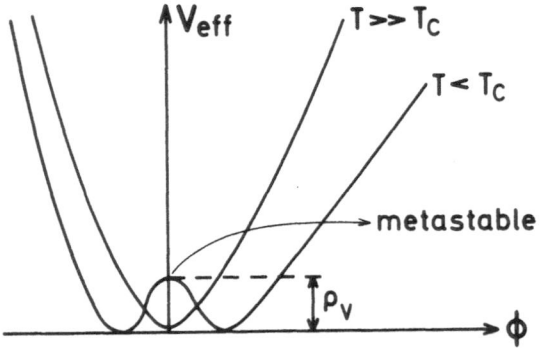

Fig.19a: Effective poten-
tial for $T \gg T_c$, and
$T \leq T_c$.

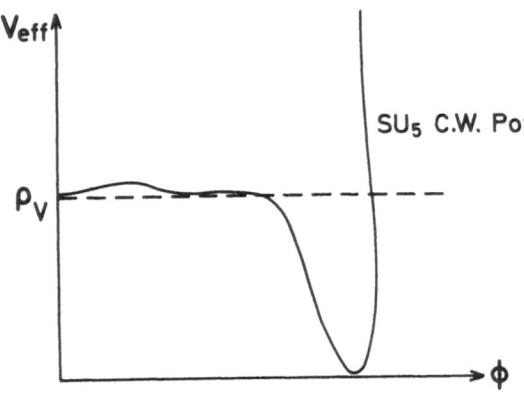

Fig.19b: SU_5 Coleman-
Weinberg potential.

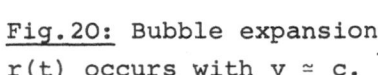

Fig.20: Bubble expansion
$r(t)$ occurs with $v \simeq c$.

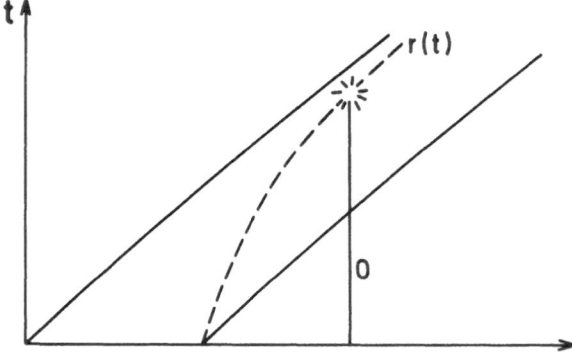

a bubble.) At the same time the part of the universe inside the Higgs is reheated to the temperature T_c (by the coupling of the Higgs field to all kinds of particles which are annihilated into radiation). This reheating is an irreversibel process which produces a large entropy $\propto z^3$, where

$$z = \frac{T_c}{T_b} e^{\chi \Delta t} \quad . \tag{96}$$

Linde (1982) and Albrecht and Steinhardt (1982) assume $T_b \simeq 10^6$ to $10^8\ ^\circ K$, then $z >> 10^{28}$, i.e.

$$R(t) >> R(t_c)\ 10^{28} \quad . \tag{97}$$

A small region at $t = 10^{-35}$ sec has grown by at least a factor of 10^{28} during the phase transition, i.e.

$$\text{during} \qquad \chi \Delta t \simeq 3260 \quad . \tag{98}$$

(The perceptive reader may note that Linde (1982) and Albrecht and Steinhardt (1982) actually give an expansion

$$\frac{R(t)}{R(t_c)} = \exp(3260)\ \ ) .$$

This rapid exponential growth of $R(t)$ reduces our whole observable universe to a tiny part of the initial bubble of the stable phase. There is therefore no longer a "horizon problem" nor a "flatness problem" ($\Omega=1$). The problems are not swept under the rug, but pushed out of sight by the exponential de Sitter phase at t_c.

5.3 A Few Critical Remarks

This is a very new subject, so new questions appear about as fast as new inflationary bubbles. Let me list a few comments:

i) The models start with a homogeneous and isotropic FRW universe at $t < 10^{-35}$ sec. It seems to me that the inflationary scenario should be applied to a less smooth initial state (anisotropic, inhomogeneous).

ii) The high temperature state expectation value is set $<T_{\mu\nu}>_T = \rho_\nu g_{\mu\nu}$ with constant ρ_ν. However, systems in thermodynamics equilibrium define a preferred rest system, where there is no net motion. So, why not put

$$<T_{\mu\nu}>_T = \rho_\nu \delta_{\mu 0} \delta_{\nu 0} \tag{99}$$

(E. Seiler, private communication), or of perfect-fluid type!

iii) If $\rho_\nu(t)$ is not constant, then the de Sitter phase may not occur.

iv) There is a danger in relying on one-loop or 2-loop effective potentials, when describing the phase structure. It can be proved that V_{eff} has to be strictly convex at $\phi=0$, and that in Higgs models $V'_{eff}(\phi=0)=0$ (Elitzur 1975; Kogut 1979; cf. also Wightman 1979;,Symanzik 1970). Then there is no double hump structure as in Fig.19. (E. Seiler private communication).

v) The event horizons in a de Sitter spacetime should radiate with a temperature $T_H \simeq 10^{11}$ GeV (Gibbons and Hawking 1977). If this is true (doubts are permitted!) then only temperatures $T>T_H$ can occur, and a "supercooling"only down to 10^{11} GeV makes the whole scenario more difficult.

CONCLUSION

We might agree that the conclusion in the following poem of C.Morgenstern applies to this (and perhaps to other?) talks in this meeting:

    "Korf liest gerne schnell und viel:

    darum widert ihn das Spiel

    all des zwölfmal unerbetnen

    Ausgewalzten, Breitgetretnen.

    Meistens ist in sechs bis acht

    Wörtern völlig abgemacht

    und in ebensoviel Sätzen

läßt sich Bandwurmweisheit schwätzen.

Es erfindet drum sein Geist
etwas, was ihn dem entreißt:
Brillen, deren Energieen
ihm den Text - zusammenziehen!
Beispielsweise dies Gedicht
läse, so bebrillt man - nicht!
Dreiunddreißig seinesgleichen
gäben erst - Ein - Fragezeichen!!"

ACKNOWLEDGEMENT

I want to thank Jürgen Ehlers for many helpful
suggestions, and for a careful reading of the manuscript.

REFERENCES

1. A. Albrecht, P. Steinhardt, Phys. Rev. Lett $\underline{48}$ (1982) 1220

2. C.H. Bender, F. Cooper, "Failure of the naive loop
    expansion ...", Los Alamos preprint 1983.

3. B. Cabrera, Phys. Rev. Lett $\underline{48}$ (1982) 1378.

4. J. Ehlers, P. Geren, R.K. Sachs, J. Math. Phys. $\underline{8}$ (1968) 1344.

5. J. Ehlers, Proc. "Enrico Fermi" Summer School, Varenna 1969
    (R.K. Sachs ed., Academic Press: 1971), p.1.

6. J. Ehlers, AG Mitt. $\underline{38}$ (1976) 41.

7. J. Einasto, M. Jôeveer, E. Saar, Mon. Not. R. Astr. Soc $\underline{193}$
    (1980) 353.

8. S. Elitzur, Phys. Rev. $\underline{D12}$ (1975) 3978.

9. G.F.R. Ellis, Proc. "Enrico Fermi" Summer School, Varenna
    1969 (R.K. Sachs ed., Acad. Press: 1971), p.104.

10. G.F.R. Ellis, R. Maartens, S.D. Nel, Mon. Not. R. Astr. Soc.
    $\underline{184}$ (1978) 439.

11. G.F.R. Ellis, Proc. IX Texas Symp. Ann. N.Y. Acad. Sci. 336 (1980) 130.

12. I.M.H. Etherington, Phil. Mag. 15 (1933) 761.

13. A. Fabian, Proc. X Texas Symp. Ann N.Y. Acad. of Sci. (1982).

14. H.C. Ford, R.J. Harms, F. Bartko, R. Ciardullo, E. Eason, Bull Ann. Astron. Soc. 12 (1980) 472.

15. W.A. Fowler, F. Hoyle, Ann. Phys. 10 (1960) 280.

16. A. Friedmann, Zs. f. Phys. 10 (1922) 377.

17. J. Fry, K. Olive, M. Turner, Phys. Rev. D22 (1980) 2953.

18. G.W. Gibbons, S.W. Hawking, Phys. Rev. D15 (1977) 2752.

19. A.H. Guth, Phys. Rev. D23 (1981) 347.

20. S.W. Hawking, R. Penrose, Proc. Roy. Soc. London A314 (1970) 529.

21. S.W. Hawking, G.F.R. Ellis, 'The large scale structure of space-time' (Cambridge Univ. Press: 1973).

22. E. Hubble, Astrophys. J. 64 (1926) 321.

23. E. Hubble, Proc. NAS 15 (1929) 168.

24. F. Jüttner, Zs. f. Physik 47 (1928) 542.

25. T. Kirsten, 'The Origin of the Solar System' (S.F. Dermott ed., Wiley & Sons : 1978) p. 267.

26. J. Kogut, Rev. Mod. Phys. 51 (1979) 659.

27. E.W. Kolb, S. Wolfram, Astrophys. J. 239 (1980) 428.

28. E.W. Kolb, preprint - Santa Barbara (1981).

29. J. Kristian, R.K. Sachs, Astrophys. J. 143 (1966) 379.

30. A.D. Linde, Phys. Lett. 108B (1982) 389.

31. W. Mattig, Astr. Nach. 284 (1958) 109.

32. C.W. Misner, K.S. Thorne, J.A. Wheeler, 'Gravitation' (W.H. Freeman & Co, San Franc.: 1973)

33. R.A. Muller, Sci. Amer. 238 (1978) 64.

34. R.A. Muller, Proc. IX Texas Sym. Ann. of N.Y. Acad. Sci. 336 (1980) 116

35. P.E. Nissen, ESO-preprint 1983.

36. P.E. Nissen, ESO Messenger 28 (1982) 4.

37. K.A. Olive, D.N. Schramm, G. Steigman, Phys. Rev. Lett. 43 (1979) 239.

38. J.P. Ostriker, P.J.E. Peebles, Astrophys. J. 186 (1973) 467.

39. R. Penrose, 'Perspectives in Geometry and Relativity' (B.Hoffman ed., Indiana Univ. Press: 1966), p.271.

40. A.A. Penzias, R.W. Wilson, Astrophys. J. 142 (1965) 419.

41. P.J.E. Peebles, 'Physical Cosmology' (Princeton Univ. Press: 1971).

42. P.J.E. Peebles, M. Seldner, Astrophys. J. 225 (1978) 7.

43. P.J.E. Peebles, 'The Large Scale Structure of the Universe' (Princeton Univ. Press: 1980).

44. P.J.E. Peebles, Proc. X Texas Symp. N.Y. Acad. Sci.(1982).

45. J. Rogerson, D.G. York, Astrophys. J. 186 (1973) L95.

46. V.C. Rubin, Comments on Astrophys. 8 (1979) 79.

47. A. Sandage, G.A. Tammann, Astrophys. J. 190 (1974) 525.

48. A. Sandage, G.A. Tammann, Astrophys. J. 207 (1976) L1

49. A. Sandage, Astrophys. J. 252 (1982) 553.

50. A. Sandage, G.A. Tammann, Astrophys. J. 256 (1982) 339.

51. J. Schmid-Burgk, Nucl. Astrophys. (1981) p. 295.

52. D.N. Schramm, R.V. Wagoner, Ann. Rev. Nucl. Sci. 27 (1977) 37.

53. D.N. Schramm, Proc. X Texas Symp. N.Y. Acad. Sci. (1982).

54. G. Smoot, F. Gorenstein, R.A. Muller, Phys. Rev. Lett. 39 (1977) 898.

55. G. Steigman, Ann Rev. Nucl. Part Sci. 29 (1979) 313.

56. R.A. Sunyaev, Ya.B. Zel'dovich, Ann. Rev. Astron. Astrophys. 18 (1980) 537.

57. K. Symanzik, Commun. Math. Phys. 16 (1970) 48.

58. F.K. Thielemann, J. Metzinger, H.V. Klapdor, preprint (1983) (Astron. Astrophys. in preparation).

59. B.M. Tinsley, Astrophys. J. 198 (1975) 145.

60. G. de aucouleurs, Nature 266 (1977) 126.

61. R.V. Wagoner, W.A. Fowler, F. Hoyle, Science 155 (1967) 1369.

62. S. Weinberg, "Gravitation and Cosmology" (Wiley, New York: 1972).

63. S. Weinberg, 'The First 3 Minutes'(Basic, New York:1977).

64. R. Weiss, Ann. Rev. Astron. Astrophys. 18 (1980) 489.

65. A.S. Wightman, Introduction to 'Convexity in the Theory of Lattice Gases' (R.B. Israel, Princeton Univ.Press: 1979).

66. C.M. Will, Annals N.Y. Acad. Sci. 336 (1980) 307.

67. D.P. Woody, P.L. Richards, Phys. Rev. Lett. 42 (1979) 925.

68. J.M. Uson, D.T. Wilkinson, Phys. Rev. Lett. 49 (1982) 1463.

69. J. Yang, D.N. Schramm, G. Steigman, R.T. Rood, Astrophys. J. 227 (1979) 697.

70. Ya.B. Zel'dovich, J. Einasto, S.F. Shandarin, Nature 300 (1982) 407.

Acta Physica Austriaca, Suppl. XXV, 71–100 (1983)

INTRODUCTION TO SUPERSYMMETRY AND SUPERGRAVITY[+]

by

H. NICOLAI

CERN

I. INTRODUCTION

Over the past years, we have witnessed a dramatic
increase in "public interest" in supersymmetry and supergravity. This interest cannot as yet be based on any
solid experimental facts, but it has nevertheless become
clear that supersymmetry has many attractive features to
offer and that it will probably play an important role in
the ultimate unification of the fundamental particle interactions. This unification constitutes one of the most
ambitious endeavors in theoretical physics, because it
necessarily involves the extrapolation in energy over many
orders of magnitude where no experimental information is
expected to become available. The main reason why supersymmetry and supergravity are so attractive is that, at the
present time, there appear to be no other candidate theories

[+]Lectures given at the XXII. Internationale Universitätswochen
für Kernphysik,Schladming,Austria,February 23 - March 5,1983.

that may enable us to simultaneously solve the three outstanding problems of modern elementary particle physics.
These are

(1) The unification of gravity with the other fundamental
    interactions and the construction of a consistent, i.e.
    finite or renormalizable, theory of quantum gravity,

(2) the explanation of the current proliferation of funda-
    mental fields and coupling constants ("who ordered the
    muon?"), and

(3) the hierarchy problem in its most fundamental form: why
    are the relevant mass scales in nature so tiny in compari-
    son with the Planck mass of 10^{19} GeV, or why is the
    gravitational force so much weaker than the other forces
    in nature?

Needless to say that we are still far from realizing this
ambitious goal but there are hints that we may be on the
right track.

In these lectures, I do not intend to give a complete
review of the subject since that would fill several volumes
in view of the rapid progress that has been and is being
made. Nor do I want to discuss any particular topic in exhaustive depth because I would end up hunting down minus
signs in that case. I will rather concentrate on a few representative topics to stimulate the interest of those
readers who are not very familiar with what supersymmetry is
all about and to provide them with an "appetizer". For those
who want to study it more thoroughly, I recommend J. Wess'
Princeton Lectures [1] and P. van Nieuwenhuizen's Physics
Report [2] where the reader may find many technical details
which are indispensable for doing actual calculations as well
as many relevant references.

Nothing will be said here about possible physical
applications although this is currently a very popular subject. My main reason for this omission is that none of the

models proposed so far has come close to any kind of experimental test[+]. Since no trace of supersymmetry in nature has been detected so far all we can presently say is that the scale at which supersymmetry becomes visible may lie anywhere between 20 GeV and 10^{19} GeV.

II. WHAT IS SUPERSYMMETRY?

It is well-known that there exist two types of particles in nature, bosons and fermions. Supersymmetry [3,4] is a new kind of symmetry that relates them in the sense that bosons are transformed into fermions and vice versa. Since fermionic operators anticommute one is naturally led to extend the concept of Lie algebra in such a way that the basic algebra contains anticommutators as well as commutators. One then speaks of a "graded Lie algebra" because each generator of the algebra is graded according to whether it is bosonic or fermionic. The grading is defined to be

$$|B| = 0 \qquad \text{if B is bosonic} ,$$

$$|F| = 1 \qquad \text{if F is fermionic} . \qquad (2.1)$$

Similarly, one introduces a graded commutator by

$$[A_1, A_2\} \equiv A_1 A_2 - (-)^{|A_1||A_2|} A_2 A_1 \quad . \qquad (2.2)$$

These graded commutators satisfy "graded Jacobi identities", namely

---

[+]A possible refutation to this statement may be provided by the dominant decay mode of the proton [26].

$$\sum_{\text{cycl.}} (-)^{|A_1|(|A_2|+|A_3|)} [A_1,[A_2,A_3\}\} = 0 \ . \tag{2.3}$$

One easily sees that the cyclic sum in (2.3) differs from the usual one only if two of the operators are fermionic. The full algebraic structure is then given by the following set of commutators and anticommutators

$$[B_i, B_j] = f_{ijk} B_k \ ,$$

$$[B_i, F_j] = g_{ijk} B_k \ ,$$

$$\{F_i, F_j\} = h_{ijk} F_k \ , \tag{2.4}$$

where f,g and h are generalized structrue constants. A new feature is that, in order to integrate the graded Lie algebra to a group, one must introduce anticommuting quantities ζ, ζ' ,.. (Grassmann numbers). These obey

$$\{\zeta,\zeta'\}= \{\zeta,F_i\} = \zeta^2 = 0 \tag{2.5}$$

but commute with bosonic quantities. By use of Grassmann numbers one can rewrite the algebra (2.4) in terms of commutators only, because

$$\zeta_i\{F_i,F_j\}\zeta_j' = [\zeta_i F_i,F_j\zeta_j'] \ . \tag{2.6}$$

Thus, any element g of the "supergroup" can be represented as

$$g(\omega,\zeta) = \exp[\omega_i B_i + \zeta_j F_j] \tag{2.7}$$

with commuting ω_i and anticommuting ζ_j. In complete analogy with the case of ordinary Lie algebras, one can classify graded Lie algebras following Cartan's ideas [5]. Their study has led to extensive mathematical investigations, but it turns out that the algebras which may actually be relevant in physics are of a quite definite type, see following section.

To have a concrete (and utterly simple) example which is also a little closer to physics, let us consider a non-interacting system of a harmonic oscillator and a spin. The first is described by creation and annihilation operators a^+ and a, which obey canonical commutation relations $[a,a^+]=1$; the spin is likewise described by creation and annihilition operators ψ^+ and ψ which satisfy canonical anticommutation relations $\{\psi,\psi^+\} = 1$. The Hamiltonian of the system is given by

$$H = a^+a + \psi^+\psi .\qquad(2.8)$$

The Hilbert space of this system is, of course, obtained by applying bosonic and fermionic creation operators to the vacuum state, i.e.

$$|n_b,0> = \frac{(a^+)^{n_b}}{\sqrt{n_b!}}|0,0>, \; n_b \in N \; ,$$

$$|n_b,1> = \psi^+|n_b,0> .\qquad(2.9)$$

We next introduce a fermionic charge Q by

$$Q = a^+\psi \rightarrow Q^+ = a\psi^+\qquad(2.10)$$

(of course $[a,\psi] = \ldots = 0$). Clearly,

$$H = \{Q, Q^+\} \; ,$$

$$[H,Q] = [H, Q^+] = 0 \; ,\qquad(2.11)$$

and $Q(Q^+)$ generates a symmetry of the system. What is its physical meaning? Since the action of Q on a state removes one fermion (or "flips the spin from up to down") and adds one boson, and since Q commutes with the Hamiltonian, it follows that

$$\langle n_b, 1 | H | n_b, 1 \rangle =$$

$$= \frac{1}{\sqrt{n_b+1}} \langle n_b, 1 | HQ^+ | n_b+1, 0 \rangle =$$

$$= \frac{1}{\sqrt{n_b+1}} \langle n_b, 1 | Q^+ H | n_b+1, 0 \rangle =$$

$$= \frac{1}{\sqrt{n_b+1}} (\langle n_b+1, 0 | HQ | n_b, 1 \rangle)^* =$$

$$= \langle n_b+1, 0 | H | n_b+1, 0 \rangle \quad . \tag{2.12}$$

Here, in going from the second to the third line, we have
used that the fermionic charge commutes with the Hamiltonian.
Therefore, the spectrum of this model exhibits a degeneracy
between bosonic and fermionic states: to each bosonic state
of a certain energy there corresponds a fermionic state of
the same energy. The only nondegenerate state is the vacuum
state $|0,0\rangle$, for which

$$Q|0,0\rangle = Q^+|0,0\rangle = 0 \quad . \tag{2.13}$$

Consequently, this is also the only supersymmetric state.
Unless the supersymmetry is spontaneously broken, in which
case no state exists which is annihilated by both Q and Q^+,
the supersymmetric state always coincides with the state of
lowest energy. This follows from the positivity of the
Hamiltonian which, in turn, is a consequence of (2.11)

$$H = QQ^+ + Q^+Q \geq 0 \quad . \tag{2.14}$$

The reader will certainly agree that this example is indeed
very trivial. But nonetheless, it does exhibit some of the
general features which will recur in more complicated
situations, namely the boson-fermion degeneracy, which is
ultimately responsible for the cancellation of ultraviolet
divergences between boson and fermion loops in supersymmetric
theories, and the lower boundedness of the Hamilton operator.

In fact, the step from (2.8) and (2.10) to the case of a
free field theory only requires appending momentum labels,
spinor indices and, if desired, internal symmetry indices
to the various operators. If one introduces interactions, the
basic expressions for H and Q are, of course, no longer
quadratic, but the underlying algebraic structure remains
unaffected.

Let us finally note the consequences of (2.14) for
spontaneous symmetry breaking. It is useful to introduce a
"statistics operator" $(-)^F$ whose action on bosonic and
fermionic states is defined by[+]

$$(-)^F |boson> = |boson> \quad ,$$

$$(-)^F |fermion> = -|fermion> \quad . \tag{2.15}$$

One can easily convince oneself that the "supertrace"

$$n \equiv \sum_i <i| (-)^F |i> \tag{2.16}$$

only receives contributions from the groundstates since each
state of higher energy has a partner of the opposite statistics.
Hence, assuming that all groundstates are bosonic, the
"Witten index" n [6] is equal to the number of supersymmetric
groundstates. From the previous arguments, it is not difficult to see that a necessary (but not always sufficient)
criterion for spontaneous breaking of supersymmetry is the
vanishing of n [6]. In the case of an harmonic oscillator and
a spin without interactions, of course, n=1. For more
sophisticated models the actual calculation of n may turn
out to be quite difficult [6].

[+] Of course, F is just the fermion number.

III. THE BASIC SUPERALGEBRA AND ITS REPRESENTATIONS

In this and the following section of these lectures I will follow the notation and conventions of ref. [1] where further references may be found. The graded Lie algebra that turns out to be the relevant one in the context of four-dimensional field theory contains the Poincaré algebra with the translation generator P_μ and the rotation generator $M_{\mu\nu}$ as a subalgebra; besides, it contains 2N fermionic charges Q_α^i and their hermitean conjugates $\bar{Q}_{\dot\alpha i} \equiv (Q_\alpha^i)^+$ and, last not least, the generators of an internal (S)U(N). That this is essentially the most general algebra which may be relevant for elementary particle physics has been shown in [7] (the most general algebra also contains "central charges" and the generators of the conformal group [7]). The generators P_μ and $M_{\mu\nu}$ obey the usual commutation relations. The new relations are

$$\{Q_\alpha^i, \bar{Q}_{\dot\beta j}\} = 2\delta^i{}_j \sigma^\mu_{\alpha\dot\beta} P_\mu \quad ,$$

$$\{Q_\alpha^i, Q_\beta^j\} = \{\bar{Q}_{\dot\alpha i}, \bar{Q}_{\dot\beta j}\} = 0 \quad ,$$

$$[P_\mu, Q_\alpha^i] = [P_\mu, \bar{Q}_{\dot\alpha i}] = 0 \quad , \qquad\qquad (3.1)$$

where we have defined[+]

$$\sigma^\mu_{\alpha\dot\beta} \equiv (1, \sigma^k) \qquad\qquad\qquad\qquad (3.2)$$

(σ^k are the usual Pauli-matrices). The indices $\alpha, \beta, \ldots = 1,2$ are SL(2,C) spinor indices, and dotted indices refer to the complex conjugate representations. The indices $i, j, \ldots$ run from 1 to N, and for a given N, one speaks of "N-extended supersymmetry". There are a few more commutation relations

---

[+]One similarly defines $\bar\sigma^{\mu\dot\alpha\beta} \equiv (1, -\sigma^k)$, see [1].

that have not been written out: the commutator $[M_{\mu\nu},Q]$ simply specifies that Q_α^i ($\bar{Q}_{\dot\alpha i}$) transforms as a two-component spinor (its complex conjugate) under the Lorentz group SL(2,C), while the commutator between Q_α^i and the internal symmetry generator indicates that Q_α^i transforms as the fundamental representation of (S)U(N).

One of the most striking features of the algebra (3.1) is that it mixes space-time symmetries with internal symmetries: the anticommutator of Q and $\bar{Q}$ is the translation operator P_μ. In this way, it becomes possible to fuse these two kinds of symmetries, and this was, in fact, one of the original motivations for introducing supersymmetry [4]. The beauty of this unification of symmetries will only become fully apparent in the context of supergravity, where the transformation parameters become space-time dependent.

The algebra (3.1) has been given in terms of two-component spinors. One can, of course, also switch to four-component notation by introducing a four-component Majorana spinor

$$Q^i = \begin{pmatrix} Q_\alpha^i \\ \bar{Q}_i^{\dot\alpha} \end{pmatrix} \tag{3.3}$$

(SL(2,C) indices are raised and lowered with the "metric" $\varepsilon_{\alpha\beta}$). The γ-matrices are given by [1]

$$\gamma^\mu = \begin{pmatrix} 0 & \sigma^\mu \\ \bar{\sigma}^\mu & 0 \end{pmatrix}, \qquad \gamma^5 = \begin{pmatrix} 1 & 0 \\ 0 & -1 \end{pmatrix}, \tag{3.4}$$

and the first equation in (2.1) may also be written as

$$\{Q^i, \bar{Q}_j\} = 2\delta^i{}_j \gamma^\mu P_\mu . \tag{3.5}$$

From (3.3) and (3.4), it is obvious that the (S)U(N) group is a <u>chiral</u> group since Q_α^i and $\bar{Q}_{\dot\alpha i}$ are just the chiral projections in (3.3) (in N=1 supersymmetry, the associated chiral invariance is sometimes called "R-invariance").

To analyze the irreducible multiplets of the algebra (3.1) [8], we have to distinguish between the massive and the massless case. Here, only the latter will be considered. If $M^2 = 0$, one can choose a frame of reference such that $P_\mu = (P,0,0,P)$; without loss of generality, we may assume that $P>0$. In this frame, we get from (3.1)

$$\{Q_1^i, \bar{Q}_{1j}\} = 4P ,$$

$$\text{all others} = 0 .$$

$$(3.6)$$

In particular Q_2^i and $\bar{Q}_{2i}$ create only unphysical (i.e. zero-norm) states and may therefore be discarded in the analysis of the particle multiplets. We next observe that

$$[M_{12}, Q_1^i] = -\frac{1}{2} Q_1^i .$$

$$(3.7)$$

Since we are quantizing about the z-axis, this means that application of the operator Q_1^i decreases the helicity of a given state by one half. One therefore proceeds as in ordinary quantum mechanics when one constructs irreducible multiplets of the rotation group by use of raising and lowering operators. We thus apply the operator $(2\sqrt{P})^{-1}Q_1^i$ to a state of given helicity λ_{max}; the result is a state of helicity $\lambda_{max} - \frac{1}{2}$, i.e.

$$\frac{Q_1^i}{2\sqrt{P}}|\lambda_{max}\rangle = |\lambda_{max} - \frac{1}{2}, i\rangle .$$

$$(3.8)$$

Continuing and descending in helicity by steps of one-half, we obtain after k steps

$$\frac{Q_1^{i_1}}{2\sqrt{P}} \cdots \frac{Q_1^{i_k}}{2\sqrt{P}} |\lambda_{max}\rangle = |\lambda_{max} - \frac{k}{2}, [i_1 \ldots i_k]\rangle .$$

$$(3.9)$$

Since the Q's anticommute, the index k-tuple $[i_1 \ldots i_k]$ which labels the state of helicity $\lambda_{max} - \frac{k}{2}$ is antisymmetric and

consequently, there are $\binom{N}{k}$ such states. Because of the
anticommutation property, this procedure terminates after
N steps. We also have to take into account the CPT conjugate
states of reversed helicity. It immediately follows that

$$|\lambda_{max}| \geq \frac{N}{4} \left(\frac{N+1}{4} \text{ if N is odd}\right) \quad .$$ (3.10)

For renormalizable interactions, we have $|\lambda_{max}|=1$ and the
upper limit is N = 4; if gravity is included, one has
$|\lambda_{max}| = 2$ and the maximal N is N = 8. The respective tables
are given below (s = spin, N = number of supersymmetries).

N \ S	1	$\frac{1}{2}$	0
1	1	1	–
2	1	2	1+1
3	1	3+1	3+3
4	1	4	6

Table 1 : Massless multiplets with $|\lambda_{max}| = 1$

One sees from these tables that the multiplets of highest
N are CPT-self-conjugate. Note also that these multiplets
are unique in the sense that no other multiplets exist
with lower maximum helicity. Another feature is that as one

N \ S	2	$\frac{3}{2}$	1	$\frac{1}{2}$	0
1	1	1	-	-	-
2	1	2	1	-	-
3	1	3	3	1	
4	1	4	6	4	1+1
5	1	5	10	10+1	5+5
6	1	6	15+1	20+6	15+15
7	1	7+1	21+7	35+21	35+35
8	1	8	28	56	70

Table 2: Massless multiplets with $|\lambda_{max}| = 2$

proceeds to higher N, the number of possible models that
realize this supersymmetry decreases. For instance, there is
only one model having global N = 4 supersymmetry, namely
the celebrated N = 4 Yang-Mills theory [9]. Beyond N = 4,
supersymmetry can only be realized as a local symmetry. The
presence of spin-2 particles in the multiplet makes it
exceedingly plausible that gravity must be an integral part
of locally supersymmetric theories, and this is indeed true.
Thus, the lowest line in table 2 gives the particle content
of the maximally extended N = 8 supergravity [10]: one graviton, eight gravitinos, 28 vector particles, 56 spin-$\frac{1}{2}$
particles and 70 scalars. This theory is the most interesting
candidate for the ultimate unification of all particle interactions since it is (almost) unique; particle multiplet and
the various couplings are entirely fixed and there is only
one free parameter, namely the gravitational coupling
constant κ.

Finally, it should be mentioned that the analysis presented in this section may also be carried out for algebras which contain "central charges" [11].

IV. SUPERSYMMETRIC MODELS

In the foregoing section, I have derived the (massless) representations and the irreducible particle multiplets of the basic superalgebra (3.1). It now remains to actually construct Lagrangian field theories which realize this algebra and are thus supersymmetric. I will here briefly discuss Wess-Zumino type models, that is models only containing spin-0 and spin-$\frac{1}{2}$ fields. Their construction is most easily achieved by the use of superfields although, historically, this was not the method by which the Wess-Zumino model was originally found [4].Since this topic has received extensive treatment in the existing literature [1] and since there is no point in unnecessary repetition, this chapter will be rather cursory. The basic idea is to extend space-time to superspace, i.e. a manifold that is, at least locally, parametrized by the space-time coordinates x_μ and anticommuting variables θ_α and $\bar{\theta}_{\dot{\alpha}}$, and to consider functions on this superspace. These functions are called superfields and given by

$$\Phi(x,\theta,\bar{\theta}) = A(x) + \theta^\alpha \phi_\alpha(x) + \bar{\theta}_{\dot{\alpha}} \bar{\psi}^{\dot{\alpha}}(x) + \theta^\alpha \sigma^\mu_{\alpha\dot{\beta}} \bar{\theta}^{\dot{\beta}} V_\mu(x)$$

$$+ \theta^\alpha \theta_\alpha F(x) + \bar{\theta}_{\dot{\alpha}} \bar{\theta}^{\dot{\alpha}} G(x) + \theta^\alpha \theta_\alpha \bar{\theta}_{\dot{\alpha}} \bar{\lambda}^{\dot{\alpha}}(x) + \bar{\theta}_{\dot{\alpha}} \bar{\theta}^{\dot{\alpha}} \theta^\alpha \rho_\alpha(x)$$

$$+ \theta^\alpha \theta_\alpha \bar{\theta}_{\dot{\alpha}} \bar{\theta}^{\dot{\alpha}} D(x) \quad . \tag{4.1}$$

Here, it is important that, owing to the anticommutation properties of θ_α and $\bar{\theta}_{\dot{\alpha}}$, the Taylor-expansion of $\phi(x,\theta,\bar{\theta})$ in θ_α and $\bar{\theta}_{\dot{\alpha}}$ terminates after a finite number of steps. The coefficient functions in this expansion are called component fields; for instance, A(x) is a complex scalar field, $\phi_\alpha(x)$ is an SL(2,C)-spinor, $V_\mu(x)$ is a complex vector field, ect. The next step in our construction is to realize the operators P_μ, Q_α and $\bar{Q}_{\dot\alpha}$ of the abstract algebra (3.1) in terms of differential operators which act on the function $\Phi(x,\theta,\bar\theta)$. Explicitly, one finds [1]

$$P_\mu = -i\partial_\mu,$$

$$Q_\alpha = \frac{\partial}{\partial\theta^\alpha} - i\sigma^\mu_{\alpha\dot\alpha}\bar\theta^{\dot\alpha}\partial_\mu \quad,$$

$$\bar{Q}_{\dot\alpha} = \frac{\partial}{\partial\bar\theta^{\dot\alpha}} - i\theta^\alpha\sigma^\mu_{\alpha\dot\alpha}\partial_\mu \quad. \tag{4.2}$$

The reader is invited to check that, upon (anti)commuting these operators, the algebra (3.1) is indeed reproduced. An infinitesimal supersymmetry transformation is now defined by

$$\delta\Phi(x,\theta,\bar\theta) = (\xi^\alpha Q_\alpha + \bar\xi_{\dot\alpha}\bar{Q}^{\dot\alpha})\,\Phi(x,\theta,\bar\theta) \quad. \tag{4.3}$$

To obtain the transformation laws of the component fields, one simply expands both sides of (4.3) in terms of θ_α and $\bar\theta_{\dot\alpha}$, and compares coefficients. In this way, one finds

$$\delta\Lambda(x) = \xi^\alpha\phi_\alpha(x) + \bar\xi_{\dot\alpha}\bar\psi^{\dot\alpha}(x) \qquad\qquad \text{etc.} \tag{4.4}$$

In particular, the anticommutation properties of θ_α and $\bar\theta_{\dot\alpha}$ imply that

$$\delta D = \theta^\alpha\theta_\alpha\bar\theta_{\dot\alpha}(-i\xi^\beta\sigma^\mu_{\beta\dot\beta}\bar\theta^{\dot\beta}\partial_\mu)\bar\lambda^{\dot\alpha}(x) + \bar\theta_{\dot\alpha}\bar\theta^{\dot\alpha}\theta^\alpha(-i\theta^\beta\sigma^\mu_{\beta\dot\beta}\bar\xi^{\dot\beta}\partial_\mu)\rho_\alpha(x) \tag{4.5}$$

and therefore the <u>highest component of a superfield always transforms as a total derivative</u>. This is very important for the construction of invariants since, from (4.3),

$$\delta(\Phi_1\Phi_2) = \delta\Phi_1 \cdot \Phi_2 + \Phi_1\delta\Phi_2 \quad, \tag{4.6}$$

the same property also holds for products of superfields. Thus,

$$\delta\int d^4x \left\{ \begin{array}{l} \text{any product} \\ \text{of superfields} \end{array}\right\}\Big|_{\text{highest comp.}} =$$

$$= \int d^4x \partial^\mu K_\mu \quad \sim 0 \tag{4.7}$$

for suitable K_μ. Thus, supersymmetric Lagrangians are not invariant, transforming as total derivatives, but the associated actions are invariant if the fields fall of sufficiently rapidly at infinity (in the quantum theory this can be rigorously justified in the presence of a mass gap).

From (4.3), it is evident that superfields are representations of supersymmetry. One can thus regard superfields as a convenient book-keeping device to describe the supermultiplet, which consists of the various component fields, in a compact way. In this perspective, one does not have to attach any intrinsic meaning to superspace as such or to worry about the problem of how to rigorously justify formal manipulations with anticommuting objects.

The representation of supersymmetry which corresponds to the superfield (4.1) is, however, not irreducible. One can reduce it by imposing suitable covariant constraints. One possibility is to require $\Phi(x,\theta,\bar\theta)$ to be real, i.e.[+]

$$\Phi(x,\theta,\bar\theta) = [\Phi(x,\theta,\bar\theta)]^* \quad . \tag{4.8}$$

From the expansion (4.1), we can directly read off what (4.8) means in terms of the component fields, namely

---

[+]Complex (or rather hermitean) conjugation also inverts the order of fermionic quantities.

$$A(x) = A^*(x) \quad ,$$

$$\bar{\psi}_{\dot{\alpha}}(x) = [\phi_\alpha(x)]^* \quad ,$$

$$V_\mu(x) = V_\mu^*(x) \quad , \text{ etc.} \tag{4.9}$$

It can be shown that the constraint (4.8) indeed leads to an irreducible representation of supersymmetry. If one adds a gauge invariance, the multiplet (4.8) describes spin-1 and spin-$\frac{1}{2}$; it is therefore called the vector-multiplet. There is no room here to describe this construction in any more detail, and we refer the reader to ref. [1] for an exhaustive treatment.

The other possibility is to impose a differential constraint by means of the operator [1]

$$D_\alpha = \frac{\partial}{\partial\theta^\alpha} + i\sigma^\mu_{\alpha\dot{\alpha}}\bar{\theta}^{\dot{\alpha}}\partial_\mu \quad ,$$

$$\bar{D}_{\dot{\alpha}} = -\frac{\partial}{\partial\bar{\theta}^{\dot{\alpha}}} - i\theta^\alpha\sigma^\mu_{\alpha\dot{\alpha}}\partial_\mu \quad . \tag{4.10}$$

These are almost like the Q-operators of eq. (4.2), but a little calculation shows that

$$\{D_\alpha, Q_\beta\} = \{D_\alpha, \bar{Q}_{\dot{\beta}}\} = \ldots = 0 \tag{4.11}$$

(in the language of differential geometry in superspace, D_α and $\bar{D}_{\dot{\alpha}}$ are invariant vector fields under supersymmetry transformations). Hence, the constraint

$$\bar{D}_{\dot{\alpha}}\Phi(x,\theta,\bar{\theta}) = 0 \tag{4.12}$$

is covariant: the variation $\delta\Phi(x,\theta,\bar{\theta})$ also obeys it because of (4.11). Working out the details, one finds that (4.12) implies [1]

$$\Phi(x,\theta,\bar{\theta}) = A(x) + \theta^{\alpha}\psi_{\alpha}(x) + \theta^{\alpha}\theta_{\alpha}F(x) + i\theta^{\alpha}\sigma^{\mu}_{\alpha\dot{\alpha}}\cdot\bar{\theta}^{\dot{\alpha}}\partial_{\mu}A(x)$$

$$- \frac{i}{2}\theta^{\alpha}\theta_{\alpha}\partial_{\mu}\psi^{\beta}(x)\sigma^{\mu}_{\beta\dot{\beta}}\bar{\theta}^{\dot{\beta}} + \frac{1}{4}\theta^{\alpha}\theta_{\alpha}\bar{\theta}_{\dot{\alpha}}\cdot\bar{\theta}^{\dot{\alpha}}\Box A(x) \quad . \tag{4.13}$$

This multiplet, which is called <u>chiral multiplet</u>, describes spin-O and spin-$\frac{1}{2}$ particles which are represented by the complex scalar field A(x) and the SL(2,C)-spinor field $\psi_{\alpha}(x)$, respectively. The complex field F propagates no physical degrees of freedom; it is an auxiliary field which can be eliminated by its equation of motion.

The chiral multiplet and the vector multiplet are the basic building blocks of rigidly N = 1 supersymmetric theories. Their derivation also illustrates the general procedure which is expected to yield irreducible representations of extended supersymmetry. In extended supersymmetry, the number of component fields contained in a superfield rapidly increases with N because there are 4N spinorial quantities θ^{i}_{α} and $\bar{\theta}_{\dot{\alpha}i}$. Therefore, unconstrained extended superfields are highly reducible. There are two ways of getting rid of superfluous components. One is to impose algebraic and/or differential constraints; the other is to introduce additional gauge invariances such that most of the remaining components become unphysical gauge degrees of freedom. If the latter are gauged away, one obtains the theory in a special gauge (so-called "Wess-Zumino gauge") which contains only the physical degrees of freedom. Of course, it requires some ingenuity to arrange things in such a way that one is left precisely with the right number of physical degrees of freedom after this procedure. The algebraic structure is still rather easily recognized in N = 1 supersymmetric models but the example of N = 4 supergravity which is the maximally extended theory for which this program has been partially carried out [12] shows the intricacies and complications that arise in extended supersymmetry: in order to embed the physical theory into a superspace formutation, one must add more than 100 000 auxiliary degrees of freedom! Thus, the completion of the general program remains one of the outstanding challenges.

One can now construct interacting field theories based on the two irreducible multiplets. Since the details of this construction can be looked up in ref. [1], I only quote the result for chiral multiplets. For a set of n chiral multiplets $\Phi_i(x,\theta,\bar\theta)$, the most general supersymmetric and renormalizable Lagrangian is given by

$$L = A_i \Box A_i^* + i\partial^\mu \bar\psi_{i\dot\alpha} \sigma_\mu^{\dot\alpha\alpha} \psi_{i\alpha} - \{\frac{1}{2} \frac{\partial^2 W}{\partial A_i \partial A_j} \psi_i^\alpha \psi_{j\alpha} + c.c.\}$$

$$- \frac{\partial W}{\partial A_i} (\frac{\partial W}{\partial A_i})^* \tag{4.14}$$

(after elimination of the auxiliary fields F_i and F_i^*). The complex function W: $C^n \to C$

$$W(A_i) = \lambda_i A_i + \frac{1}{2} m_{ij} A_i A_j + \frac{1}{3} g_{ijk} A_i A_j A_k \tag{4.15}$$

is called superpotential. Its relation with the scalar potential $P(A_i)$ follows from (4.14)

$$P(A_i) = \sum_{i=1}^{n} | \frac{\partial W}{\partial A_i} |^2 . \tag{4.16}$$

Obviously, $P(A_i) \geq 0$ in accordance with the general consideration leading to (2.14). The original Wess-Zumino model is re-obtained by taking one chiral multiplet $\Phi(x,\theta,\bar\theta)$ and by putting

$$W(A) = \frac{1}{2} m A^2 + \frac{1}{3} g A^3 . \tag{4.17}$$

From the positivity of (4.16), it follows that the model (4.14) has a supersymmetric groundstate whenever there exists an $A_i^{(o)}$ such that

$$\frac{\partial W(A^{(o)})}{\partial A_i} = 0 \text{ for all i .} \tag{4.18}$$

If no such $A_i^{(o)}$ exist, the supersymmetry is spontaneously broken. For spontaneous supersymmetry breaking, one needs at least three chiral multiplets, and an example, where it actually occurs is [13]

$$W(A_1, A_2, A_3) = \mu A_1 A_2 + \lambda A_3 (A_1^2 - \mu^2) \ . \tag{4.19}$$

Finally, the use of superfields greatly simplifies quantum calculations [1], [14]. One finds that the WZ-type models discussed here have the remarkable feature that all quadratic divergences cancel and that only one wave function renormalization is required to cancel the remaining logarithmic divergence and to render these models finite to all orders in perturbation theory [15]. This property has been invoked recently to provide a solution to the technical part of the hierarchy problem: a fine-tuning at the tree-level is not upset by higher order corrections [16]. However unobjectionable such a philosophy may be as far as the perturbation expansion is concerned, one should nevertheless be careful in subscribing to it: one can argue, that the absence of any other genuine renormalization in these models is precisely what is needed to establish their inconsistency [17]!

V. SUPERSYMMETRY WITHOUT ANTICOMMUTATION

As we have seen, superfields contain more degrees of freedom than just those of the physical particles. This approach is based on the assumption that the underlying structure becomes more transparent upon the addition of suitable auxiliary fields. In this section, I want to describe another approach [18] which goes precisely in the opposite direction in that one not only eliminates the auxiliary fields but also the fermionic ones. One can then attempt to understand supersymmetry in terms of purely bosonic functional integration measures, i.e. without the use of anticommuting quantities.

Since the basic idea is best illustrated by super-symmetric quantum mechanics[+], let us go back to the simple example of section 2, where we discussed the supersymmetric system of one spin and one harmonic oscillator. We will now add anharmonic interactions. Since this is most easily accomplished in a Lagrangian formulation, let us replace the creation and annihilation operators a^+ and a by the the con-figuration variable $q(t)$ and the canonical momentum $p(t)$ and let us pass from there to the Lagrangian formulation. The Lagrangian which generalizes the simple model of section 2 is given by

$$L = \frac{1}{2} \dot{q}(t)^2 - \frac{1}{2}[V(q(t))]^2 - i\bar{\psi}(t)\dot{\psi}(t) + \bar{\psi}(t)\psi(t)V'(q(t)) \quad (5.1)$$

The noninteracting case is recovered by putting $V(q) = q$. The Lagrangian (5.1) is invariant up to a total derivative under the supersymmetry transformations

$$\delta q(t) = \bar{\varepsilon}\psi(t) + \bar{\psi}(t)\varepsilon \quad,$$

$$\delta\psi(t) = (i\dot{q}(t) + V(q(t)))\varepsilon \quad. \quad\quad (5.2)$$

Going from real to imaginary times amounts to replacing (5.1) and (5.2) by

$$L = \frac{1}{2} \dot{q}^2 + \frac{1}{2}V(q)^2 + \bar{\psi}\dot{\psi} - V'(q)\bar{\psi}\psi \quad\quad (5.1')$$

and

$$\delta q = \bar{\varepsilon}\psi + \bar{\psi}\varepsilon \quad,$$

$$\delta\psi = [\dot{q} + V(q)]\varepsilon \quad, \quad\quad \delta\bar{\psi} = [-\dot{q} + V(q)]\bar{\varepsilon} \quad\quad (5.2')$$

[+]Supersymmetric Quantum mechanics is discussed in [19].

so $\bar{\psi}$ is no longer the complex conjugate of ψ. The subsequent considerations will be based on (5.1') and (5.2').

The (Euclidean) correlation functions of this quantum mechanical model are given by

$$<q(t_1)\ldots q(t_n)> =$$

$$= \frac{1}{Z}\int \prod_t dq(t)d\psi(t)d\bar{\psi}(t)q(t_1)\ldots q(t_n)e^{-\int L(t')dt'} \tag{5.3}$$

where[+]

$$Z = \int \prod_t dq(t)d\psi(t)d\bar{\psi}(t)e^{-\int L(t')dt'} \quad . \tag{5.4}$$

Since the Lagrangian (5.1') is quadratic in the fermionic variables one can integrate them out using Berezin's prescription [20]. The result is

$$\int \prod d\psi(t)d\bar{\psi}(t)\exp\{-\int\bar{\psi}(\frac{d}{dt} + V'(q))\psi dt\} =$$

$$= \det\{(\frac{d}{dt} - V'(q(t)))\delta(t-t')\} =$$

$$= \det\{(\frac{d}{dt} + V'(q(t)))\delta(t-t')\} \quad . \tag{5.5}$$

The analog of this determinant in quantum field theory is just the Matthews-Salam determinant [21]. The functional measure (5.3), from which one computes the various correlation functions has now become a purely bosonic measure, i.e.

$$<q(t_1)\ldots q(t_n)> = \frac{1}{Z}\int d\mu(q)q(t_1)\ldots q(t_n) \tag{5.6}$$

with

[+]It will shortly become obvious that Z=1.

$$d\mu(q) \equiv \prod_t dq(t) \cdot \det\{(\frac{d}{dt} + V'(q(t)))\delta(t-t')\}$$

$$\times \exp\{-\frac{1}{2}\int(\dot{q}(t)^2 + V(q(t))^2)dt\} \quad . \tag{5.7}$$

Closer inspection now shows that the supersymmetry which is
hidden in $d\mu(q)$ may be characterized by the following statement [18]: there exists a field redefinition such that the
measure (5.7) becomes Gaussian and such that the Jacobian
of the field transformation equals the Matthews-Salam determinant (5.5). It is now obvious why I have chosen supersymmetric quantum mechanics to illustrate this theorem:
the field transformation can be constructed explicitly. It
is [22],[23][+]

$$\xi(t) = \dot{q}(t) \pm V(q(t)) \quad . \tag{5.8}$$

Indeed,

$$\det \frac{\delta\xi(t)}{\delta q(t')} = \det\{(\frac{d}{dt} \pm V'(q(t)))\delta(t-t')\} \tag{5.9}$$

and

$$\frac{1}{2}\int_{-\infty}^{+\infty}\xi(t)^2dt = \frac{1}{2}\int_{-\infty}^{+\infty}\{\dot{q}(t)^2 + V(q(t))^2\}dt \pm \int_{-\infty}^{+\infty}\dot{q}(t)V(q(t))dt. \tag{5.10}$$

Since the last term in (5.10) is a topological invariant, we
have

$$d\mu(q) = \exp\{-\frac{1}{2}\int_{-\infty}^{+\infty}\xi^2dt - |Q|\}\prod_t d\xi(t) \tag{5.11}$$

where Q is the topological charge of the configuration $q(t)$
(it can take only three values). We note that (5.8) is a
stochastic differential equation for $q(t)$ in terms of the

[+]See also the third reference in [19].

"white noise" $\xi(t)$ [23] and that the winding number of the mapping (5.8) is equal to the number of nontrivial topological sectors, for, whenever $\xi(t) = \dot{q}(t) \pm V(q(t)) = 0$, $q(t)$ describes an instanten solution with a definite topological charge Q.

It can be argued that the characterization of supersymmetry in terms of purely bosonic functional integration measures which I have just explained by the example of supersymmetric quantum mechanics is generally valid [18]. Of course, in the general case, one cannot expect the result to be as simple as (5.8); sometimes nonlocal field redefinitions are required and the transformation can only be obtained iteratively. For example, in the Wess-Zumino model, the transformation up to second order is given by[+]

$$A'(x) = A(x) + mg\int dy \ C(x-y)A^2(y) - g^2\int dydz \partial_\mu C(x-y)A^*(y)\partial_\mu C(y-z)A^2(z)$$

(5.12)

where g is the coupling constant and

$$C(x) = \int \frac{e^{ikx}}{k^2+m^2} \ \frac{d^4k}{(2\pi)^4}$$

(5.13)

the usual propagator.

There are indications that the locality of the field redefinition is related to the finiteness of the supersymmetric theory under consideration. This is borne out by the example of N = 2 WZ-type models in two dimensions [22,23] which are finite and for which the field redefinition can be shown to be local. A more intriguing example would be the N = 4 super Yang-Mills theory in four dimensions, but the existence of the mapping is still an open problem in this case [18].

What have we learnt from these considerations? First of

[+]This result has been derived independently in [24].

all, anticommuting variables are not strictly necessary to understand supersymmetry. Secondly, we have discovered a fascinating new facet of the boson fermion-symmetry present in supersymmetric theories. Thirdly, we have supplemented the superfield method which allows us to understand the perturbative aspects of supersymmetric theories by an independent characterization that will eventually allow us to gain a better understanding of the nonperturbative aspects as well.

VI. WHY LOCAL SUPERSYMMETRY IMPLIES GRAVITY

From the basic supersymmetry algebra (3.1), it is almost obvious why local supersymmetry implies gravity. The commutator of two local supersymmetry transformations with parameters $\varepsilon_1(x)$ and $\varepsilon_2(x)$, respectively, yields a translation with space-time dependent parameter $\bar{\varepsilon}_1(x)\gamma_\mu\varepsilon_2(x)$ which can be interpreted as a general coordinate transformation (Although this argument is not quite rigorous, it can be made so). But this means that gravity must be included, and the actual calculations confirm this conclusion. In the preceding sections we have described models which possess a global supersymmetry. A rather general feature was that the Lagrangians were invariant under supersymmetry transformations up to a total derivative such that the action was invariant if the fields decay sufficiently rapidly at infinity. Thus, one invariably finds that

$$\delta L = \bar{\varepsilon}\partial^\mu K_\mu \tag{6.1}$$

where ε is the supersymmetry transformation parameter and K_μ some model dependent vector spinor. One can now ask what happens if we make the transformation parameter ε local i.e. space-time dependent[+]. Since the given action is invariant

---

[+]The following discussion has been adapted from [2].

for constant ε, it is not difficult to see that the non-invariant part is proportional to $\partial_\mu \varepsilon(x)$. Indeed, one finds

$$\delta S = \delta \int L d^4 x = \int \partial_\mu \bar{\varepsilon}(x) \cdot j^\mu(x) d^4 x \qquad (6.2)$$

where $j^\mu(x)$ is the Noether current

$$\bar{\varepsilon}(x) j^\mu(x) = \frac{\delta L}{\delta \partial_\mu \phi(x)} \, \delta\phi(x) - K^\mu(x) \qquad (6.3)$$

(remember that it is a vector spinor). Clearly, the action is no longer invariant for local $\varepsilon(x)$. Therefore, one must now add new terms to Lagrangian and transformation rules by an iterative procedure ("Noether method") such as to restore full invariance. As in any gauge theory, one must introduce a gauge fields which carries a vector index μ in addition to the indices of the gauge group. Since the latter are spinor indices, the gauge field of supersymmetry is a vector spinor $\psi_{\mu\alpha}(x)$ which is called the gravitino. (It is perhaps useful at this point to keep in mind the analogy with ordinary electrodynamics. The Dirac Lagrangian $L = i\bar{\psi}\gamma^\mu \partial_\mu \psi$ is invariant under rigid phase rotations $\psi \to e^{i\alpha}\psi$. In order to promote this to a local invariance, one must introduce a gauge field A_μ and a Noether coupling $A^\mu j_\mu$ where $j_\mu = \bar{\psi}\gamma_\mu \psi$ is the electromagnetic current). We then add the Noether coupling

$$S^N = -\kappa \int \bar{\psi}_\mu(x) j^\mu(x) d^4 x \qquad (6.4)$$

to the original Lagrangian. If we postulate that $\psi_\mu(x)$ transform as

$$\delta\psi_\mu(x) = \frac{1}{\kappa} \partial_\mu \varepsilon(x) + \ldots \qquad (6.5)$$

we readily see that the unwanted term (6.2) is cancelled by the variation of the gravitino field in (6.4), and we have thus achieved the desired cancellation to lowest order. The dimensionful parameter κ is needed because ε has dimension $-\frac{1}{2}$ whereas ψ_μ has dimension $+\frac{3}{2}$; its appearance is another hint that local supersymmetry leads to supergravity. The transformation rule (3.5) ensures that ψ_μ is pure spin-$\frac{3}{2}$ in the same way that general coordinate invariance in general relativity ensures that the graviton is pure spin-2, and therefore local supersymmetry fills the gap between general relativity with a spin-2 gauge field and ordinary gauge theories with a spin-1 gauge field.

Continuing with the iterative procedure that has led us to include (6.4) and varying the Noether current $j^\mu(x)$ in (6.4), one arrives at a new term

$$\delta(S + S^N) = i\kappa \int T^{\mu\nu}(\phi)\,\bar{\psi}_\mu^{(x)}\gamma_\nu \varepsilon(x)\,d^4x \tag{6.6}$$

where $T^{\mu\nu}$ is precisely the energy-momentum tensor that can be derived from the original Lagrangian (it is a general result that the "supercurrent" always varies into the energy momentum tensor). To cancel (6.6), one introduces a new field g which transforms as

$$\delta g_{\mu\nu} = -i\kappa \bar{\psi}_\mu \gamma_\nu \varepsilon - i\kappa \bar{\psi}_\nu \gamma_\mu \varepsilon \quad . \tag{6.7}$$

But $T_{\mu\nu}$ is precisely what we get when we make the original Lagrangian invariant under general coordinate transformations by replacing derivatives by covariant derivatives including a factor $\sqrt{g}$ and then taking the variational derivative $\frac{\delta(\sqrt{g}L)}{\delta g^{\mu\nu}}$. Hence, $g_{\mu\nu}$ must be identified with the metric tensor, and we arrive once more at the conclusion that local supersymmetry necessitates gravity. Once can now continue the Noether procedure which will terminate after a finite number of steps once one has introduced the principle of general coordinate covariance. However the result is rather involved and not very illuminating, and I will therefore not pursue this line of argument here.

The requirement of local supersymmetry has forced us to introduce two new fields, the gauge field of local supersymmetry which is the analog of ordinary gauge fields in Yang-

Mills theories, and the graviton field which has no analog
in ordinary Yang-Mills theories. Just as for Yang-Milld theories
one can construct an invariant action which only contains
the gauge fields and no (super) matter fields. This, then,
is the action of pure N = 1 supergravtiy. Simple N = 1
supergravity was first constructed in [25]. Although it had
been clear for some time that one would end up with an
extension of general relativity, the actual construction was
quite laborious and simplifications were only later found. The
key to these results is the use of first order formalism,
i.e. the treatment of the Lorentz-connection $\omega_{\mu ab}$ as an independent variable and the use of vierbeine e_μ^a instead of the
metric $g_{\mu\nu}$. In the full theory (6.5) becomes

$$\delta\psi_\mu(x) = \frac{1}{\kappa} D_\mu(\omega)\epsilon(x) =$$

$$= \frac{1}{\kappa}(\partial_\mu + \frac{1}{4}\omega_{\mu ab}\sigma^{ab})\epsilon(x) \quad , \tag{6.8}$$

i.e. the derivative in (6.5) has to be covariantized with
respect to local Lorentz-transformations. The analog of (6.7)
in terms of vierbein fields is

$$\delta e_\mu^a(x) = -i\kappa\bar\psi_\mu(x)\gamma^a\epsilon(x) \tag{6.9}$$

whence (6.7) follows by use of $g_{\mu\nu} = e_\mu^a e_{\nu a}$. The complete
action of N = 1 supergravity which is invariant under the
combined transformation laws (6.8) and (6.9) is just the sum
of the usual Einstein Lagrangian and the properly covariantized
Rarita-Schwinger Lagrangian

$$L = -\frac{1}{4\kappa^2} e\, e_a^\mu e_b^\nu R_{\mu\nu}^{\ \ ab}(\omega) - \frac{1}{2}\epsilon^{\mu\nu\rho\sigma}\bar\psi_\mu\gamma_5\gamma_\nu D_\rho(\omega)\psi_\sigma \tag{6.10}$$

where

$$R_{\mu\nu}^{\ \ ab} = \partial_\mu\omega_\nu^{\ ab} + \omega_\mu^{\ ac}\omega_{\nu c}^{\ \ b} - (\mu\leftrightarrow\nu) \tag{6.11}$$

is the usual curvature tensor. The proof that (6.10) is indeed invariant under the local supersymmetry transformations (6.8) and (6.9) may be found in [2]. In addition, (6.10) is invariant under general coordinate transformations and under local Lorentz-transformations.

We thus see that the requirement of local supersymmetry forces us to go all the way to gravity; there is no local supersymmetry without gravity. This is certainly one of the most attractive features of supersymmetry, because a unification of matter interactions can be naturally achieved in this way. In the case of $N = 1$ supergravity, one still has considerable freedom in adding matter interactions. In higher extended supergravities, there is less and less freedom to choose the matter multiplets and matter interactions, and for $N \geq 5$, one is only left with the pure supergravity theories. These models are therefore (almost) unique since the field multiplets, the form of the interactions and the internal symmetries are (almost) entirely fixed. Consequently, there appears to be some chance that supergravity may realize the (philosophical) principle that the "theory of the world" should be uniquely determined by the requirement of mathematical consistency, and thereby · provide an answer to the three questions cited in the introduction.

VII. CONCLUSION

At the end of these lectures, it is evident that in order to go beyond what I have said so far, I would have to start afresh with a whole new set of lectures and I would no longer be able to avoid discussing minus signs. Since it was not my intention to belabor the reader with all the technical details that are necessary to master supergravity but rather to generate interest in this new and exciting branch of theoretical physics, I refer the reader once more to refs. [1] and [2] for a more detailed exposition.

REFERENCES

1. J. Wess and J. Bagger, "Supersy-metry and Supergravity" (Princeton University Press, 1982).

2. P. van Nieuwenhuizen, Phys. Rep. 68 (1981) 189.

3. D. Volkov and V.P. Akulov, Phys. Lett. 46B (1973) 109.

4. J. Wess and B. Zumino, Nucl. Phys. B70 (1974) 39.

5. W. Nahm, V. Rittenberg and M. Scheunert, Phys. Lett. 61B (1976) 383.

6. E. Witten, Nucl. Phys. B188 (1981) 513.

7. R. Haag, J. Lopuszanski and M. Sohnius, Nucl. Phys. B88(1975)61.

8. D.Z. Freedman, Irreducible Representations of Super-symmetry, in: "Recent Developments in Gravitation", Cargèse 1978, eds. M. Levy and S. Deser (Plenum Press, 1979).

9. F. Gliozzi, J. Scherk and D. Olive, Nucl. Phys. B122 (1977) 256. L. Brink, J. Schwarz and J. Scherk, Nucl. Phys. B121 (1977) 77.

10. D.Z. Freedman and B.de Wit, Nucl. Phys. B130 (1977) 105. B.de Wit, Nucl. Phys. B158 (1979) 189. E.Cremmer and B. Julia, Phys. Lett. 80B (1978) 48; Nucl. Phys. B159 (1979) 141. B.de Wit and H. Nicolai, Phys. Lett. 108B (1981) 285; Nucl. Phys. B208 (1982) 323.

11. S. Ferrara, C.A. Savoy and B. Zumino, Phys. Lett. 100B(1981)393.

12 P. Howe, H. Nicolai and A. Van Proeyen, Phys. Lett. 112B (1982) 446. E. Bergshoeff, M.de Roo and B.de Wit, Nucl.Phys. B217 (1983)489.

13. L.O'Raifeartaigh, Nucl. Phys. B96 (1975) 331.

14. E.g. M. Grisaru, W. Siegel and M. Roček, Nucl. Phys. B159 (1979) 420.

15. J. Wess and B. Zumino, Phys. Lett. 49B (1974) 52. J. Iliopoulos and B. Zumino, Nucl. Phys. B76 (1974) 310.

16. E. Witten, Nucl. Phys. B188 (1981) 513. S. Dimopoulos and H. Georgi, Nucl. Phys. B193 (1981) 150. N. Sakai, Z. Phys. C11 (1982) 153.

17. N.V. Krasnikov and H. Nicolai, Phys. Lett. 121B (1983) 259.

18. H. Nicolai, Phys. Lett. 89B (1980) 341, 117B (1982) 408;

Nucl. Phys. B176 (1980) 419.

19. H. Nicolai, J. Phys. A9 (1976) 1497. E. Witten, Nucl. Phys. B188 (1981) 513. P. Salomonsen and J.W. van Holten, Nucl. Phys. B196 (1982) 509.

20. F.A. Berezin, The Method of Second Quantization (Academic Press, N.Y., 1966).

21. T. Matthews and A. Salam, Nuovo Cim. 12 (1954) 563; 21 (1955) 120.

22. S. Cecotti and L. Girardello, Phys. Lett. 110B (1982) 39; Harvard preprint HUTP-82/A013 (1982).

23. G. Parisi and N. Sourlas, preprint LPTENS 82/6 (1982).

24. M. Goltermann, Utrecht preprint (1982).

25. D.Z. Freedman, S. Ferrara and P. van Nieuwenhuizen, Phys. Rev. D13 (1976) 3214. S. Deser and B. Zumino, Phys. Lett. 62B (1976) 335.

26. J. Ellis, D.V. Nanopoulos and S. Rudaz, Nucl. Phys. B202 (1982) 43.

Acta Physica Austriaca, Suppl. XXV, 101–141 (1983)
© by Springer-Verlag 1983

MOSTLY MAGNETIC MONOPOLES[+]

by

W.P. TROWER
Virginia Polytechnic Institute and State University
Blacksburg,Virginia 24061,USA

INTRODUCTION

In these lectures I will discuss the exerimental aspects of
some of the particle physics which is being conducted out of
the mainstream and, for the most part, independent of the
programmatic research at the great accelerators. My principal
topic will be the search for the magnetic monopole to which
I will devote the first two lectures. What I will present will
be more of a "trail guide" for the occasional hiker rather
than an "atlas" for the geographer. I will emphasize the art
of the hunt more than the theoretical ideas motivating it. I
will also organize some of the speculations as to where monopoles might be found.

In the final lecture I will discuss some of the typical
experiments in four categories defined by their relationship
to the current orthodoxy, gauge theory. In doing so I will
attempt to answer the question: Is there life beyond gauge theory?

[+] Lectures given at the XXII. Internationale Universitätswochen
für Kernphysik,Schladming,Austria,February 23-March 5,1983.

MAGNETIC MONOPOLES

The Monopole Idea

Five important ideas have defined the monopole for which
we search and so have direct impact on our experiments.

The first is contained in a letter written in 1269 AD
by Petrus Peregrinus de Maricourt, a French military engineer
[2].Here Peregrinus suggests the possible existence of lines
of magnetic force and isolated magnetic poles. His motivation
for putting forward this thesis was more pecuniary than pure.
It seems that while participating in the siege of a small
rebellious principality in southern Italy, Peregrinus' thoughts
turned to his post military employment. This, in turn, apparently caused him to consider the commercialization of a per-
petual motion machine he had invented which was based on the
then little understood principles of magnetism. Thus, he pre-
pared his digression on lodestones, as a modern inventor would
prepare an offering prospectus, to attract potential investors.
Nonetheless, he was the first to articulate the possiblity of
free magnetic poles.

The great James Clerk Maxwell contributed the second
important idea about magnetic monopoles by omitting them from
his famous unification of electricity and magnetism. He had
explicitly considered magnetic monopoles in early formulations
of his theory [4]. However, he finally abandoned them as there
was no physical phenomenon which required magnetic poles for
its explanation". The lack of orthographic symmetry, which
has sometimes teased rotation which is, ironically, bereft of
any new physics [5].

The third idea concerning magnetic poles came from
P.A.M. Dirac [6] in 1931. Fresh from performing the triumphant
marriage of quantum mechanics and special relativity on the
body of the hydrogen atom Dirac studied quantum-mechanically
the motion of an electron in the field of a magnetic monopole.
There Dirac found his now famous quantization condition,

$$g = n \ (e/2\alpha) \ , \tag{1}$$

where e and g are, respectively, the fundamental electric and magnetic charges; n is the principal quantum number; and α is the fine structure constant which characterizes the strength of electromagnetic interactions. He demonstrated that if a single magnetic pole exists anywhere in the universe, electric charge quantization would result. It was both the strength, 68 times that of the electric charge, and the definite value of the magnetic charge, which Dirac theory predicted, that were essential for experiment.

In the mid-sixties Schwinger contributed the fourth idea, the possibility of a particle combining both electric and magnetic charge which he called a dyon[7]. Because of the large electromagnetic field strength of dyons, Schwinger constructed an argument which would explain the strong nuclear force by having prosaic particles composed of dyons.

The most recent idea occurred in 1974 when Polyakov and 't Hooft [8]independently showed that monopoles are an inevitable consequence of a class of elementary particle theories, spontaneously broken non-Abelian gauge theories. However, the gauge theory that successfully unified the weak, SU(2), and electromagnetic, U(1), interactions and the theory that described the nuclear force, Quantum Chromo Dynamics, QCD were not one of these. Non-Abelian gauge theories are being used in the attempted unification of the electroweak and strong interactions by the so-called Grand Unification Theories, GUTs. An intrinsic property of Polyakov-'t Hooft monopoles is a mass enormously larger than that of any known particle.

The properties of magnetic monopoles before and after the advent of GUTs are summarized in Table I.

Table I: Summary of Magnetic Monopole Properties.

Property	Pre-GUTs	Post-GUTs
Magnetic Charge (g)	$e/2\alpha$ (+,−,±??)	$e/2\alpha$(+,−,±??)
Electric Charge (e)	??	??
Kind	(lepton, hadron,??)	Composite
Spin	??	Not Applicable
Statistics	??	Not Applicable
Parity	??	Not Applicable
Size	??	Like Hadrons
Structure	??	Onion-like
Mass	??	$\sim 10^{16}$ GeV/c^2
Stability	??	Absolute (except annihilation)
Excited States	??	Yes
Sources and Sinks	Accelerators, Matter Cosmic Rays	Cosmic Rays Astronomical Objects
Cross Section	??	??
Flux	??	Undecided
Origin	??	Big Bang

Electromagentic Interactions of Monopoles

The detection of magnetic monopoles rests on its unique electromagnetic interactions. Further, many of the techniques and devices by which electrically charged particles are measured have the potential of being applied to monopoles searches.

A monopole in a magnetic field, B, will be accelerated by a force

$$F = gB, \tag{2}$$

and gain energy at the rate

$$\Delta E \sim 2 \text{ MeV } B \cdot L \quad, \tag{3}$$

where B is in gauss and L is in meters. A monopole will lose energy in matter to ionization at the rate, calculated in the impulse approximation, of [9],

$$\left(\frac{dE}{dx}\right)_I^M \sim \left(\frac{\beta}{2\alpha}\right)^2 \left(\frac{dE}{dx}\right)_I^{E \, min} \quad, \text{ where } \beta = \frac{v}{c} \sim 1 \quad ; \tag{4}$$

by bremsstrahlung [9],

$$\left(\frac{dE}{dx}\right)_B^M \sim 2\left(\frac{1}{2\alpha}\right)^4 \left(\frac{\beta m_e}{M_m}\right)^2 \left(\frac{dE}{dx}\right)_M^E \quad, \tag{5}$$

but which is only appreciable for $\gamma = (1-\beta^2)^{-1/2} > 20$; and by Cherenkov radiation [10],

$$\left(\frac{dE}{dx}\right)_C^M \sim \left(\frac{1}{2\alpha} \eta(\omega)\right)^2 \left(\frac{dE}{dx}\right)_C^E \quad, \tag{6}$$

where $\eta(\omega)$ is the index of refraction. $(\frac{dE}{dx})^E$ is the energy loss of an electrically charged particle under the same conditions for the same process.

A monopole induces eddy currents in a conductor and loses energy at the rate [11]

$$(\frac{dE}{dx})_E^M \sim 6 \times 10^{-3} \beta\sigma \frac{GeV}{cm} \quad , \tag{7}$$

where σ is the electrical conductivity.

In transition radiation[12], photons are produced when a charged particle traverses the interface of two dissimilar materials. It has an intensity in the visible region $\sim\log\gamma$ and in the x-ray region, $\sim\gamma$. This radiation is confined to a narrow forward cone.

Magnetic monopoles can be trapped microscopically on nuclei, atoms, molecules; or macroscopically in bulk matter [13].The paramagnetic monopole binding energies in atoms and molecules are comparable to the chemical bond energies, $\sim$ 1eV. Monopoles bind to nuclei with a typical energy of a few keV. The monopole binding energy in bulk paramagnetic and ferromagnetic materials are $\sim$ 10 eV and $\sim$100 eV, respectively.

Alvarez [14] noted that a magnetic monopole passing through a ring would induce a current. The value of this current would only depend on the strength of the magnetic charge and the properties of the ring. The current would be independent of all other properties of the monopole and its circumstances. The induced ring current is quickly extinguished by the ring material's resistance unless it is superconducting.

The supercurrent density in a superconducting ring is

$$\vec{j}_e = \frac{he^*}{2im^*} (\psi^*\vec{\nabla}\psi - \psi\vec{\nabla}^*\psi) - \frac{e^{*2}}{m^*c} \psi^* \psi \vec{A} , \tag{8}$$

where the particles involved are Cooper pairs whose mass and electric charge are twice that of the electron. $\vec{A}$ is the electromagnetic vector potential. ψ, the coherent many-body state of Cooper pairs, has a local pair density, $\psi^*\psi = n_s/2$, half the superelectron density, n_s. Solving this equation [5] in conjunction with Maxwell's equations for a closed ring, gives the magentic flux through the ring to be twice the magnetic flux quantum,

$$\phi_o = (hc/2e) = 207 \ nG \cdot cm^2 \ . \tag{9}$$

For a magnetic monopole passing through the ring, shown in Fig.1,

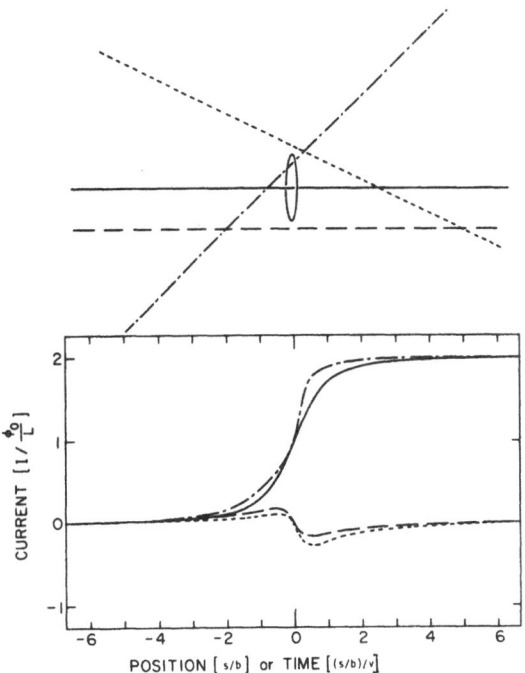

Fig.1: The monopole induced current in a superconducting ring for ring-passing and near-miss trajectories. s is the trajectory distance from the ring plane. (ref.5)

$$\phi(t) = \phi_s(t) + \phi_g(t) = - 4\pi g\theta(t) , \tag{10}$$

where $\phi = 0$ at $t = -\infty$ is assumed and θ is the solid angle subtended by the ring at the monopole. The first term is the self induced ring supercurrent

$$\phi_s(t) = -I(t)L , \tag{11}$$

where L is the ring self inductance. The second term is the monopole flux coupling the ring

$$\phi_g(t) = 2\pi g(1- 2\theta(t) + \gamma vt((\gamma vt)^2 + b^2)^{-1/2}) . \tag{12}$$

The induced ring current is

$$I(t) = (\phi_o/L)(1 + \gamma vt((\gamma vt)^2 + b^2)^{-1/2}) . \tag{13}$$

Ring current changes for several monopole trajectories are shown in Fig.1 while a schematic of the flux line behavior of a ring-passing monopole is shown in Fig.2. The characteristic time, $b/\gamma v$, is ~ 1 μs for $\beta \sim 10^{-4}$ and b $\sim$ few centimeters. A ring-passing magnetic monopole will induce a persistent current change. A near-miss monopole, a ring-passing electric charge and a ring-passing magnetic dipole will cause only small, transient current changes whose maximum excursion is that for a ring-passing monopole.

GUT Independent Experiments

Induction; Static: Alvarez and his colleagues did the first experiments looking for magnetic monopoles using electromagnetic induction[15]. Their experimental setup, sketched in Fig.3, consisted of a ring with its attendant readout electronics through which they circulated a sample of material. The original ring was operated at room temperature and thus required a high circulation speed or the monopole induced

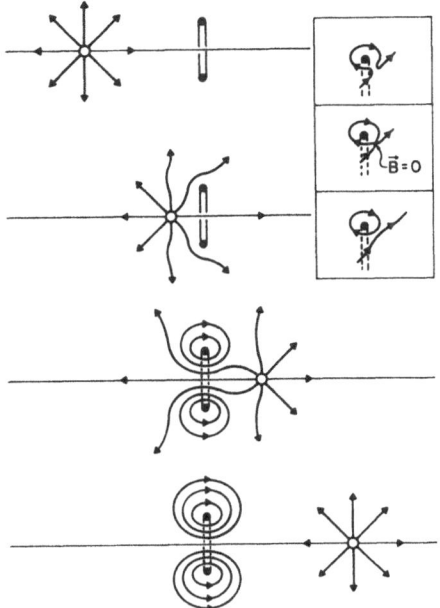

Fig.2: A monopole passing through a current ring pulls flux
with it until the lines break and form a residual toroidal field around the ring. (Cabrera in ref.56)

current change would be degraded by resistive effects before
the sensitivity of the conventional electronics was reached.
With the advent of superconducting rings, the circulation speed
became irrelevant as a monopole induced current change would
persist. The addition of a Superconducting Quantum Interfero-
metric Device, SQUID, enormously improved their readout sensitivity and decreased the number of passes required for detection.
They examined many kilograms of various materials which included meteorites, deep sea nodules, snow from the polar regions and lunar dust. The final experiments using lunar samples placed the most stringent limit on the monopole content

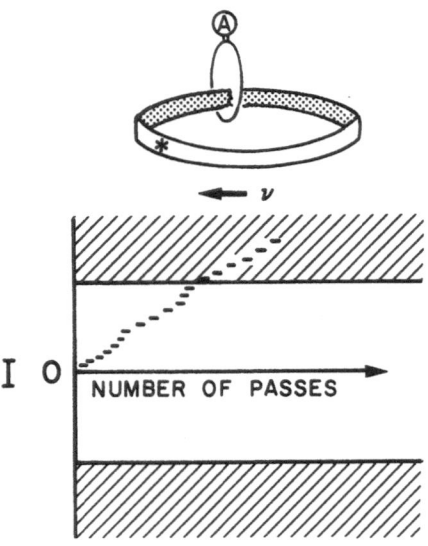

NUMBER OF PASSES

<u>Fig.3:</u> A schematic of the Berkeley static multiple pass
induction monopole detector (top). A graph of the
current incrementation per pass if a monopoles were
present (bottom). Current changes in the shaded regions are detectable.

in matter, $\rho_m < 2 \times 10^{-4}$ monopoles/gram for >.05 g. Here,
about seven passes of the sample were required to detect a
single Dirac charge.

The development of a superconductive technique to
achieve ultra-low magnetic field regions by the Fairbank group
at Stanford, increased the sensitivity of magnetically charged
particle detectors a thousand fold[16]. Cabrera obtained
these low field regions using cylinders constructed of 70 μm
lead, fan-folded like a grocery-bag. As the bag cools through
its transition temperature, the ambient magnetic flux should
be spontaneously expelled by the Meissner effect. In practice,
pinning forces prevent the trapped flux vortices from migra-

ting. However, once superconducting, changes in the external magnetic field are exactly cancelled by the bag's surface supercurrents. Now if the bag is mechanically expanded, its interior magnetic field is reduced. A new folded bag can then be inserted and cooled inside the expanded one. Repeating this procedure, sketched in Fig.4, four times can produce interior field of <100 nG. All field components will decrease exponentially with the distance from the bag's opening. The residual trapped fields will be entirely horizontal.

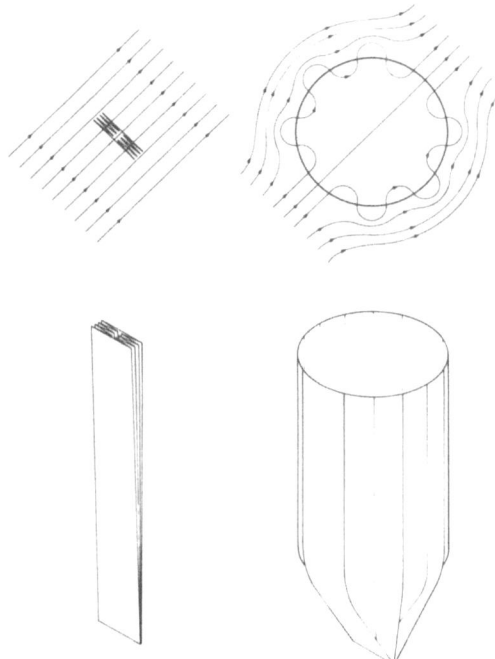

Fig.4: Schematic of the field exclusion process during one cycle which Cabrera uses to achieve ultra-low magnetic regions (ref.5).

Before the above procedure is begun, the dewar is wrapped in a mumetal shield which reduces the earth's field about a hundred fold. A detection and a calibration ring are then placed near the bottom of the bag and maintained at liquid helium temperature.

This assembly, shown in Fig.5, was used by Cabrera to measure the net magnetic charge of eleven materials ranging from titanium to teflon. He passed the samples through the ring in a holder suspended by a quartz fiber. Restricted to a single pass through the ring, he measured decigram size samples with a sensitivity of >.02 g to established ρ_m < .5 monopoles/gram.

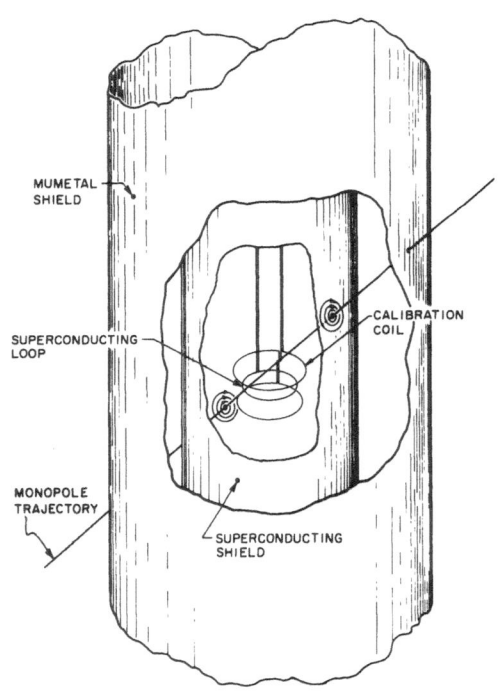

MUMETAL SHIELD

CALIBRATION COIL

SUPERCONDUCTING LOOP

MONOPOLE TRAJECTORY

SUPERCONDUCTING SHIELD

Fig.5: Schematic of Cabrera's monopole detector of the kind used to measure the net magnetic charge of various materials and later detected the candidate cosmic ray monopole event. This ultra low field superconducting magnetometer is a four-turn 5-cm diameter ring made of niobium wire positioned with its axis vertical. The ring is connected to a SQUID. The SQUID and ring are mounted inside an ultralow field shield and, in turn, are mounted inside a single mumetal cylinder to provide 180 db isolation from external magnetic field changes. (ref.5)

His most interesting samples were the niobium spheres from the quark experiment[17]. When these spheres were passed through the ring, they induced an appreciable transient current implying that the trapped magnetic flux gave the spheres a substantial dipole moment. Over the past decade experiments have been made on twelve spheres, two of which had previously been found to have a fractional electric charge. No magnetic charge was detected. These experiments continue with the purpose of bracketing a magnetic charge measurement with two unchanged non-zero fractional electric charge measurements.

Now let's summarize what can be learned from static induction experiments. If no monopoles are detected in a sample, we can infer that monopoles don't bind to that particular substance, have been dislodged in handling, or don't exist in our environment. The latter can be due to their inability to be stopped here or their not being here. The null measurements on niobium spheres which were previously measured to have fractional electric charge imply that the fractional charge was lost prior to the magnetic measurement, or that it was not due to a dyon. If a net magnetic charge is detected on a sample, then its source and abundance are determined and its binding and mass can be studied. The direction of the induced current offset and the sample circulation direction uniquely determine the polarity of the magnetic pole. Once the polarity of g is known, interesting questions such as, "Are monopoles, like protons, preponderantly of one polarity on our vicinity?", "Are there multiply charged monopoles?", and so on, can be investigated.

Induction; Dynamic: The instrument just described was used in the first dynamic induction experiment. Here the passage of a monopole with charge g through the ring would produce an $8\phi_o$ flux change in the ring-SQUID circuit. The device was principally employed in an experiment to measure h/m_e and only saw service as a monopole detector by default. Unexpectedly, an event was recorded on February 14, 1982 at 1:52 PM which was consistent with the passage of a particle with magnetic charge g

to ± 5%[19]. This datum, reproduced in Fig.6, is not easily
attributable to any other probable cause and is unique among
all the other observed events as can be seen in Fig.7. However,
it's possible, but highly improbable that some internal
release of stress could have produced the offset. The experiment ran continuously from February until October 1982
with improvements to the instrumentation being made without
seeing a second event.

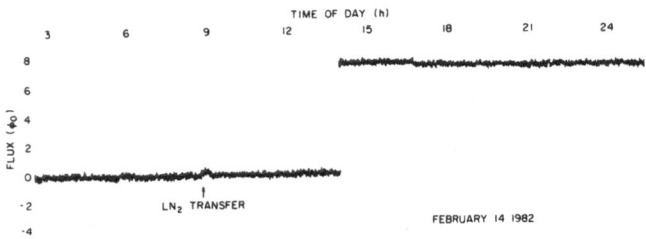

Fig.6: The candidate monopole event induced current change.
(ref. 18)

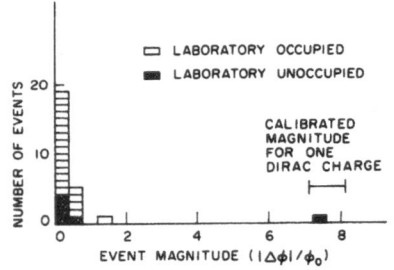

Fig.7: Histogram of persistent
current changes in the single
ring detector during the
period when the candidate
event was found. (ref. 18)

Since August 1982 the Cabrera group has operated a
larger cross section coincidence induction detector with improved mechanical stability. This device, sketched in Fig.8,
consists of three mutually orthogonal rings and has a cross

section integrated over solid angle 7 times that of the single ring detector. This increased area and the fact that 80% of all monopole trajectories would intersect at least two rings will provide a valuable crosscheck on any future event. No event answering to the monopole's description has yet been seen in this detector.

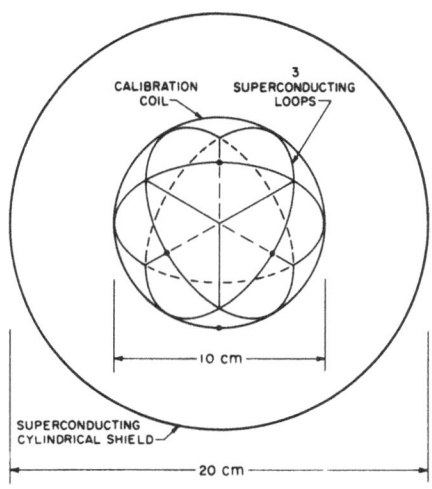

Fig.8: Top view of Cabrera's triaxial detector. (Cabrera in ref.56)

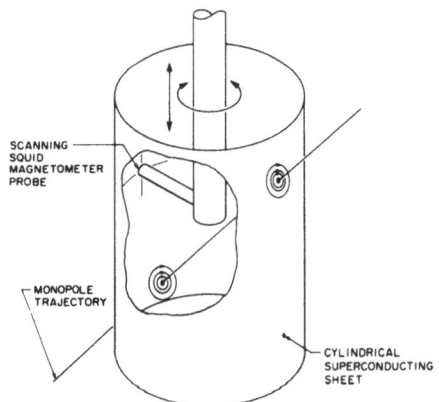

Fig.9: Sketch of Cabrera's scanning SQUID detector. (Cabrera in ref.56)

A third detector now being built by the Cabrera group, shown in Fig.9, will have a one square meter area. This device, a thin superconducting cylindrical shell, will permanently record any doubly quantized vortices produced by the passage of a magnetically charged particle. While the shell remains superconducting, flux pinning due to lattice and surface defects will prevent vortex motion. The trapped background flux pattern, about $1/cm^2$ in a 100 nG field, will be periodically scanned using a small ring-SQUID detector. Subtraction techniques will be used to enhance the sensitivity. There are a variety of other ideas for induction experiments currently being implemented throughout the world that should provide further understanding of the phenomenon that caused Cabrera's candidate event[19].

Now let's summarize what can be learned from dynamic induction experiments. If there are no further cosmic ray events, the flux limit on cosmic ray monopoles in this part of the solar system can be established. However, its value will probably not be reduced much more than 10^3 below that quoted by Cabrera, 6×10^{-10} monopoles/cm^2/s/sr[19]. If more events are found, then not only can the local cosmic ray flux be determined, but so can its variation in time. The absolute value of g, the signal in these experiments, might be found in integer multiples, but the polarity of g cannot be determined by dynamic induction experiments alone. Because of noise restrictions, the SQUID can only be sampled at intervals of tenths of seconds; so coincidence measurements will provide redundancy, but not the monopole's speed. The direction of the current offset will ambiguously determine the monopole direction: polarity as (down: +g or up: -g) or (up: +g or down: -g). A detector like the one Cabrera currently has under construction would determine the monopole's orbit with little information on its time of arrival.

Ionization and Other Techniques: Since the monopole magnetic charge produces a field about seventy times stronger than that of an electric charge, a rapidly moving monopole should have enormously larger electromagnetic energy losses. Implicit in all the null experiments until the mid-seventies was the assumption that the monopole mass was comparable with that of other particles [20] and could attain relativistic velocities. Optical detection of bremsstrahlung, de-excitation, Cherenkov and transition photons which such monopoles would produce in abundance should be a simple matter. Only one experiment looked for the unique polarization from Cherenkov light [21] while all the others sought the photons resulting from ionization[5]. These monopoles should be present in cosmic rays and could be produced by accelerators.

In these ionization searches these experimental components were employed: detector (emulsion, plastics, scintillators, wire chambers, bubble chambers), source (cosmic rays, accelerator, both ancient and new matter) and device (concentrator, extractor). All experiments mixed and matched these generic components. To demonstrate the rich variety of these searches, I will describe three. In 1951 Malkus [22] conducted the first monopole search. He used an electromagnet to concentrate cosmic ray monopoles into a stack of photographic emulsion. Finding no dark thick monopole tracks when he developed his plates, he concluded that the monopole flux was $< 2 \times 10^{-11}$ monopoles/cm^2/s/sr for charge 1-3 g.

A search conducted at Fermilab [23] used a block of metal into which the primary proton beam was dumped as a source. The idea was that monopoles created in the target would lose energy and come to rest there. This target was then subjected to a magnetic field with sufficient strength to extract and accelerate any resident monopoles. A scintillator hodoscope would then detect these monopoles by the anomalously large light output they produced. No monopole signals were seen which indicated that the production cross section of monopoles by 300 and

400 GeV protons was $<6 \times 10^{-42}$ cm^2/nucleon for $M_m < 12$ GeV/c^2 and charge <24 g. This is one of the smallest cross sections of any kind ever measured at an accelerator. The world survey of accelerator searches is presented in Fig.10.

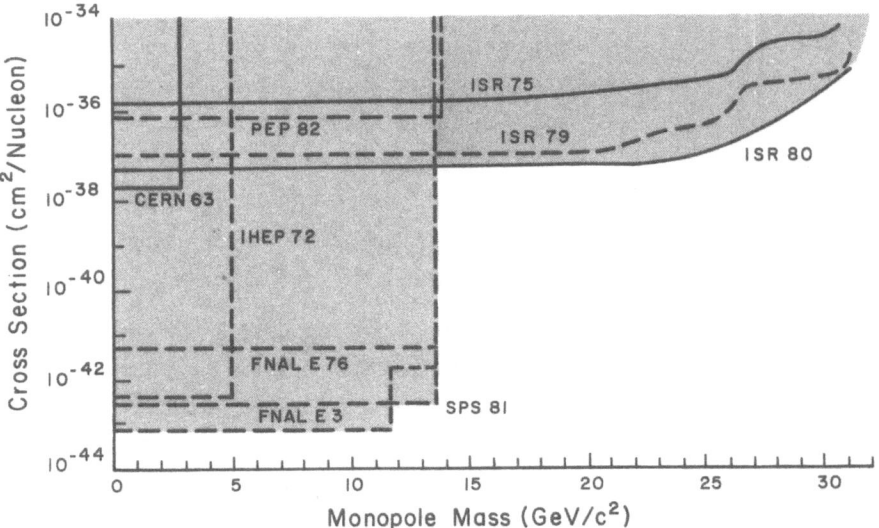

Fig.10: Summary of the limits accelerator searches have placed on the mass and production cross section of magnetic monopoles. Solid lines indicate direct measurements and dashed lines indirect. The shaded region has been explored. (Data is from ref.57)

The third experiment looked at ancient minerals, mica and obsidian, by a beautiful etching technique developed at the General Electric Corporation[24]. Samples were placed in an acid bath which etched the radiation damaged sections of the minerals to produce conical pits. The cone axis orientation indicated the producing particle's direction, the cone angle its charge and the cone depth its energy. By this method and geologically dating the mineral's age, the cosmic ray flux was determined to be $<10^{-19}$ monopoles/cm^2/s/sr for charge

<2 g, independent of the monopole mass.

Experimental searches for magnetic monopoles to date include 18 at accelerators, 15 using cosmic rays as a source and 9 searching in matter [57].Monopoles, however, have been the subject of over 2,400 papers [25].the vast majority of them concerned with the theoretical aspects. Thus, we can calculate a "reality ratio", R,

$$R = \frac{\text{Number of Experimental Papers}}{\text{Number of Theoretical Papers}} < .02 \ , \tag{14}$$

which for monopoles is small.

Experimental With the GUT Idea

The idea that the monopole could be enormously more massive than any of the known particles had profound implications for experiment. Although this possibility was introduced by 't Hooft and Polyakov, in hindsight it could have been deduced from other arguments[20]. Bogomol'yni showed that gauge theories with monopoles could put limits on the monopole mass[26]. The monopole mass must be greater than the mass of the gauge boson associated with the theory divided by the appropriate coupling constant. For an electroweak theory, with 80 GeV/c^2 intermediate bosons and the fine structure constant, $M_m > 10^4$ GeV/c^2. For GUT monopoles, $M_m > 10^{16}$ GeV/c^2.

The obvious consequence of such massive monopoles is that they will not be produced by accelerators. Since they are so heavy, it is unlikely that they can be relativistic. Cosmic ray monopoles would be accelerated by the small vast cosmical magnetic fields to very high energies without approaching the speed of light. Thus, monopoles would not produce Cherenkov, bremsstrahlung or transition radiation.

Matter searches for massive monopoles are unpromising. Those monopoles which were in the solar system at its formation would have been collected by gravity in the interior of the sun, earth and other planets. Monopoles present in lunar sam-

ples and meteors would not survive the journey to earth. The
accelerations they would experience would be sufficient to
dislodge them even from nuclei. Massive cosmic ray monopoles
would have huge penetrating power and so would not be found
in the earth's crust.

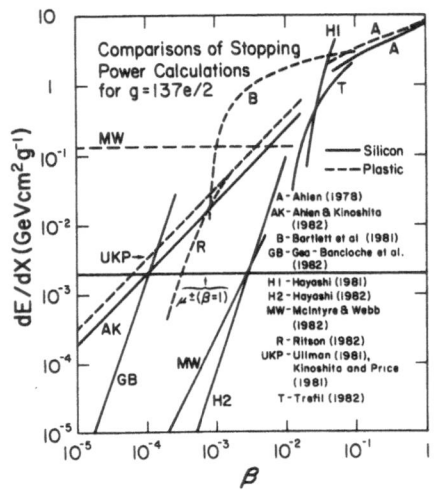

Fig.11: Predicted rate of
energy loss to silicon for protons and magnetic monopoles.
The lines are the predictions
by various authors.(Ahlen in
ref.56)

Atomic de-excitation may hold promise for the detection
of monopoles, but the situation is still unclear. The half
dozen calculations of energy loss[27], shown in Fig.11, at
β of 10^{-3} for a GUT monopole differ by almost five orders of
magnitude. This uncertainity not only bears on the possibility
of detecting monopoles in various media, but also relates to
the possibility of their being trapped in astronomical bodies
since their range will depend on the rate at which they lose
energy to this process.

The origin of this uncertainity in energy loss estimates
lies in the fact that atomic collisions are complicated processes. For relativistic charged particles, these inherent
complications are simplified by arbitrarily bifurcating the
interaction into two soluble classes: close and distant collisions. In close collisions the energy transfers are so much

larger than the electron's binding energy that the electron
is considered free and the impulse approximation is satisfactory. The energy loss is then easily calculated from simple
kinematics and scattering cross sections. In distant collisions, the atom is considered to be excited by the electric
field perturbation of the glancing projectile and the dipole
approximation is used.

These calculations break down when the velocity is
sufficiently small so that most of the collisions no longer
fall into one distinct class. Then a real model for the atom
must be used to understand the dynamics. Further, the actual
calculation requires approximations be made in order to produce
a result. Between models, approximations and human error,
it is understandable that no two theoretical predictions agree.

Recently, energy loss to Zeeman splitting at low velocities in hydrogen has been studied from a fundamental basis
[28].

Even if the energy loss for a slow monopole were opti-
mistically that of a minimum ionizing electrically charged
particle, there would still be problems with its detection.
The typical monopole would produce its photons in a time of
∿ microseconds which would frustrate the standard fast electronics techniques used in particle physics which operates
in the nanoseconds time regime. This difficulty may be overcome by using integrating amplifiers, but not without further
complications. The longer you leave the electronic gate open
the more random noise you also collect. Besides backgrounds
the trigger that is selected to detect monopoles is based
upon assumptions about the energy loss, velocity, direction of
incidence, and so on. This caveat stated, I show a summary of
the cosmic ray monopole experimental results in Fig.12.

An interesting possibility is an induction detector
surrounded by, and providing the trigger for, an ionization detector. Such an arrangement has been proposed to be used with
Cabrera's detectors. If an induction event occurs, there are

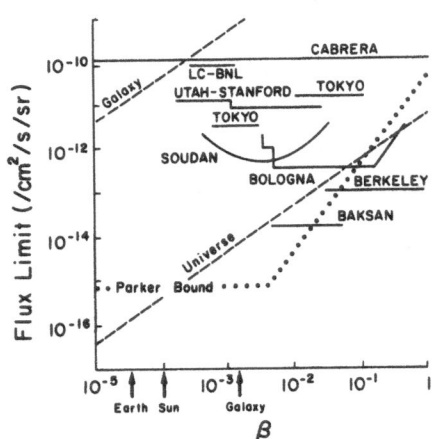

Fig.12: Published monopole cos-
mic-ray flux limits.
The Stanford result corresponds
to one candidate event. All other
detectors relied on ionization.
(ref.1)

two possible outcomes. In the case that no ionization is detected within a few seconds of the induction event, the cre-
dibility of the ionization loss estimates and/or induction
signal are called into question. Perhaps the monopole velocity is much slower than previously thought, implying an even
heavier monopole. If an attendant ionization signal is discovered, the monopole speed and direction could be determined
to give the magnetic charge polarity and general source location. Even one such ionization event would go far in resolving the confusion about the monopole ionization energy loss.

Eddy Currents: A dozen years ago Hofstader[29] suggested and
demonstrated experimentally that electron beams could produce
mechanical oscillations. Despite considerable interest in the
possibilities, no particle detector has been developed using
this principle. The problem has been the poor signal-to-noise
ratio. Only many particles, each with an energy deposition of
approximately 2 MeV/cm can produce a detectable thermoacoustic
wave. Encouraged by early estimates of monopole eddy current
energy losses of 2 GeV/cm in metals, interest in this effect

has revived for use in a monopole detector. However, calculations of the thermal fluctuation pressure suggest that this severely limits the thermoacoustic detection of monopoles for conductive media with temperatures above a few millidegrees [30]. Noise sources attendant to actual acoustic measurements are also discouragingly large. Experimental work exploring acoustical detection is being pursued [31], realizing that the models and approximations in complex problems involving condensed matter are often wrong.

Where Are Monopoles To Be Found?

The sources and sinks of heavy monopoles, unlike ordinary particles, are limited and for the most part inaccesible to experiment and even observation.

Let's start with the earth. Cosmic rays which fall on our planet at the rate of about one per square centimeter per minute can have energies as large as 10^{11} GeV. So cosmic rays can not produce massive monopoles. We know experimentally that monopoles comprise only a tiny part of the cosmic ray flux.

If a sufficiently large amount of material was exposed to cosmic rays long enough; and if that material had a propensity to collect magnetic objects; and if this material were concentrated: then, just maybe, the monopoles it might harbor could be detected. However, this material would have to be handled gently lest or the monopoles become dislodged. Sedimentary iron ore deposits may be such a source. Since millions of tons of surface iron are raised above the Curie point each day to make steel, monopoles falling out of smelters might be detected by induction coils made of sufficient size and sensitivity [32].

Monopoles which could have accreted in the formation of the earth would probably have collected near the core. There they would form in two separated groups by magnetic charge polarity. The earth's periodic magnetic field reversals begin a migration in which these groups would pass through each other. Collisions would result in annihilation contributing to the earth's heat flux [33].Thus, the effects of monopoles could be observed if magnetic field reversals were followed in a constant time by the recession of ice ages.

Iron meteors in their wanderings might act as monopole collectors. A careful capture of such a meteor in space where an unaccelerated search could be conducted might yield monopoles[34]. Besides the obvious technical problems involved in such a search, meteors could well routinely experience forces sufficient to dislodge any hitch hiking monopoles.

The sun may harbor monopoles[35]. If the sun contains 10^{26} monopoles and emits 10^9 monopoles/second over its lifetime, then Cabrera's published flux limit[10] would be plausible. However, the sun must possess a 1 kG interior magnetic field so that monopoles won't fall to the center as they undergo ohmic losses. In this picture the solar flare fields expel monopoles with velocities similar to the earth's orbital velocity, 30 km/s, so that they propitiously form a cloud in the earth's orbit. On the average, these monopoles would have velocities five times smaller than galactic monopoles at the earth's surface. Large temporal flux variations as seen in the solar wind could be present. This model predicts a 22 year mean solar magnetic lifetime, the only suggested mechanism which explains the sunspot cycle. The sun, because of its gravitational potential, would seem to be a good monopole concentrator. But, such does not appear to be the case[36]. Thus, any monopoles in the sun must be primordial.

A few observations of the sun suggests that it might have a magnetic monopole moment[37]. At face value, these measurements are consistent with a net north monopoles abundance of 1.5×10^{29}[38]. The presence of solar monopoles could catalyze fusion of magnetically dipolar nuclei, that is, the $^3He + ^3He$ (but not the $^3He + ^4He$) reaction. This in turn could explain the low solar flux of certain neutrinos[39]. A recent calculation claims that fusion catalysis is not possible[40].

Probably other kinds of stars than our own would be more likely to sequester monopoles. A calculation for neutron stars does not make this prospect very attractive for 10^5 GeV/c^2 monopoles[41].

Monopoles, as all mythological particles, have been suggested as the source of the dark matter which is needed the galaxy from flying apart. A recent calculation shows that a monopole halo could also prevent certain instability to flat galaxies [42].

Central to the question of galactic magnetic monopole abundance is the so-called "Parker bound" [43]. Free monopoles in a magnetic field will neutralize the field since the electric currents that generate the field have to do work on the monopoles, thereby dissipating the currents. The lifetime in seconds of a cosmical magnetical field in the presence of monopoles is,

$$\tau = \frac{B}{8 N_g g v} = \frac{B}{8 g F} \tag{15}$$

where B is the magnetic field, typically 3 microgauss, N_g is the free monopole density in cm^{-3}, and F is the monopole flux. The regeneration time of the galactic field is approximately 30 million years. This implies that the flux limit at the earth is $F < 3 \times 10^{-15}$ monopoles/cm^2/s, or roughly one hundred thousand times smaller than the Cabrera flux. This is, of course, the flux for free monopoles not bound in some way or near a monopole source.

Recently this limit has been reexamined in detail as a function of monopole mass and velocity[44]. Even for monopoles with masses near the Planck mass, 10^{19} GeV, it is difficult to achieve fluxes higher than approximately 10^{-11} monopoles/cm^2/s.

One escape from the Parker bound could be that the galactic magnetic field is due to monopoles[42]. In this picture, clouds of north and south monopoles would oscillate through each other to produce an alternating field or magnetic plasma oscillations. Rather than depleting the magnetic field, the monopoles would transfer energy back and forth from kinetic energy to magnetic field energy. However, if the galactic field were produced by magnetic monopoles, it should have a vanishing curl which contradicts observations. Indeed, the magnetic field configuration and magnitude generally fits the picture of a dynamo. The relatively short field oscillation time for a monopole galactic field would prevent dust grain alignment[45], a process known to exist in the galaxy from star light polarization measurements. The short oscillation time also leads to difficulties in confining cosmic rays. Some of these problems can be ameliorated by different choices of monopoles masses, but these solutions do not seem to fit the general picture of the galactic magnetic field distribution.

Parker bound arguments can be extended to the impact of monopoles on extragalactic magnetic fields. There is evidence for intragalactic cluster fields of the order of 10^{-7}-10^{-8} gauss. Other observations, such as a possible enhancement of high-energy cosmic rays from the Virgo cluster, indicate that extragalactic fields cannot be much larger than this. These low fields can be used to set flux limits in our galaxy 10^3 times lower than the galactic Parker bound[46].

Massive 't Hooft-Polyakov monopoles could only have been produced in the first instants after the creation of the universe in the Big Bang. Standard cosmology and reasonable GUTs suggests that the number of monopoles be roughly the same as the proton number in the universe. This is in clear contradiction to experience[47]. The estimated total protonic mass in the universe is not far from that needed to close the universe. If an equivalent number of monopoles with their enormously larger mass is added, the universe would't be

here now! Various scenarios have been suggested, the inflationary universe aomong them, which look like they will eventually do the trick[48]. However, more theoretical work and
hopefully some observational tests are needed to clarify the
creation and survivability of monopoles in the early universe.

Experiments Depending On GUT Properties

My friends who are practiced in the art of theory tell
me that all of GUT particle physics is rolled up in the 't Hooft
monopole, which is depicted in Fig.13. Near the center (10^{-29} cm)
there is a GUTS symmetric vacuum populated with virtual
grand unification particles. Much further out (10^{-16} cm) the
field is color-electroweak with Z and W bosons, with the
latter being at the edge, while at 10^{-13} cm it is color magnetic with photons and gluons. Beyond nuclear distances it
behaves a usual magnetically charged pole.

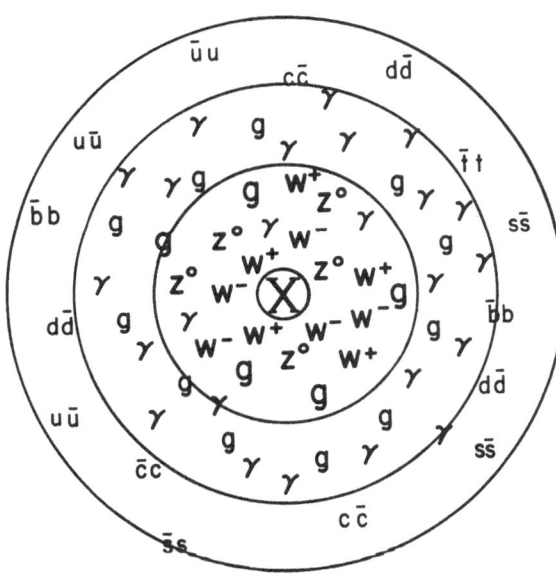

Fig.13: The onion-struc-
ture of the GUT monopole
is illustrated by regions
defined by the degree of
unification and their virtual inhabitants. From
the inside out is the
grand unification core,
the electroweak unification
region, the confinement
region, and fermion-
antifermion condensate.
(ref.1)

10^{-29} 10^{-16} 10^{-15} 10^{-13}

Radius (cm)

This view of the GUT monopole raises the possibility that when it passes through matter, a collision with a proton may occur in which the core region is overlapped causing the proton to spontaneously decay into a pion and a positron. In other words, proton decay, which must ordinarily be exceedingly rare, may be catalyzed in the presence of a monopole. The possibility of observing this effect, called baryon catalysis or the Rubakov-Callan effect[49], has attracted much interest. Theoretical opinions on the possibility of catalysis are widely divergent.

Monopole baryon catalysis offers an interesting possibility for terrestrial as well as astronomical observations. In grand unification, the proton lifetime is expected to be about 10^{31} years so that its detection requires large amounts of matter. On the other hand, a monopole passing through a proton decay detector could catalyze with a strong interaction cross section, inducing some tens of decays. Since monopoles are expected to be slow, all the detector electronic gates must be kept open during the monopole passage, typically for fifty microseconds. The upper limits on baryon catalysis are now being produced by large proton decay detectors on the basis of non-observance of that decay[50].

Neutron stars should be relatively good monopole collectors[51]. Once inside a star, catalyzing monopoles could transform nucleons at such a rate that the resulting x-ray luminosity would exceed by many orders of magnitude the measured upper limits of neutron star x-ray luminosities. This implies a monopole flux, assuming hadronic cross sections for catalysis, less than 5×10^{-22} monopoles/cm^2/s/sr, much lower than any other flux limit. A similar analysis has been made for a peculiar subclass of A stars [52].

The GUT monopole-antimonopole system, monopolonium, would produce a spectacular display[53]. Its lifetime would range from days to a hundred billion years. At a separation of an angstrom the monopoles' orbital velocity would be 1 cm/sec, the principal quantum number about a trillion and the binding energy ∿40 keV.

Monopolonium decays by classical Larmor radiation for all but the last 10 seconds of its life. It de-excites first by emitting radio, then successively light, x-rays, γ-rays, quarks and gluons, intermediate bosons, and ultimately 10^{14} GeV/c^2 X and Y particles are emitted altogether. Typically some tens of millions of particles are emitted altogether.

Monopolonium radiation at a wavelength of 1 cm is expected to give a flux of 10^{-24} eV/cm^2/s/Hz. Unfortunately, the current observational limits are 3×10^{-16} eV/cm^2/s/Hz corresponding to 50 microJanskies.

MONOPOLES: THE CODA

MONOPOLE

Fig.14: Zerelov's magnetic monopole. (ref.55)

The magnetic monopole is a fascinating object which has obviously captivated the imaginations of many species of physicists. Perphaps this poem found in the Schladming Music School by Professor Mitter best catches the essence of the monopole for those of us who search for it:

> O Menschenherz, was ist Dein Glück?
> Ein rätselhaft geborner
> und, kaum gegrüßt, verlorner
> unwiederholbarer Augenblick!
>
> N. Lenau[54]

Then again perhaps the monopole Zerelov[55] envisions in
Fig.14 is closer to reality.

EXPERIMENT AND GAUGE THEORY

The success of gauge theory has been overpowering. This
has been both good and bad for the practice of particle
physics. Lets take a look at the good first.

Electroweak gauge theory, $SU(2)_L \times U_{em}(1)$, has had a
profound impact on experiment since the prediction and experimental discovery of the weak neutral current. The constancy
of the Weinberg angle determined by a variety of different
experiments and the calculation of the $W^\pm$ and Z masses have
been a major motivation in the marshalling of enormous resources, both material and manpower, specifically aimed at
detecting these messengers of the weak force. The recent an-
nouncement of the discovery of a few $W^\pm$ bosons at CERN has
further strengthened the domination of this gauge theory on
weak interaction experiments.

The chaos of elementary objects which was organized by
$SU(3)_c$, the gauge theory of strong interaction, brought relief
to a decade of confusion in the bubble chamber groups all over
the world. The prediction of the Ω^- by this theory only
further convinced us of its power. The association of the
parton with the quark and the display of jets further solidified
the hold which this theory had on experiments. The successive
appearance of phenomena attributable to new quarks, first
the $J/\psi(c\bar{c})$, then the $\Psi(b\bar{b})$, and now a hint of the "t"
at CERN has given this theory a virtual monopoly over the

strong interaction physics experiments performed at accelerators.

With spirits thus renewed after long years in the wilderness of uncertain calculation, theorists naturally have tried to effect a unification of these two successful approaches in GUTs, the simplest form of which is SU(5). Although no prediction of this theory has been experimentally demonstrated, it has encouraged a migration of particle experimentalists underground, to the ocean and even back to nuclear reactors.

Now the world might really be as our gauge theorist friends tell us and at this time we have no reason to doubt them. However, the increasing dominance by these theories of our professional life has produced profound changes in the practice of particle physics. With the emphasis on rare phenomena, more specialized facilities are being built at larger costs and involve immense groups of physicists in experiments which take ever longer to complete. Thus, in this increasingly industrialized environment there is less variety, flexibility and spontaneity in the choice of physics to study. The machinery of funding agencies and program committees further reduce the possibility of looking for the unexpected; and, if the unexpected is seen, to encourage that it be ignored if it doesn't rest easily in the framework of gauge theory.

It is against this background that I define four classes of experiments by their relation to the current gauge theory. By examining a couple of typical phenomena in each class I will then attempt to show that there is life in physics beyond gauge theory.

Experiments Which Are Important Inspite Of Gauge Theory: Two experiments on which a gauge theory, in these cases GUTs, makes very specific predictions are the existence of magnetic monopoles and the decay of the proton, the search for which Professor Fiorini has discussed eloquently in his lectures[50]. But the existence of the monopole and nucleon instability are fundamentally interesting independent of gauge theory. Experiments into these matters were being conducted long before gauge theory had metastasized from its humble beginnings in QED to claim all of particle physics. So although the orthodoxy has placed its benediction on these experiments they would be done, albeit not with the present urgency, in the absence of gauge theory.

Experiments On Which Gauge Theory Is Indifferent: The experiments I use as examples here, double beta decay and neutron-antineutron oscillation have again been described in detail by Fiorini[50]. These Experiments, like those in the previous class, have a long creative past independent of gauge theory and will continue to survive the shifting winds of theoretical fashion again because they are of fundamental importance and therefore merit study.

Experiments At Odds With Gauge Theory: The credo of confinement, that quarks can not exist outside the safe harbor of a particle whose pedigree has been registered in the Particle Data Book, [57],has yet to be proved theoretically. A single, albeit difficult, experiment has persistently and publically proported to have measured fractional electric charge in the amounts thought to be carried by quarks[17]. This experiment has occasioned bewilderment in the best, scoffing among the rest but has inspired few to become agnostics. Indeed funding to continue this experiment has been perilious and new proposed searches for fractional charge have received, if any, grudging support.

Let me review briefly the Fairbank experiment and bring you up to date on their continuing efforts to confirm or impune their previous results. The device, sketched in Fig.15. levitates a superconducting niobium ball which it then oscillates horizontally. From the oscillation frequency of the resident ball its electric charge is derived. This procceedure requires such technical virtuosity to implement that it unfortunately has no immitators. The sum total of all the Fair-

bank group's results to date are shown in Fig.16. For the
past two years the experiment has continuously modified and
should start its next measurement cycle this summer enhanced
by two significant new features. One is the use of niobium-
iron balls which it is hoped can also be measured in the room
temperature experiments of Ziock and Morpurgo. The second
is a measurement proceedure suggested by Alvarez in which a
randomly chosen arbitrary offset is added to the oscillation
pattern data by an uninvolved third party. After all the experimental corrections are made to these data, the ball's
charge is determined when the unknown constant is supplied.

Currently several new experiments searching for uncon-
fined fractional charge are being built. At Argonne[59] an experiment is under way using ink drops spewed horizontally and
falling under gravity while being deflected by an electric
field in the direction of their initial motion. The electric
field will sort the drops out by their net electric charge. As
many very similar drops can be made, a great deal of material
can be examined with considerable accuracy.

At Atom Sciences in Oak Ridge[60] a single atom counting
experiment is beginning to look for quarked sodium atoms extracted from sea water.

Another possibility is finding evidence for quarked
helium molecules. Helium is produced by three processes. Alpha
decay of heavy nuclei, the first process, produces the commer-
cially available helium in well gas. A quarked alpha particle
would soon acquire two or three electrons and the resulting
atom might form an exotic molecule if it didn't stick to the
well walls. Since accelerator experiments have been conspicu-
ously unsuccessful in producing quarks in collisions, alpha
decay is not a promising source of quarked helium. Quarked
helium produced in fusion reactions in stars, the second process, would probably not survive since the matter density is
so great that it would soon be neutralized in a collision with
a complimentarily charged nucleus. Most primordial quarked

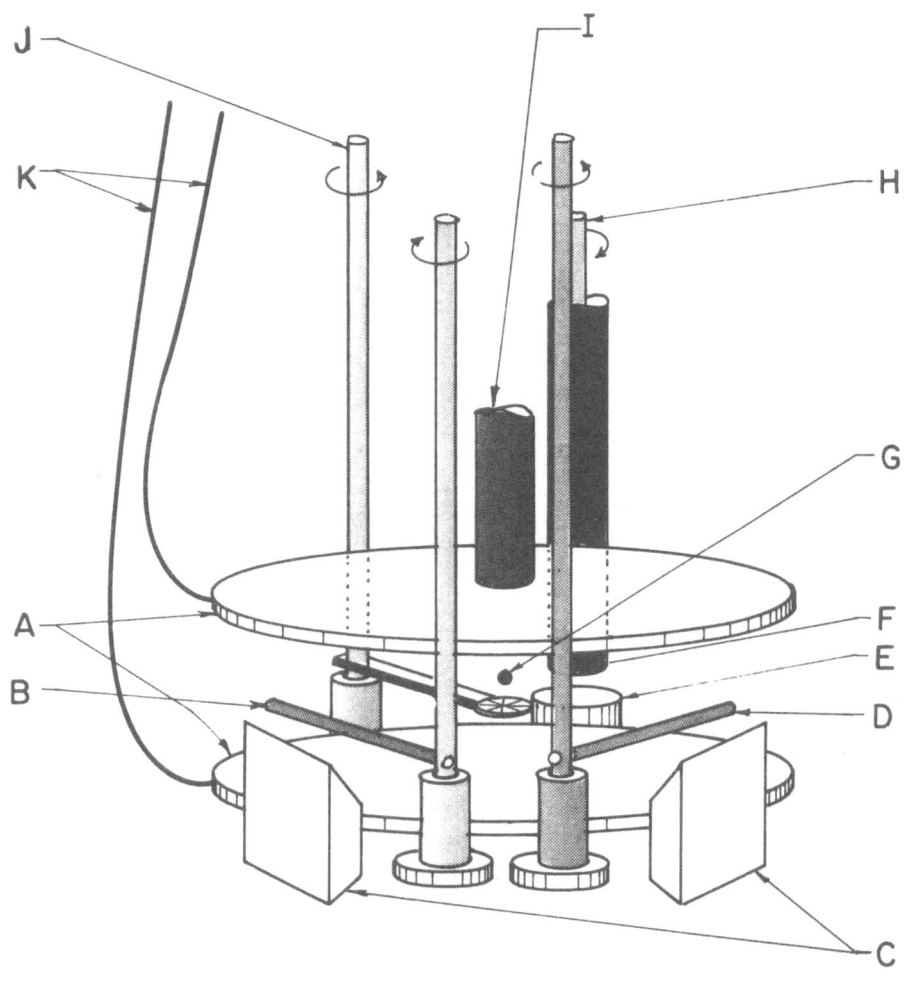

Fig.15: Schematic of the Stanford fractional electric charge
experimental apparatus showing the capacitor plates
(A), the electron (B) and positron (D) emitters, the
observation optics (C), the ball loading system (E,F,
H,J), the niobium ball (G), the SQUID (I) and the
voltage leads (K), (ref.58)

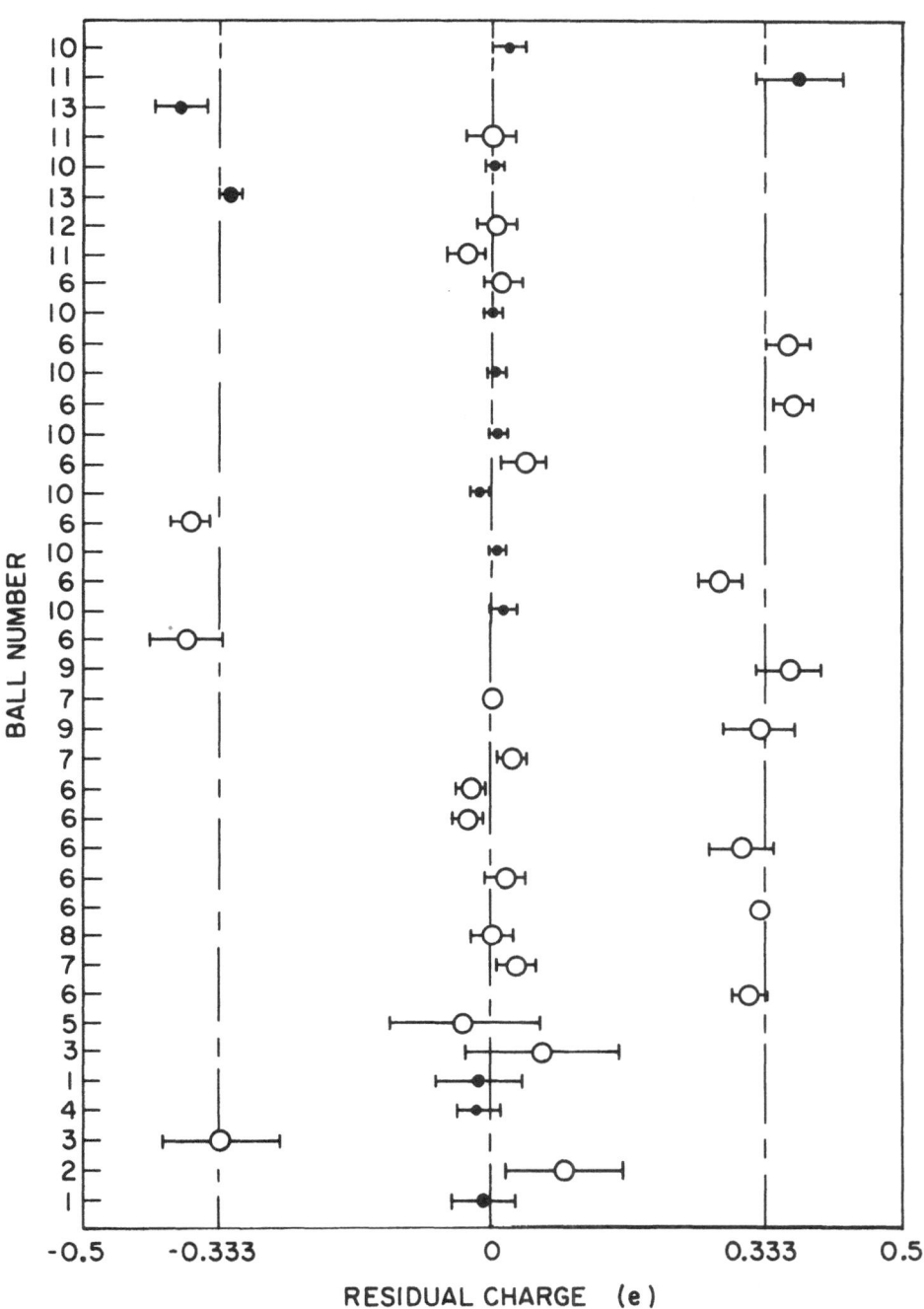

Fig.16: A summary of the Stanford measurements of the net electric charge on niobium balls. (ref.17)

helium produced in the Big Bang, the third process, would have been neutralized as it cycled through supernovas. However, some pristine intergalactic clouds may have survived. If so, then radio astronomy techniques could be used to search quarked helium spectra if the abundance was a few parts in 10^4 and if the line frequencies were known.

A second example, an experiment whose positive results would be at odds with gauge theory, is underway in Pavia[61]. Here we are looking for beta decay of medium mass nuclei far from the stability line with a few nanosecond half lives. No reasonable theory of weak interactions permits such decays. The super allowed ($0^+ \to 0^+$) beta decay of mirror nuclei in Fermi theory is characterized by the product of the half-life and a function of the spectrums end point energy which has been shown to be constant[62]. The largest measured end point energy is 7.2 keV and the shortest lifetime is .2 seconds.

In the present experiment a Californium 252 source is used to produce neutron rich fission fragments [63] whose energy and mass can be determined by their time-of-flight and energy disposition in a silicon detector. The fragment flight from the target to the silicon detector is viewed by a silicon-germanium detector telescope located at a right angle to the path. If the fragment decays and if the decay product enters the telescope, its nature and energy will be determined. Crude preliminary results shows counts in the telescope whose origins are currently being investigated.

Experiments On Which Gauge Theory Is Moot

During the nineteen fifties five events were seen by three different groups in emulsion stacks flown in ballons which have defied explanation in terms of known physics[64]. They were unusual in that only electron-positron pairs were present. These pairs appeared in great numbers in a small cone and each pair had a small opening angle indicating a very energetic process, certainly above a few TeV. Finally, each event began deep in the emulsion with no causal particle track.

A decade and a half later it was proposed that these unusual events could have been the results of the photoproduction of a virtual monopole-antimonopole pair[65]. The photon shower which pair produces electrons was the consequence of bremsstrahlung and annihilation radiation. Accelerator searches have so far failed to reproduce these pure multiphoton showers [57] and a satisfactory explanantion has not yet been found for them. The complimentary process, the so-called Centauro events[66], produce pure hadronic showers free of electrons and neutral pions. The energy of the primary is estimated to be about 1000 TeV. Centauro events have not been reproduced at accelerators, satisfactorily explained or discredited.

The gravitational interaction of antimatter with matter is widely believed to be the same as for the matter-matter interaction. However, there is no experimental evidence to support this dogma nor does gauge theory address this question. A single very difficult experiment, being conducted at Stanford to look at how positrons and electrons fall in the earth's field[67], has not produced a definitive result. Because of their greater mass protons and antiprotons would be a better subject for this study. With the advent of strong antiproton sources such as LEAR at CERN this very important measurement may be realized before the decade is out[68].

My conclusion from his idiosyncratic survey of particle physics experiments in terms of gauge theory, is that we must contrive to pursue fundamental measurements and not give the current gauge theoretic imperatives exclusive use of our resources. In this way experiment can best contribute to the understanding of the universe as it is, which indeed may be just as gauge theory tells us. It will also make possible an alternate professional life style to that present at accelerator laboratories which should enrich our practice of particle physics by giving it diversity.

ACKNOWLEDGEMENTS

I have benifited much from discussing monopoles with Qaisar Shafi, John Ficenec, Dick Carrigan and Blas Cabrera. I thank Luis Alvarez for introducing me to this subject when I was a student. Discussions with Andy Pickering have influenced my awareness of the impact of gauge theory on experiment. This work is supported in part by grants from the Jeffress Trust and the National Science Foundation.

REFERENCES

1. For a comprehensive status report see R.A. Carrigan, Jr. and W.P. Trower, FERMILAB-83/31 (1983).

2. The Letter of Petrus Peregrinus on the Magnet, A.D. 1269 translated by Brother Arnold (McGraw, New York, 1904).

3. C.W.F. Everitt, private communication.

4. J.D. Jackson, CERN-77-17 (1979) shows that this is still the case.

5. B. Cabrera and W.P. Trower, Found. Phys. 13 (1983) 195.

6. P.A.M. Dirac, Proc. Roy. Soc. London, Ser. A, 133 (1931) 60 and Phys. Rev. 74 (1948) 817.

7. J. Schwinger, Phys. Rev. 144 (1966) 1087; 173 (1968) 1536; and Science 166 (1969) 797.

8. G. 't Hooft, Nucl. Phys. B79 (1974) 276 and 105 (1976) 538; and A.M. Polyakov, JETP Lett. 20 (1974) 194.

9. E. Amaldi, in Old and New Problems in Elementary Particles, edited by G. Puppi (Academic, New York, 1978), p. 1, contains a good account of these effects.

10. S.A. Bunyatov, Sov. J. Part. Nucl. 10 (1979) 259.

11. V.P. Martem'yanov and S.Kh. Khakimov, Sov. Phys. JETP 35 (1972) 20.

12. I. Frank and V. Ginsburg, J. Phy. (Moscow) 9 (1945) 353; G.M. Garibian, Sov. Phys. JETP 6 (1958) 1079; 10 (1960) 372.

13. E. Goto, Prog. J. Phys. Soc. Japan. 13 (1963) 1413; D.

Sivers, Phys. Rev. $\underline{D2}$ (1970) 2048; and L. Bracci and G. Fiorentini, Pisa Preprint IFUP TH-83/2 (1983).

14. L.W. Alvarez, Lawrence Radiation Laboratory Physics Note 470, (1963)(unpublished); L.J. Tassie, Nuovo Cimento $\underline{38}$ (1965) 1935; and L. Vant-Hull, Phys. Rev. $\underline{173}$ (1968) 1412.

15. L.W. Alvarez, P.H. Eberhard, R.R. Ross and R.D. Watt, Science $\underline{167}$ (1970) 7C1; Phys. Rev. $\underline{D4}$ (1971) 3260; Phys. Rev. $\underline{D8}$ (1973) 698; and P.H. Eberhard, R.R. Ross, J.D. Taylor, L.W. Alvarez and H. Oberlack, Phys. Rev. $\underline{D11}$ (1975) 3099.

16. B. Cabrera (Ph. D Thesis, Stanford, 1975) unpublished.

17. G.S. LaRue, J.D. Phillips, and W.M. Fairbank, Phys. Rev. Lett. $\underline{46}$ (1981) 967 and references therein.

18. B. Cabrera, Phys. Rev. Lett. $\underline{48}$ (1982) 1378.

19. C.C. Tsuei in ref. 56.

20. R.A. Carrigan, Jr., Nuovo Cimento $\underline{38}$ (1965) 638 implied and A.S. Goldhaber, private communication, demonstrates that this assumption could have been seen to be wrong from classical arguments.

21. V.P. Zrelov, L. Kollarova, D. Kollar, V.P. Lupiltsev, P. Pavlovic, J. Ruzicka, V.I. Sidorova, M.F. Shabashov and R. Janik, Czech. J. Phys. $\underline{B26}$ (1976) 1306.

22. W.V.R. Malkus, Phys. Rev. $\underline{83}$ (1951) 899.

23. R.A. Carrigan, F.A. Nezrick and B.P. Strauss, Phys. Rev. $\underline{D8}$ (1973) 3717; and Phys. Rev. $\underline{D10}$ (1974) 3867.

24. R.L. Fleischer, P.B. Price and R.T. Woods, Phys. Rev. $\underline{184}$ (1969) 1398.

25. From bibliographic compilations of D.M. Stevens, VPI-EPP-73-5 (1973); R.A. Carrigan, Jr., FERMILAB-77/42 (1977); R.E. Craven, W.P. Trower and R.A. Carrigan, Jr., FERMILAB-81/37 (1981); and R.E. Craven and W.P. Trower, FERMILAB-82/96 (1982).

26. E.B. Bogomol'yni, Sov. J. Nucl. Phys. $\underline{24}$ (1976) 449.

27. S.P. Ahlen in ref. 56; and S.P. Ahlen and K. Kinoshita Phys. Rev. $\underline{D26}$ (1982) 2347.

28. S.D. Drell, N.M. Kroll, M.T. Mueller, S.J. Parke and M.A. Ruderman, Phys. Rev. Lett. $\underline{50}$ (1983) 644.

29. B.L. Beron and R. Hofstader, Phys. Rev. Lett. 23 (1969) 184.

30. C.W. Akerlof, Phys. Rev. D26 (1982) 1116; and 27 (1983) 1675.

31. B.C. Barish, in ref. 56.

32. D. Cline in ref. 56.

33. R.A. Carrigan, Nature (London) 288 (1980) 348.

34. G. 't Hooft, private communication.

35. S. Dimopoulos, S.L. Glashow, E.M. Purcell and F. Wilczek, Nature (London) 298 (1982) 824.

36. K. Freese and M.S. Turner, Chicago-EFI-82-56 (1982).

37. J.M. Wilcox, Comments Astrophys. Space Phys. 4 (1972) 141.

38. L.W. Alvarez, private communication.

39. R. Davis, Jr., Proc. Brookhaven Solar Neutrino Conf., BNL 50879 (1978) p.1.

40. J.S. Trefil, H.P. Kelly and R.T. Rood, Nature 302 (1983) 111.

41. M. Bonnardeau, Phys. Rev. D23 (1981) 323.

42. E.E. Saltpeter, S.L. Shapiro, and I. Wasserman, Phys. Rev. Lett. 49 (1982) 1114.

43. E.N. Parker, Astrophys. J. 160 (1970) 383; E.N. Parker, Cosmical Magnetic Fields (Clarendon, Oxford, 1979).

44. M.S. Turner, E.N. Parker and T.J. Bogdan, Phys. Rev. D26 (1982) 1296.

45. E.M. Purcell, in ref. 56.

46. Y. Rephaeli and M.S. Turner, Chicago -EFI-82-37 (1982).

47. J.P. Preskill, Phys. Rev. Lett. 43 (1979) 1365.

48. Q. Shafi, lectures in this volume.

49. V. Rubakov JETP Lett. 33 (1981) 644 and Nucl. Phys. B203 (1982)311; and C.G. Callan, Phys. Rev. D25 (1982) 2141 and Nucl. Phys. B203 (1982) 311.

50. E. Fiorini, lectures in this volume.

51. E.W. Kolb, S.A. Colgate and J.A. Harvey, Phys. Rev. Lett. 49 (1982) 1373.

52. D.M. Ritson, SLAC-PUB-2977 (1982).

53. C. Hill, in ref.56.

54. An english translation is:

Human heart, what is your happiness?

A short moment

born out of an enigma

lost already, when just welcomed

and unrepeatable.

55. J. Ruzick and V.P. Zrelov, JINR 61-2-80-850 (1980).

56. Magnetic Monopoles, edited by R.A. Carrigan, Jr. and W.P. Trower (Plenum New York, 1983).

57. M. Roos et al., Phys. Lett. B111 (1983) 111.

58. S. Felsh, Am. J. Phys. to be published.

59. R. Hagestrom, private communication.

60. W.M. Fairbank, Jr., private communication.

61. T. Pinelli, private communication.

62. D.H. Wilkinson and D.E. Alburger, Phys. Rev. C13 (1976) 2517.

63. H.W. Schmitt, W.E. Kiker and C.W. Williams, Phys. Rev. B137 (1965) 837.

64. G.B. Collins, J.R. Ficenec, D.M. Stevens, W.P. Trower and J. Fischer, Phys. Rev. D8 (1973) 982 summarize these events.

65. M.A. Ruderman and D. Zwanziger, Phys. Rev. Lett. 22 (1969) 146.

66. Brazil-Japan Collaboration, in Proc. Plovdiv International Conf. Cosmic Rays. 7 (1977) 208.

67. E.C. Whitteboron and W.M. Fairbank, Phys. Rev. Lett. 17 (1967) 1049.

68. T. Goldman and M.M. Nieto, Phys. Lett. B122 (1982) 437.

Acta Physica Austriaca, Suppl. XXV, 143–144 (1983)

EXPERIMENTS ON BARYON AND LEPTON NON-CONSERVATION[+]

by

E. FIORINI

Dipartimento di Fisica dell'Università and

I.N.F.N. - Milano

ABSTRACT

The recent theories which try to unify electroweak and
strong interactions imply in most cases non conservation of
the baryon and lepton numbers. Baryon non conservation can
be investigated in a model independent way by searching for
radioactive remnants or for abnormal isotopic abundance con-
sequent to the decay of a proton or a neutron inside the nucleus. Since the sensitivity of these methods is for the moment
quite low (up to 10^{27} years as lower lifetime for nucleon
decay) more emphasis is presently devoted to "direct" experiments where nucleon stability is searched with huge detectors placed deep underground to reduce strongly the background
due to cosmic rays. Two experiments are at present in operation where the detector is essentially a calorimeter made
by plates of Iron interleaved with planes of detecting tubes
(proportional and limited streamer tubes). In the former of
these experiments, running since two years in southern India,
three nucleon decay candidates have been found, corresponding
to a nucleon decay lifetime of about 10^{31} years. In the latter,
carried out by an Italian collaboration in the Mont Blanc
tunnel, many events totally confined inside the detector have

---

[+] Lectures given at the XXII. Internationale Universitätswochen
für Kernphysik,Schladming,Austria,February 23–March 5,1983.

been observed. One of them, hard to be interpreted, unlike
the others, as a neutrino interaction, could be due to a proton decay. Its lifetime would range between 2 and 3×10^{31} years.
A very large detector based on the detection of the Cerenkov
light emitted by the charged secondaries of nucleon decays
has up to now provided no evidence for proton decay into a
positron and a neutral pion. An alternative way to investigate
nucleon stability is to search n-n̄ oscillations with thermal,
cold or ultracold neutrons from a reactor. The only running
experiment, at the Grenoble reactor, shows no evidence for
such oscillations and yields a limit of 10^6 sec for the free
neutron oscillation time.

The problem of non conservation of lepton number is
connected with the existence of a massive neutrino, as suggested by a recent russian experiment. There is however no
evidence at nuclear reactors or at particle accelerators of
neutrino oscillations which one would like to assume as a natural consequence of a finite neutrino mass. Moreover electron
and muon leptonic numbers seems to be separately conserved.
The more sensitive tool to investigate non conservation of
the total lepton number, double beta decay in the neutrino-
less mode, has also indicated no violation. The obtained experimental limits seem hardly compatible with the results of
the russian group on the non vanishing mass of the electron
neutrino.

REFERENCES

1. H.H. Williams, Grand unification, proton decay and neutrino
   oscillations-Talk presented at the SLAC Summer Institute on
   particle physics, August 16-24, 1982.
2. E. Fiorini, Phil. Trans. R. Soc. London A304 (1982) 105.
3. H. Primakoff and S.P. Rosen, Ann. Rev. of Nuclear Sci. 31
   (1981) 145.
4. L. Zanotti, Double beta decay - Invited paper given to
   ICOMAN 83, Frascati January 17-21, 1983.

Acta Physica Austriaca, Suppl. XXV, 145–248 (1983)
© by Springer-Verlag 1983

REVIEW OF GUTs AND SUSY-GUTs[+]

by

G. ROSS
Rutherford Lab.

1. INTRODUCTION

Over the last decade remarkable progress has been made
in understanding the strong, weak and electromagnetic
interactions. The most successful theory we have is quantum
electrodynamics, a theory for electromagnetism. It has been
tested to great accuracy. For example the prediction for the
magnetic moment of the electron is 1.0011596553 in natural
units and experiment gives a value of 1.0011596524, both
with uncertainties ±0.0000000030. Quantum electrodynamics
is a gauge field theory and may be derived from the requirement that electric charge is locally conserved.

It now seems likely that the theories for the strong
and the weak forces may also be gauge field theories derived from the requirement of local invariances under new
symmetries. For the strong interactions the symmetry group
is SU(3) acting in the colour quantum number of quarks. For
the weak and electromagnetic interactions the symmetry group
is that introduced by Glashow, Salam and Weinberg [1] namely
SU(2)×U(1) which contains charge conservation. Together the
gauge field theory built on the symmetry SU(3)×SU(2)×U(1)
has come to be known as the "standard" model and its predic-

[+]Lectures given at the XXII.Internationale Universitätswochen
 für Kernphysik,Schladming,Austria,February 23-March 5,1983.

tions are entirely consistent with all experimental data.

If the strong, weak and electromagnetic interactions
are separately described by gauge field theories, it is natural to ask whether they are all related. In grand unified
theories this idea is explained to the full by embedding
$SU(3) \times SU(2) \times U(1)$ in a semisimple group G (e.g. $SU(5)$) with
a single coupling constant g_{GU}. The strong, weak and electromagnetic interactions are then seen to be different facets of
the same fundamental interaction based on a field theory with
local gauge invariance under G. The group G is spontaneously
broken at a scale M_X to $SU(3) \times SU(2) \times U(1)$ and the strong weak
and electromagnetic couplings are related to g_{GU} by radiative
corrections. Remarkably the predictions for these couplings
agree with experiment provided the scale M_X is very large
($\simeq 10^{15}$ GeV in $SU(5)$); the reason such a large scale arises
is that the radiative corrections needed to get agreement
depend only logarithmically on M_X. However a large mass scale
turns out to be essential for in $SU(5)$, and in most GUTs, the
new gauge bosons mediate novel processes, in particular proton decay. To avoid violating the current lower bound on
the proton lifetime of $O(10^{30}$ years) the mass M_X of these
new gauge bosons must be $\gtrsim O(10^{15}$ GeV).

The appearance of such a large scale gives rise to a
serious problem for GUTs, the "hierarchy" problem. In these
theories the natural size for the weak interaction breaking
scale M_W is $O(M_X)$ yet the actual value needed for M_W/M_X is
$O(10^{-13})$. To achieve this parameters in the theory must be
tuned to an accuracy of one part in 10^{13}, and no one has given
a reason why this should be so within the framework of $SU(5)$.
Recently there has been much interest in a generalisation of
GUTs which avoids this problem through the introduction of
a new symmetry, supersymmetry, which guarantees that M_W
should be small. Grand unified theories with supersymmetry,
SUSY-GUTs, have been constructed and provide the first self-
consistent GUTs. A particularly interesting feature of these
models is that they require a new set of states, supersymmetric

partners of the observed states, which must be relatively light ($O(M_W)$) and should be observable with the new machines such as LEP.

In these lectures I will discuss the status of GUTs and SUSY-GUTs. In section 2 the successes and failures of the standard model are reviewed. Section 3 introduces GUTs and discusses the minimal SU(5) version. Section 4 discusses the classic predictions of SU(5) and gives a critique of its achievements. In section 5 some of the possible generalisations of SU(5) are introduced. Finally section 6 and 7 discusses supersymmetric grand unification in the globally supersymmetric case (SUSY GUTs) and in the locally supersymmetric case (SUGRA GUTs) respectively.

2.1 The Standard Model

Building on the local gauge principle, gauge theories for the weak, electromagnetic and strong interactions have been constructed. QCD, the theory for the strong interactions is based on the gauge group SU(3), which transforms the colour quantum number carried by all strongly interacting particles. The Glashow, Salam and Weinberg model [1] of the weak and electromagnetic interactions is based on the gauge group SU(2)×U(1) which transform weak isospin and hypercharge. Together the SU(3)$_c$×SU(2)×U(1) model, or (3,2,1) model, provides a potentially complete description of the strong, weak and electromagnetic interactions. It already has much experimental evidence in favour of it.

The structure of the model is given by the Lagrangian density

$$L_{(3,2,1)} = L_{kin} + L_{Yuk} + L_{scalar}$$

where the kinetic term L_{kin}, describes the kinetic energy of the gauge and matter fields and through the local gauge gauge principle the coupling of the gauge bosons.

$$L_{kin} = \sum_{fermions\ j} i\bar{\psi}_j (\partial_\mu - ig_3 A^a_\mu \frac{\lambda^a}{2} - ig_2 W^b_\mu \frac{\sigma^b}{2} - ig_1 B_\mu Y)\gamma^\mu \psi_j$$

$$+ \sum_{scalars\ k} | (\partial_\mu - ig_2 W^b_\mu \frac{\sigma^b}{2} - ig_1 B_\mu Y)\phi_k |^2$$

$$- \frac{1}{4} F^a_{\mu\nu} F^{\mu\nu}_a - \frac{1}{4} W^j_{\mu\nu} W^{\mu\nu}_j - \frac{1}{4} B_{\mu\nu} B^{\mu\nu} \quad , \tag{2.1}$$

where $\frac{\lambda^a}{2}$ and $\frac{\sigma^b}{2}$ represent the generators of SU(3) and SU(2) respectively, and the kinetic terms for the gauge bosons involve the gluon field strength

$$F^a_{\mu\nu} = \delta_\mu A^a_\nu - \delta_\nu A^a_\mu + g_3 f^{abc} A^b_\mu A^c_\nu \quad , \qquad a = 1 \ldots 8$$

and the W and B field strengths

$$W^j_{\mu\nu} = \partial_\mu W^j_\nu - \partial_\nu W^j_\mu + g_2 \epsilon^{jkm} W^k_\mu W^m_\nu \quad ,$$

$$B_{\mu\nu} = \partial_\mu B_\nu - \partial_\nu B_\mu \quad . \tag{2.2}$$

Y is chosen to satisfy

$$Q = \frac{\sigma_3}{2} + Y \quad . \tag{2.3}$$

Fermions appear to be grouped in three families (the e, μ and τ families) with SU(3)×SU(2) content

$$(3,2) + 2(\bar{3},1) + (1,2) + (1,1) \tag{2.4}$$

as shown in Fig. 1.

L_{Yuk} describes the coupling of matter fermions to scalars in the theory. It is needed to introduce masses to the quarks and leptons, for the gauge interactions of L_{kin} preserve the chirality of quarks and leptons, while mass terms mix chirality. For the minimal (3,2,1) model it is possible to give quarks and leptons mass with a single doublet ϕ.

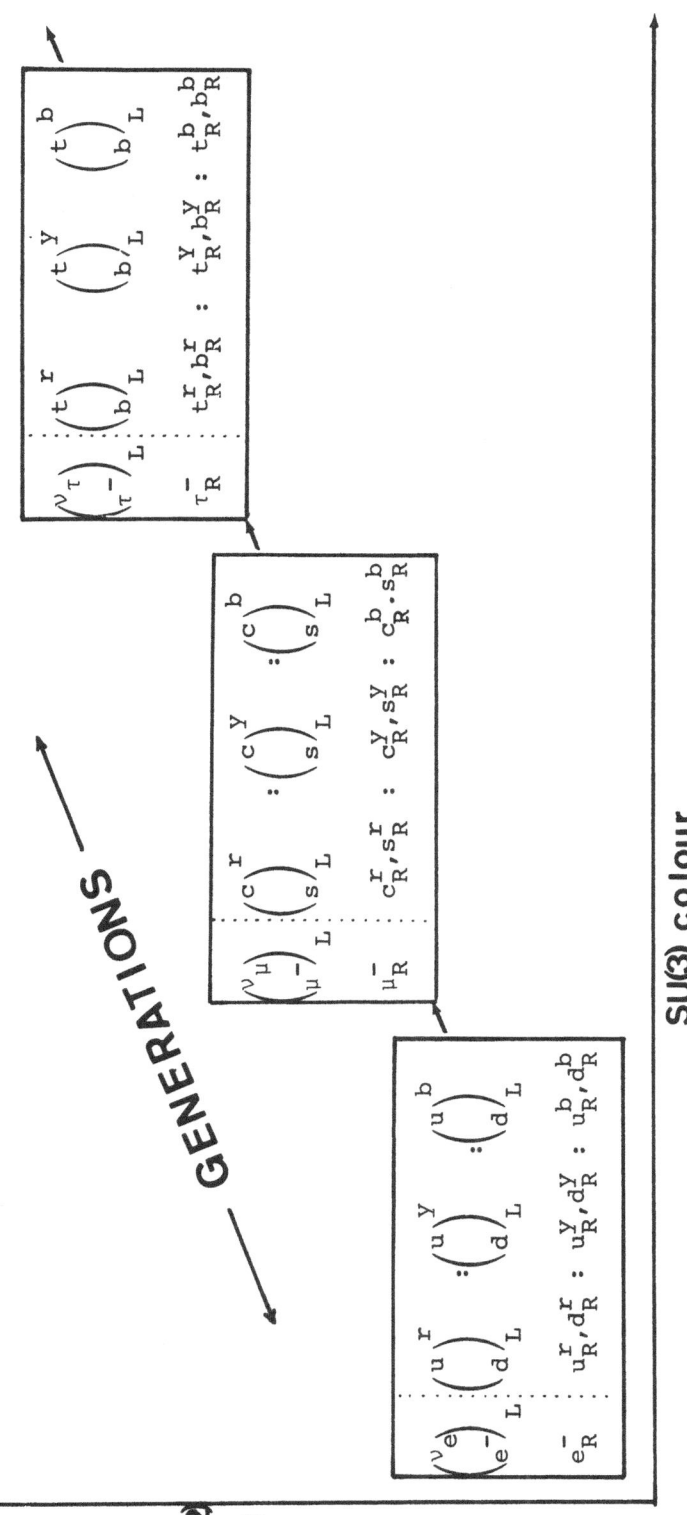

Fig. 1

$$L_{Yuk} = \sum_{\substack{\text{quarks} \\ \text{generations } j,k}} c_{jk}(\bar{u}'_j, \bar{d}'_j)_L \begin{pmatrix} \phi^{o*} \\ -\phi^- \end{pmatrix} u_{k'R} + \tilde{c}_{jk}(\bar{u}'_j, \bar{d}'_j)_L \begin{pmatrix} \phi^+ \\ \phi^o \end{pmatrix} d'_{kI}$$

$$+ \sum_{\substack{\text{leptons} \\ \text{generations } i}} \tilde{d}_{ij}(\bar{\nu}_i, \bar{\ell}_i)_L \begin{pmatrix} \phi^+ \\ \phi^o \end{pmatrix} \ell_{jR} + h.c. \quad . \qquad (2.5)$$

When ϕ^o develops a vacuum expectation value (vev), $\langle\phi\rangle$, the terms in L_{Yuk} generate quark and lepton masses. u'_i, d'_i and ℓ'_i are mixtures of mass eigenstates u_i, d_i and ℓ_i. In terms of these eigenstates the neutral currents all remain diagonal but the charged currents coupled to the W bosons, in eq.(2.1), may be written

$$L_{c.c.} = \frac{g}{2\sqrt{2}} (\bar{u}, \bar{c}, \bar{t})\gamma^\lambda (1-\gamma_5)U \begin{pmatrix} d \\ s \\ b \end{pmatrix} W^-_\lambda + h.c. \qquad (2.6)$$

where the Kobayashi-Maskawa [2] mixing matrix U may be written in the form

$$U = \begin{bmatrix} c_1 & s_1 c_3 & s_1 s_3 \\ -s_1 c_2 & c_1 c_2 c_3 - s_2 s_3 e^{i\delta} & c_1 c_2 s_3 + s_2 c_3 e^{i\delta} \\ s_1 s_2 & -c_1 s_2 c_3 - c_2 s_3 e^{i\delta} & -c_1 s_2 s_3 + c_2 c_3 e^{i\delta} \end{bmatrix}$$

$$(2.7)$$

where $c_i = \cos \theta_i$, $s_i = \sin \theta_i$ and θ_i and δ are arbitrary.

Finally there is the lagrangian density describing the interaction of scalar fields to trigger spontaneous symmetry breakdown.

$$L_{scalar} = -V(\phi) = -\frac{1}{2}\lambda^2|\phi|^4 + \frac{1}{2}\mu^2|\phi|^2 \tag{2.8}$$

giving

$$|<\phi>|^2 = \frac{\mu^2}{2\lambda^2} \quad . \tag{2.9}$$

After spontaneous symmetry breakdown, $W_\mu^{\pm}$ and Z_μ acquire masses while the photon field A_μ^γ remains massless

$$A_\mu^\gamma = B_\mu \cos\theta_W + W_\mu^3 \sin\theta_W \quad ,$$

$$Z_\mu = -B_\mu \sin\theta_W + W_\mu^3 \cos\theta_W \quad , \tag{2.10}$$

where

$$\tan\theta_W = \frac{g_1}{g_2} \quad ,$$

$$e = \frac{g_1 g_2}{(g_1^2+g_2^2)^{1/2}} \quad ,$$

$$M_W = \frac{37.3}{\sin\theta_W} \quad GeV \quad ,$$

$$M_Z = M_W/\cos\theta_W \quad ,$$

$$M_\phi^2 = 2\mu^2 \quad . \tag{2.11}$$

The neutral current coupling is given by

$$J_\mu^Z = \sum_i \bar{\psi}_L^i \gamma_\mu (T_3 - Q\sin^2\theta_W)\psi_L^i$$

$$+ \sum_j \bar{\psi}_R^i \gamma_\mu (-Q\sin^2\theta_W)\psi_R^j \tag{2.12}$$

and T_3 is the third component of weak isospin $(T_{3_L} = \pm \frac{1}{2})$.

2.2 Successes of the Standard Model

The theory described above with the multiplet structure summarized in Fig.(1) has an impressive list of successes.

(1) It is renormalisable and perturbatively unitary. As a result the theory may be used beyond the tree level to predict to arbitrary accuracy (limited by the endurance of the calculator and the convergence of the perturbation series) all but a finite number of quantities - those quantities being the fundamental parameters of the theory which require renormalisation. The perturbative unitarity of the theory means that an amplitude calculated at a given order in perturbation theory has good high energy behaviour and does not violate unitarity bounds [3]. That this is so is highly nontrivial for it comes about as a result of cancellation of graphs which separately are much larger than the final amplitude. For example in $\nu\bar{\nu} \to W^+W^-$ good high energy behaviour is arranged by a cancellation with s channel Z exchange, see Fig.2(a).

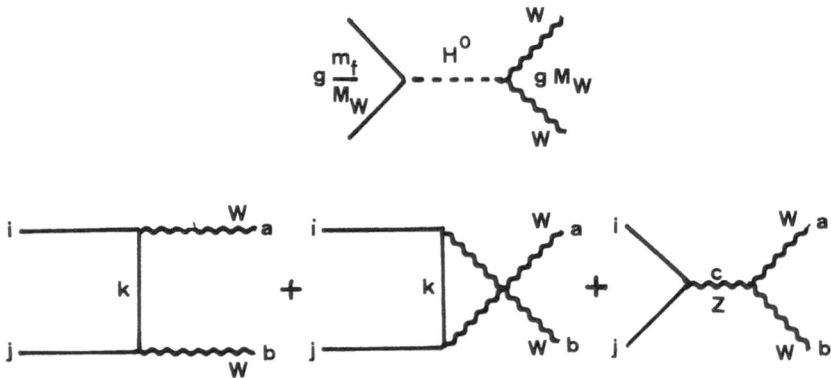

Fig.2: Graphs contributing to $f\bar{f} \to W^+W^-$.

Each term separately grows like s in cross section but there is a cancellation because their contributions are proportional to $(L^aL^b)_{ij} - (L^bL^a)_{ij} - if_{abc}L^c_{ij}$, and this term vanishes in gauge theories because the L^s form a Lie algebra with structure functions f. However there is still a residue violation of unitarity like $\sqrt{s}$ which is only cancelled by the scalar (Higgs) exchange of Fig.2(b).

Obviously this cancellation will not occur until energies above the Higgs mass, and if unitarity is not to be violated this imposes the condition on the Higgs mass m_H[3]

$$m_H \lesssim 1 \text{ TeV} \quad . \tag{2.13}$$

Of course if this condition is not satisfied unitarity does not break down, but it is achieved only through higher order perturbative terms becoming large, i.e. the alternative is a breakdown of perturbation theory.

(2) The strong interactions (QCD plus quark mass terms) automatically conserve P, C, and strangeness [4]. There is an appropriate chiral symmetry which may be realised nonlinearly. This can only occur in a theory with strong forces mediated by vector gluons, an interaction which preserves the fermion chirality. The theory is asymptotically free allowing for explanation of the observed near scaling in large momentum reactions.

(3) The low-energy weak interactions are well described by the currents following from eqs. (2.1) and (2.6) with a single parameter θ_W relating charged and neutral current phenomena [5].

$$\sin^2\theta(M_W)_{\overline{MS}} = 0.215 \pm 0.015 \quad . \tag{2.14}$$

(Here $\sin^2\theta$ is corrected [6] for the predicted radiative corrections of the model).

The recent discovery of the W is consistent with the

(radiatively corrected) predictions [6,7].

$$M_W = 83.1 \begin{smallmatrix} +3.1 \\ -3.8 \end{smallmatrix} \text{ GeV },$$

$$M_Z = 93.9 \begin{smallmatrix} +2.5 \\ -2.2 \end{smallmatrix} \text{ GeV }. \tag{2.15}$$

2.3 Limitations of the Standard Model

Although the standard model has many impressive successes it falls short of a complete theory of the strong electromagnetic and weak interactions for several reasons.

(1) There are too many parameters (mainly connected with the Higgs sector) needed to describe the standard model. The model of eqs. (2.1)-(2.7) has seventeen, six quarks and three lepton masses, 3 mixing angles and a phase parameterising CP violation, three gauge couplings and two boson mass scales M_W and M_ϕ. There is a further parameter θ which describes potential strong violation of CP which, it has been realised, must be included due to the anomaly in the axial vector current. There must be added to the (3,2,1) Lagrangian a term [8]

$$L_\theta = \frac{1}{32\pi^2} \theta_{QCD} F^a_{\mu\nu} \tilde{F}^{\mu\nu}_a \tag{2.16}$$

where

$$\tilde{F}^{\mu\nu}_a = \frac{1}{2} \epsilon^{\mu\nu\rho\tau} F^a_{\rho\tau} . \tag{2.17}$$

This term violates CP and in order to be consistent with experiment θ_{QCD} must be less than 10^{-9}. It is possible [9] to modify the standard model by adding a further Higgs doublet to replace the θ^* term in eq. (2.5) which generates up quark masses. Then one may show θ_{QCD} is zero automatically, but the model then predicts a light pseudogoldstone state, the axion, which has not been found experimentally.

(2) There is no reason why the matter multiplet structure chosen for the standard model in Fig. 1 should be as it
is. Also there is no understanding of the family replication.

(3) Charge quantisation is not explained as Y in eq.(2.3)
is arbitrary. The relation of quark to lepton charges is also
not understood. Also we do not understand why the charged
weak interactions should be left handed for both quarks and
leptons.

(4) There is no explanation of even the gross features
of the mass spectrum. Why are quarks and leptons much lighter
than the W and Z? Why are families different in mass, and
what relates quark and lepton masses? Neutrinos are massless
because one excludes right handed neutrinos, but why are
neutrinos different in this respect?

For many people these limitations suggest that the
(3,2,1) model is only a step towards a more fundamental theory
and that at best it is an effective theory valid up to a
scale M_X at which the underlying theory that will answer the
above questions appears.

There are two main possibilities for this underlying
theory, if is exists. The first is that some or all of the
fields of the standard model may be composite and there is
some more fundamental level of structure. The second is
that the fields of the standard model are themselves fundamental, but they are related by further symmetries, broken
at the scale M_X. The latter approach leads to grand unified
theories (GUTs) and to supersymmetric theories (SUSY-GUTs),
and are the subject of these lectures. In GUTs the additional
symmetries are gauge symmetries based on larger Lie algebra
than SU(3)×SU(2)×U(1) which may relate particles of the same
spin. In the ideal GUT all the fundamental fields of a given
spin will belong to a single irreducible representation of
a gauge group G and hence their interactions will also be related by the (gauge) transformations of G. In SUSY-GUTs the
additional symmetry is based on graded Lie algebra which may

relate particles of different spin and ideally may relate all
matter particles and all interactions to the fundamental
gauge bosons and gauge interactions. How far along this road
it is possible to proceed we will discuss in the following
sections.

3. GRAND UNIFIED THEORIES

Grand unified theories seek to embed the standard model SU(3)×SU(2)×U(1) in a group G which is either simple
with a unique gauge coupling g_{GU} or else is the product of
identical simple factors with a discrete symmetry relating
the couplings. G must have rank $\geq$ 4 to accommodate the four
operators of SU(3)×SU(2)×U(1) which may be simultaneously
diagonalised. The groups with the minimal rank 4 are SU(5)
$[SU(2)]^4$, $[O(5)]^2$, $[SU(3)]^2$, $[G_2]^2$, O(8), O(9), SP(8) and F_4.
Only the first four have complex representations capable of
accommodating the known fermions. $[SU(2)]^4$ and $[O(5)]^2$ do not
contain SU(3) as a subgroup and $[SU(3)]^2$ cannot accommodate
the quarks of Fig. 1, leaving SU(5) of Georgi and Glashow
[10] as the unique possibility of rank 4.

3.1 SU(5) - The Prototype GUT[11]

SU(5) contains 24 gauge bosons, 12 of which are those
of the standard model. The 12 new ones are called X and Y
and transform as (3,2) + ($\bar{3}$,2) under SU(3)×SU(2). We denote
charge-conjugate spinors $f^c \equiv c\bar{f}^T$ where c is the charge conjugation operator. Thus we may either choose to write our
fermion basis in terms of f_L and f_R or f_L and f_L^c. Since gauge
interactions conserve the helicity of fermions it is convenient to work with fermions of a fixed helicity, by convention
the left handed set f_L and f_L^c. In SU(5) it is found the
fermion of one generation may be grouped in a $\bar{5}$ and 10 representation of SU(5).

$$\psi_{\bar{5}} \;=\; \begin{Bmatrix} d^{c1} \\ d^{c2} \\ d^{c3} \\ e^{-} \\ \nu \end{Bmatrix}_{L} \;=\; (\bar{3},1) \,+\, (1,2) \tag{3.1}$$

where d^{ci} represents a d^{c} quark of colour i and we have shown the SU(3)×SU(2) properties of the representation.

For them the generalisation of L_{Kin} is

$$L_{Kin}^{\bar{5}} \;=\; i\bar{\psi}_{L}\gamma^{\mu}(\partial_{\mu} - \tfrac{i}{2}\, g_{GU}T^{a}V_{\mu}^{a})\,\psi_{R} \quad . \tag{3.2}$$

The matrices T^{a} representing SU(5) may be chosen conveniently as follows.

$$T^{j} \;=\; \left(\begin{array}{ccc:cc} & & & 0 & 0 \\ & \lambda^{j} & & 0 & 0 \\ & & & 0 & 0 \\ \hdashline 0 & 0 & 0 & & 0 \\ 0 & 0 & 0 & & \end{array}\right) \qquad \begin{array}{l} j = 1..8 \text{ (generators of} \\ \qquad\qquad\quad SU(3)) \end{array}$$

$$T^{9,10} \;=\; \left(\begin{array}{ccc:cc} & & & 0 & 0 \\ & 0 & & 0 & 0 \\ & & & 0 & 0 \\ \hdashline 0 & 0 & 0 & & \sigma^{\pm} \\ 0 & 0 & 0 & & \end{array}\right) \qquad \text{(charged generators of SU(2))}$$

$$T^{13} \;=\; \left(\begin{array}{ccc:cc} & & & 1 & 0 \\ & 0 & & 0 & 0 \\ & & & 0 & 0 \\ \hdashline 1 & 0 & 0 & 0 & 0 \\ 0 & 0 & 0 & 0 & 0 \end{array}\right)$$

$$T^{14} = \begin{pmatrix} & & -i & 0 \\ & 0 & 0 & 0 \\ & & 0 & 0 \\ \hline i & 0 & 0 & \\ 0 & 0 & 0 & 0 \end{pmatrix}$$

(3.3)

and others $T^{15} \ldots T^{24}$ obtained by putting 1 and $\pm i$ in the same pattern. Finally, there are two further diagonal matrices which may conveniently be chosen as

$$T^{11} = \frac{1}{\sqrt{6}} \begin{pmatrix} 1 & & & \\ & 1 & & \\ & & 1 & \\ & & & -3 \\ & & & & 0 \end{pmatrix}$$

$$T^{12} = \frac{1}{\sqrt{10}} \begin{pmatrix} 1 & & & \\ & 1 & & \\ & & 1 & \\ & & & 1 \\ & & & & -4 \end{pmatrix} .$$

(3.4)

With these assignments the $SU(3) \times SU(2)$ transformation properties of $\psi_{\bar{5}}$ given in eq.(3.1) are obvious. Since the group contains $SU(2) \times U(1)$ it necessarily contains the charge operator Q which must therefore be traceless. For the multiplet of eq.(3.1)

$$\text{Tr } Q = 3 Q_d c + Q_{\bar{\nu}} + Q_{e^-} \ , \tag{3.5}$$

i.e.

$$Q_d c = - \frac{1}{3} e \ . \tag{3.6}$$

Thus the identification of the quark states with d^c is clear and moreover SU(5) has explained one of the original puzzles, charge is quantised and the relation of quark charges to lepton charges is explained. Indeed the factor of $\frac{1}{3}$ in eq.(3.6) clearly follows from the fact there are three colours, a remarkable postdiction indeed.

From five objects a_j transforming as a $\bar{5}$ of SU(5) it is easy to construct a 10 by taking the antisymmetric product.

$$a_{jk} = \frac{1}{\sqrt{2}} (a_j a_k - a_k a_j) \ , \qquad k,j = 1..5 \ . \tag{3.7}$$

Remarkably a_{jk} may accommodate the remaining states of a family in Fig.1 as follows

$$\chi_{10} = \frac{1}{\sqrt{2}} \begin{bmatrix} 0 & \bar{u}_3 & -\bar{u}_2 & u_1 & d_1 \\ -\bar{u}_3 & 0 & \bar{u}_1 & u_2 & d_2 \\ \bar{u}_2 & -\bar{u}_1 & 0 & u_3 & d_3 \\ -\bar{u}_1 & -u_2 & -u_3 & 0 & e^+ \\ -\bar{d}_1 & -d_2 & -d_3 & -e^+ & 0 \end{bmatrix}_L$$

$$= (3,2) + (\bar{3},1) + (1,1) \ . \tag{3.8}$$

For $j,k = 1,2,3$ a_{jk} is the product of two SU(3) triplets ($3 \times 3 = 6 + \bar{3}$) the $\bar{3}$ being the antisymmetric combination. Since these objects are singlets under the SU(2) group they may be identified with $\varepsilon_{jkm} \bar{u}_{m_L}$ as shown. a_{j4} (a_{j5}) represents a colour triplet which has the third component of weak isospin $\frac{1}{2}$ ($-\frac{1}{2}$), and may be identified with u_{j_L} (d_{j_L}) as shown. a_{45} is clearly a singlet under colour and weak isospin, the e_L^+. Thus $\psi_{\bar{5}}$ and χ_{10} contain just the correct SU(3)×SU(2) content to describe a complete family as in eq.(2.4) and Fig.1. SU(5) does not explain why we need N_G generation, but they are readily included by having N_G ($\bar{5}$ + 10) representations.

We see some progress has been made towards understanding the fermion representatives. Just a $\bar{5}$ + 10 is an improvement on the multiplet structure of eq.(2.4) for a single family. Moreover there is no room for the right handed neutrino (although it could be included as a singlet). More definitely the multiplet structure demands that quarks and leptons should both couple to the charged weak current in a left handed manner;

again a non-trivial prediction. The kinetic term for the 10
dimensional representation is

$$L_{Kin}^{10} = i\bar{\chi}_{L_{jk}}\gamma^{\mu}(\partial_{\mu}\delta_{kk'} - ig_{GU}T_{kk'}^{a}\cdot V_{\mu}^{a})\chi_{jk'} \qquad (3.9)$$

The charge operator is given by $-\sqrt{\frac{2}{3}}\,T^{11}$, V_{μ}^{11} is the
photon field A_{μ} and $g_{GU} = 2\sqrt{\frac{2}{3}}\,e$. Then V_{μ}^{12} must
be identified with Z_{μ}. An immediate consequence of this is
$Tr\{Q^{\gamma}Q^{Z}\} = 0$. Since $Q^{Z} = T_{3_{L}} - Q\sin^{2}\theta_{W}$ this gives

$$Tr\{Q^{\gamma}, Q^{Z}\} \sim Tr\{T_{3_{L}}Q - \sin^{2}\theta Q^{2}\} = 0 \quad . \qquad (3.10)$$

Using $Q = T_{3_{L}} + Y$ gives

$$\sin^{2}\theta_{W} = \frac{Tr(T_{3_{L}}^{2})}{Tr(Q^{2})} \quad . \qquad (3.11)$$

The sum implied in eq.(3.11) runs over the members of
any representation of SU(5). For example with the simplest
choice the $\bar{5}$ of eq.(3.1) we find

$$\sin^{2}\theta_{W} = \frac{2.(1/4)}{1+3.(1/9)} = \frac{3}{8} \quad . \qquad (3.12)$$

Such a prediction for $\sin^{2}\theta_{W}$ is expected since it is
(cf. eq.(2.11)) related to g_{2} and g_{1} and in SU(5) these
couplings are both related to g_{GU}. However the value is
clearly inconsistent with eq.(2.14), and for this reason
SU(5) was largely ignored for some time until it was pointed
out by Georgi, Quinn and Weinberg [12] that it applied at a
scale M_{X} when SU(5) is a good symmetry and must be renormalised to low energies to compare with the experimental result of eq.(2.14). We return to this in the next section.
Using the interaction of eq.(3.2) it is easy to check that
Z does have the coupling of the standard model. For example

$$\frac{g_{GU}}{Z} \, \bar{\chi}_L \gamma_\mu T^{15} \chi_L Z^\mu = \sqrt{\frac{2}{3}} \, e\{\bar{a}_{54} T^{15}_{44} \, a_{54} + \bar{a}_{45} \, T^{15}_{55} \, a_{45}\}$$

$$= \sqrt{\frac{3}{5}} \, \bar{e} \, \bar{\psi}_{e_R} \gamma_\mu \psi_{e_R} Z^\mu \qquad (3.13)$$

in agreement with eq.(2.12) which gives (with $\sin^2\theta_W = \frac{3}{8}$)

$$\frac{e}{\sin\theta_W \cos\theta_W} \, (-\sin^2\theta_W) Q_e \, \bar{\psi}_{e_R} \gamma_\mu \psi_{e_R} Z^\mu = e\sqrt{\frac{3}{5}} \, \bar{\psi}_{e_R} \gamma^\mu \psi_{e_R} Z_\mu$$

$$(3.14)$$

The coupling of the new gauge bosons X_j, Y_j, $j = 1,2,3$ with charge $(\pm 4/3)$ and $(\pm 1/3)$ respectively may similarly be read off from eqs.(3.2) and (3.9). They give the couplings of Fig. 3.

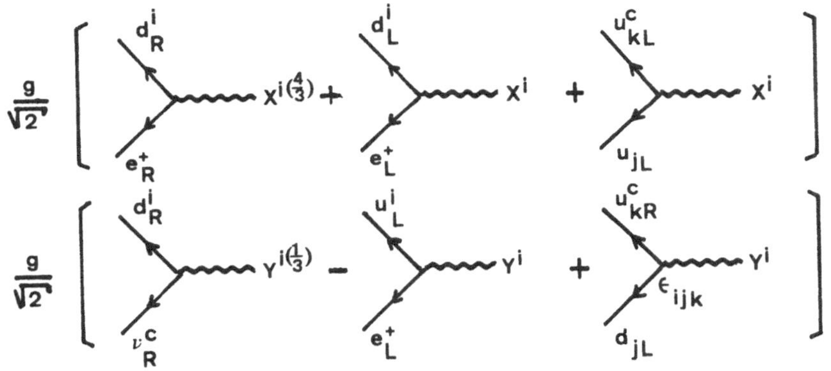

Fig. 3: Fermion couplings of X and Y in SU(5).

It is clear from Fig.3 that neither lepton number L nor baryon number B is conserved, although one may easily check that (B-L) is conserved by these interactions with $(B-L)_{Y_j, X_i} = \frac{4}{3}$. For this reason they must be very heavy, otherwise protons (and neutrinos) will decay rapidly to leptons - see below.

Of course the couplings of Fig.1 refer to the "current" eigenstates. In general these are related to the mass eigenstates by the rotations

$$\psi_{u'_L} = A^{(u,L)} \psi_{u_L} \quad , \qquad \psi_{d'_i} = A^{(d,L)} \psi_{d_L} \qquad \text{etc.} \qquad (3.15)$$

where

$$\psi_e = \begin{pmatrix} u \\ c \\ t \end{pmatrix} \quad , \qquad \psi_d = \begin{pmatrix} d \\ s \\ b \end{pmatrix} \quad , \qquad \psi_e = \begin{pmatrix} e \\ \mu \\ \tau \end{pmatrix}$$

and the primes refer to current eigenstates. The six matrices $A^{(f,L)}$, $A^{(f,R)}$, $f = u,d,e$ are unitary and, since colour is unbroken, independent of colour. The neutral currents remain diagonal (the GIM mechanism) because one meets always terms of the form

$$\bar{u}'_L \{..\} u_L = \bar{u}_L A^{(u,L)^+} \{..\} A^{(u,L)} u_L = \bar{u}_L \{..\} u_L \ .$$

However the charged currents are not invariant under this rotation. For the charged currents coupling to $W^{\pm}_{\mu}$ we find (suppressing the Lorentz structure)

$$(\bar{u}'_L d'_L + \bar{v}'_L e'_L) W = (\bar{u}_L A^{u,L^+} A^{d,L} d_L + \bar{v}'_L A^{e,L} e_L) W$$

$$= (\bar{u}_L U^{(u,d)} d_L + \bar{v}_L e_L) W \qquad (3.16)$$

where $U^{(u,d)}$ is the Kobayashi-Maskawa matrix of eq.(2.7). The reason the leptons do not have the equivalent structure is because the neutrinos are massless so we may redefine ν states $\bar{\nu}_L' A^{e,L} \equiv \bar{\nu}_L$.

In SU(5) there are further charged currents coupling to $X_j^{\pm 4/3}$ and $Y_j^{\pm (1/3)}$ so there are additional angles and phases involved in their coupling. For example

$$\overline{d_L^c} \nu'_L Y = \overline{d_L^c} A^{\bar{d},L^+} A^{e,L} \nu_L Y$$

$$\equiv \overline{d_L^c} U^{(d,\nu)} \nu_L Y . \tag{3.17}$$

The form of these (unitary) matrices $U^{(i,j)}$ will depend on the specific form of the mass matrices, which must be generated on spontaneous breakdown of the gauge theory and we turn to this question now.

3.2 Spontaneous Symmetry Breaking in SU(5)

Symmetry breaking for SU(5) must proceed in two stages

$$SU(5) \xrightarrow{M_X} SU(3) \times SU(2) \times U(1) \xrightarrow{M_X} SU(3) \times U(1)_{em} \tag{3.18}$$

where $M_X = O(10^{14}$ GeV) and $M_W \simeq 10^2$ GeV.

The first stage of breaking is achieved through an adjoint (24) of Higgs $\sum_\alpha^\beta$ (a real representation) which have the same quantum numbers as the gauge bosons. The kinetic energy term is given by

$$Tr\{|\partial_\mu \delta_\alpha^{\alpha'} \delta_{\beta'}^\beta - i \frac{g_{GU}}{2} v_\mu^a [T^a{}_\alpha^{\alpha'} \delta_{\beta'}^\beta - T^a{}_{\beta'}^\beta \delta_\alpha^{\alpha'}] \sum_{\beta'}^{\alpha'}|^2\}. \tag{3.19}$$

In order to generate SSB we construct a Higgs potential for the Σ field

$$V_\Sigma = -\mu^2 \text{Tr}(\Sigma^2) + \frac{1}{4} a[\text{Tr}(\Sigma^2)]^2 + \frac{1}{2} b \text{ Tr}(\Sigma^4) \tag{3.20}$$

where a simplifying (but inessential) symmetry has been imposed, namely $\Sigma \rightarrow -\Sigma$.

If $b > 0$, $\mu^2 > 0$ and $a > (-7/15)b$ the vev of Σ has the form [13].

$$<0|\Sigma|0> = v \begin{bmatrix} 1 & & & & \\ & 1 & & & \\ & & 1 & & \\ & & & -\frac{3}{2} & \\ & & & & -\frac{3}{2} \end{bmatrix} \tag{3.21}$$

with v given by

$$\mu^2 = \frac{15}{2} av^2 + \frac{7}{2} bv^2 . \tag{3.22}$$

Clearly the vev of eq.(3.21) leaves $SU(3) \times SU(2) \times U(1)$ unbroken and using eq.(3.19) we immediately see

$$M_X^2 = M_Y^2 = \frac{25}{8} g_{GU}^2 v^2 . \tag{3.23}$$

The 12 X and Y bosons acquire a mass and their longitudinal degrees of freedom are supplied by the twelve scalar fields of Σ carrying the same quantum numbers as X and Y. The remaining Higgs scalars are those corresponding to generators which leave the vacuum of eq.(3.21) invariant. These are clearly the remaining twelve fields carrying the same quantum numbers as the gluons (8 of them), the $W^\pm$, Z and γ.

To acchieve the second stage of SSB it is necessary to introduce a $\bar{5}$ (a complex representation) of Higgs $\bar{H}$. With a potential of the form

$$V_{\bar{H}} = -\frac{1}{2}\nu^2|\bar{H}|^2 + \frac{1}{4} \lambda (|\bar{H}|^2)^2 \tag{3.24}$$

with ν^2, $\lambda > 0$ this drives a vev for $\bar{H}$ of the form

$$\langle 0 | \bar{H} | 0 \rangle = v_o \begin{bmatrix} 0 \\ 0 \\ 0 \\ 0 \\ 1 \end{bmatrix} \tag{3.25}$$

with

$$\frac{\lambda}{2} v_o^2 = \nu^2 \ . \tag{3.26}$$

$SU(2) \times U(1)$ is broken with

$$M_W^2 = M_Z^2 \cos^2\theta_W = \frac{1}{2} g_2^2 v_o^2 \tag{3.27}$$

where g_2 is the coupling of SU(2) evaluated at a scale $O(M_W)$.

The vev of H leaves the X mass unaltered but affects the Y mass. However this is only changed to a very small extent and may be ignored.

The components H_4 and H_5 just play the role of the Higgs doublet $\begin{smallmatrix} \phi^+ \\ \phi^o \end{smallmatrix}$ (cf. (2.1)). However to order $(\frac{v_o}{v})$ the triplet components $\bar{H}_{i=1,2,3}$ remain massless and are not eaten by the Y bosons. This turns out to be a serious problem for the Higgs generate fermion masses and, as we will see, the fermion couplings introduced to do this mean that $\bar{H}_{i=1,2,3}$ mediate baryon number violation and must be very heavy. It is necessary therefore to add further terms coupling $\bar{H}$ to Σ which may give the $\bar{H}_{i=1,2,3}$ components a mass. This is done through the terms

$$V_{H,\Sigma} = \alpha |\bar{H}|^2 \ Tr(\Sigma^2) + \beta \ \bar{H}^* \ \Sigma^2 \ \bar{H} \ . \tag{3.28}$$

Notice that these terms are not unexpected for, even if absent at tree level, they would have been generated in higher order through graphs of the form of Fig.4.

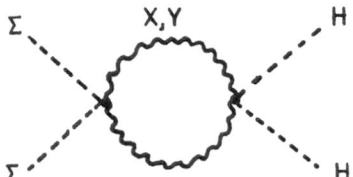

Fig. 4: Graph inducing couplings between H and Σ.

Since these graphs are divergent the terms of eq.(3.28) are needed as counter terms.

Minimising the potential of eqs.(3.20), (3.24) and (3.28) gives eq.(3.25) and

$$<0|\Sigma|0> = v \begin{bmatrix} 1 & & & & \\ & 1 & & 0 & \\ & & 1 & & \\ & & & (-\frac{3}{2} - \frac{1}{2}\varepsilon) & \\ & 0 & & & (-\frac{3}{2} + \frac{1}{2}\varepsilon) \end{bmatrix} \qquad (3.29)$$

where

$$\varepsilon = \frac{3}{20} \frac{\beta v_o^2}{bv^2} \qquad (3.30)$$

and

$$\mu^2 = \frac{15}{2} av^2 + \frac{7}{2} bv^2 + \alpha v_o^2 + \frac{9}{30} \beta v_o^2 \quad ,$$

$$\nu^2 = \frac{1}{2} \lambda v_o^2 + 15\alpha v^2 + (\frac{9}{2} - 3\varepsilon) \beta v^2 \quad . \qquad (3.31)$$

The terms involving ε in eq.(3.29) break SU(2). Since these components have weak isospin = 1 this will violate the successful formula of eq.(2.11), but as ε is so small it will not be a measurable difference. However there is an unnatural relation implicit in eq.(3.31) because it mixes two parameters v_0 and v which have vastly different sizes ($v_0/v = O(10^{-12})$). For this to work it is necessary that

$$v^2 - (15\alpha + \tfrac{9}{2}\beta)\, v^2 = O(10^{-24})v^2 \qquad (3.32)$$

We will return to a discussion of this unnatural relation, the socalled hierarchy problem, in later sections.

3.3 Fermion Masses in SU(5)

The fermion in SU(5) transform as $\bar{5} + 10$, where these representation contain LH components. Fermion mass terms involve the product $(\bar{5} \otimes 10) \otimes (\bar{5} \otimes 10)$. Now

$$\bar{5} \otimes 10 = 5 \oplus \overline{45}$$

and

$$10 \otimes 10 = \bar{5} \oplus 45 \oplus 50 \qquad (3.33)$$

Clearly there can be no bare fermion masses. However there are possible couplings of $(\bar{5} \otimes 10)$ to $\bar{5}$ and of $(10 \otimes 10)$ to 5 so we may use the $\bar{5}$ of Higgs, $\bar{H}$, to write the Yukawa couplings

$$Y = \frac{1}{\sqrt{2}}\,(\chi_i^{+})^{\alpha\beta}\,\gamma_0 M^D_{ij}[\bar{H}_\alpha \psi_\beta - \bar{H}_\beta \psi_\alpha]_j - \frac{1}{4}\varepsilon^{\alpha\beta\gamma\delta\varepsilon}\chi_{\alpha\beta i} M^u_{ij}\chi_{\delta e\; j}\bar{H}_j \qquad (3.34)$$

Here i and j are generation indices and $\alpha, \beta, \gamma, \delta, \epsilon$ SU(5) indices. Eq.(3.34) generalises the Yukawa couplings of the standard model eq.(2.5). It is possible to diagonalise M^D by unitary rotations of χ and ψ, giving a fermion mass term

$$v_o (\chi^t)^{5\beta}{}_\gamma {}^o_M M^D \psi_\beta \qquad (3.35)$$

which gives

$$m_d = m_e, \ m_s = m_\mu, \ m_b = m_t \ . \qquad (3.36)$$

These mass relations must be renormalised, and we discuss these in the next section. Up-quark masses come from the term involving M^u, but because there are no neutrinos there are no further relations between quark and lepton masses.

We may diagonalise M^u (because by eq.(3.34) it must be symmetric in generation space) by a transformation U_3 of the 10 of fermions χ.

$$M^u = U_3^T M_D^u U_3 \qquad (3.37)$$

It is convenient to remove phase factors $e^{i\phi_{ij}}$ from the first row and column of U_3 by the transformation

$U_3 = U_5 UU_4$ with

$$U_4 = \begin{bmatrix} e^{i\phi_{11}} & & 0 \\ & e^{i\phi_{21}} & \\ 0 & & e^{i\phi_{n1}} \end{bmatrix} ,$$

$$U_5 = e^{-i\phi_{11}} \begin{bmatrix} e^{i\phi_{11}} & & 0 \\ & e^{i\phi_{12}} & \\ 0 & & e^{i\phi_{1n}} \end{bmatrix} . \qquad (3.38)$$

Redefining χ to absorb U_4, and changing ψ to keep M^a real and diagonal, the term in eq.(3.34) contributing to up quark masses becomes

$$-\frac{1}{4} \, \epsilon^{\alpha\beta\gamma\delta\epsilon} \chi_{\alpha\beta} \; U^T \; U_5^2 \; M_D^u \; UH_\gamma \; \chi_{\delta\epsilon} \quad . \qquad (3.39)$$

The top quark masses come from the terms $1 \le \alpha\beta\delta\epsilon \le 4$ and it may be diagonalised by the rotation

$$U\chi_{\alpha\beta} = \chi'_{\alpha\beta} \quad , \qquad\qquad 1 \le \alpha,\beta \le 4 \quad . \qquad (3.40)$$

With the final phase transformation

$$U_5^! \; \chi'_{\alpha\beta} \;=\; \chi''_{\alpha\beta} \quad , \qquad\qquad 1 \le \alpha,\beta \le 3 \qquad (3.41)$$

we find a real diagonal mass matrix for the fields χ'', which are the mass eigenstates. The matrix U clearly is the Kobayashi-Maskawa matrix of eq.(2.7). The phase rotations of eq.(3.38) change the phases of left-handed antiquarks relative to $Q = \frac{2}{3}$ quarks and do not contribute in conventional weak interactions although they do occur in X and Y mediated interactions.

Thus we see that the mixing angles and phases of the standard model reappear in SU(5) with no constraint on their value [14]. In addition, as mentioned above, there are (N_G-1) new phases, again indetermined, which affect X and Y physics. However, as we have see there are no mixing angles relevant to X and Y exchange so that nucleon decay is predicted entirely in terms of the Kobayashi Maskawa angles.

4. PREDICTIONS OF SU(5)

4.1 Couplings and Masses

In the previous section we derived the relations

$$\sin^2\theta_W = \frac{3}{8} = \frac{g_1^2}{g_1^2 + g_1^2} \quad ,$$

$$m_d = m_e; \quad m_s = m_\mu; \quad m_b = m_\tau \quad . \tag{4.1}$$

In addition we have a relation for the strong coupling g_3:

$$g_3 = g_2 = g_{GU} \quad .$$

These results are phenomenologically unacceptable, the strong coupling g_3 being much bigger than the weak coupling constant g_2. It was the realisation however of Georgi, Quinn and Weinberg [12] that the above predictions apply at a scale $O(M_X)$ at which SU(5) is a good symmetry, and that in comparing with experiment it is necessary to include radiative corrections to continue the coupling and masses to a scale $\mu \stackrel{\sim}{\sim} O(1 \text{ GeV})$ at which laboratory measurements are made. That this is a possibility follows from the calculation of these radiative corrections which gives for the evaluation of the effective couplings of the (3,2,1) model below M_X

$$\frac{1}{\alpha_3(E)} = \frac{1}{\alpha_{GU}} + \frac{1}{6\pi} (4N_G - 33) \ln(\frac{M_X}{E}) + \ldots \quad ,$$

$$\frac{1}{\alpha_2(E)} = \frac{\sin^2\theta(E)}{\alpha(E)} = \frac{1}{\alpha_{GU}} + \frac{1}{6\pi} (4N_G - 22 + \frac{1}{2}) \ln(\frac{M_X}{E}) + \ldots \quad ,$$

$$\frac{1}{\alpha_1(E)} = \frac{3}{5}\frac{\cos^2\theta(E)}{\alpha(E)} = \frac{1}{\alpha_{GU}} + \frac{1}{6\pi} (4N_G + \frac{3}{10}) \ln(\frac{M_X}{g}) + \ldots \quad , \tag{4.2}$$

where N_G is the number of generations.

We see that the couplings of SU(3) and SU(2) decrease
with increasing energy (corresponding to the asymptotic
freedom of nonabelian gauge groups) while the U(1) coupling
increases. Moreover the negative coefficient of the log
term of eq.(4.2) is larger for SU(3) than SU(2) so that α_3
falls faster than α_2. The result is sketched in Fig.5 and
shows that even though α_3 is initially larger than α_2 and
α_1 it will eventually become equal to them at some large
scale M_X.

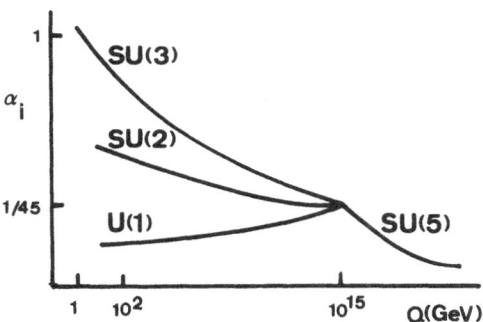

Fig. 5: Behaviour of the (3,2,1) couplings with energy

In fact eqs.(4.2) may be solved to determine M_X in
terms of α_3 and α_2 and to make a prediction for the third
coupling (or more conveniently $\sin^2\theta_W$),

$$\frac{3}{5\alpha(\mu)} - \frac{3}{5\alpha_3(\mu)} = \frac{201}{30\pi} \ln \frac{M_X}{\mu} \quad ,$$

$$\sin^2\theta(\mu) = \frac{3}{8} [1 - \frac{109}{18\pi} \alpha \ln \frac{M_X}{\mu}] \quad . \tag{4.3}$$

These lead to

$$M_X = 2 \times 10^{15} \text{ GeV}$$

and $\sin^2\theta(\mu) = 0.21$. $\hspace{4cm}$ (4.4)

$\hspace{1cm}$ The reason such a large scale emerges follows from the fact that the evolution of eq.(4.2) is only logarithmic so that M_X/μ is exponentially related to the coupling constant differences. (It is also crucial that the scale should be large to inhibit proton decay). The agreement of $\sin^2\theta(\mu)$ with experiment (eq.(2.14)) is also impressive, particularly as the initial value of 3/8 was in clear disagrement. These two facts, more than any others, encourage us to believe that GUTs have something to do with reality.

$\hspace{1cm}$ In deriving eqs.(4.3) we have been somewhat cavalier in our approach. For example the picture of Fig.5 supposes that above M_X the couplings are equal, below M_X they vary, whereas in reality they will only asymptotically be equal. Since these predictions are of central importance in GUTs I will spend a little time improving on their derivation and the inclusion of mass effects [15,16].

$\hspace{1cm}$ The renormalization group equation may be written as

$$\frac{\partial \alpha_i^{-1}}{\partial t} = F_i(\frac{p^2}{M_X^2}, \xi(t)) + \alpha_i(t)G_i(\frac{p^2}{M_X^2}, \xi(t)) + O(\alpha_i^2(t)) \hspace{1cm} (4.5)$$

where $\alpha_i(t) = g_i^2(t)/4\pi$. In an unbroken gauge theory, the functions F_i are gauge invariant constants. They are therefore independent of the gauge parameter $\xi(t)$ for $|p^2| >> M^2_X$, where G is unbroken, and independent of $\xi(t)$ for $m^2 << p^2 << M^2_X$, where there is an unbroken gauge theory G'. Therefore to a good approximation, we can freeze $\xi(p^2) = \xi(M^2_X)$ in integrating eq.(4.5). The errors this procedure introduces into α_i^{-1} when eq.(4.5) is solved are of $O(\alpha_i(M_X))$.

$\hspace{1cm}$ The F_i can be constructed by calculating the α_i^{-1} perturbatively in the modified minimal subtraction scheme (this is not necessary but it is convenient since gauge invariance is automatically respected). This will give

$$\alpha_i^{-1}(t) = \frac{4\pi}{g_\mu^2} + f_i\left(\frac{p^2}{M_X^2}, \frac{\mu^2}{M_X^2}, \xi\right) + O(\alpha_i^2) \tag{4.6}$$

where g_μ is the minimal subtraction coupling constant and μ is the mass unit introduced by going to $4-\varepsilon$ dimensions. We have

$$F_i\left(\frac{p^2}{M_X^2}, \xi\right) = \psi_i\left(\frac{p^2}{M_X^2}, 1, \xi\right)$$

where

$$\psi_i\left(\frac{p^2}{M_X^2}, \frac{\mu^2}{M_X^2}, \xi\right) = \frac{\partial f_i}{\partial t} \quad . \tag{4.7}$$

In fact ψ_i is independent of μ, which only enters f_i through

$$\lim_{\varepsilon \to 0} \left\{ \frac{1}{\varepsilon} \left(\frac{\mu^2}{\theta(p^2, M_X^2)} \right)^{\frac{\varepsilon}{2}} - \frac{1}{\varepsilon} \right\}$$

Hence

$$F_i = \frac{\partial f_i}{\partial t} \tag{4.8}$$

so the first term in eqn.(4.5) integrates to f_i, which is very easy to calculate for $p^2 \ll M_X^2$ and $p^2 \gg M_X^2$. This circumstance is peculiar to leading order.

Integrating eqn.(4.5) from $\mu_1^2 \ll M_X^2$ to $\mu_2^2 \gg M_X^2$, where α_i^{-1} and f_i are independent of i, we get

$$\alpha_i^{-1}(\mu_1) = C + f_i\left(\frac{\mu_1^2}{M_X^2}, \frac{\mu^2}{M_X^2}, \xi(M_X)\right) + O(\alpha_i) \tag{4.9}$$

(the mass unit μ is here really a dummy variable, as we shall see explicitly later). The diagrams which contribute to f_i are shown in Fig.6. The contributions of diagrams in which all the virtual particles are light may be written:

$$f_i^L = \frac{1}{6\pi}[-11\ \mathrm{Tr}(t_{iV}^L)^2 + 4\mathrm{Tr}(t_{iF}^L)^2 + \frac{1}{2}\ \mathrm{Tr}(t_{iS}^L)]\ln(\frac{\mu}{\mu_i}) + R_i(\xi(M_X)),$$

$$(4.10)$$

where, to facilitate comparison, we have a notation similar to Weinberg's [16]: t_{iV}^L, t_{iF}^L and t_{iS}^L ar the matrices which represent the generators of G on the light vector, fermion and scalar fields respectively (choosing real representations). The coefficients of $\ln(\mu/\mu_i)$ are the constants which appear in the one loop β function for unbroken gauge theories. The light scalars include the components of the Higgs fields which become the longitudinal degrees of freedom of the vector fields which gain masses when G' is broken.

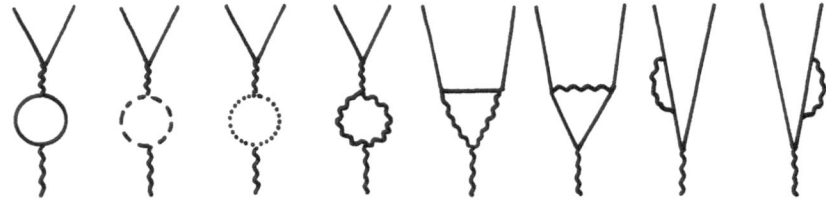

<u>Fig.6:</u> Diagrams contributing to f_i

Diagrams in which one or more internal particles are heavy contribute

$$f_i^H = \frac{1}{6\pi} (-11(t^{\cdot\cdot}_{iV})^2 + \frac{1}{2}(t^{\cdot\cdot}_{iS})^2) \ln(\frac{\mu}{M_V}) + \frac{1}{2}(t^{\cdot\cdot}_{iV})^2$$

$$+ 4(t^{\cdot\cdot}_{iF})^2 \ln(\frac{\mu}{M_F}) + \frac{1}{2}(t^{\cdot\cdot}_{iS})^2 \ln(\frac{\mu}{M_S})$$

$$+ R'_i(\xi) \tag{4.11}$$

where M_V, M_F and M_S are the masses of the superheavy vector, fermion and scalar fields respectively and t^H_{iV}, t^G_{iF} and t^H_{iS} are the matrices which represent the generators of G on the superheavy vector fields, their associated Goldstone bosons and the superheavy fermion and scalar fields with the Goldstone bosons projected out. In fact, of course, $\mathrm{Tr}(t^G_{iS})^2 = \mathrm{Tr}(t^H_{iV})^2$ and we shall use this henceforth.

When eqs.(4.8) to (4.) are substituted into eqn.(4.5), the final result is then

$$\alpha_i^{-1}(\mu) = C + \frac{1}{16} \mathrm{Tr}[(-11(t^L_{iV})^2 + 4(t^L_{iF})^2 + \frac{1}{2}(t^L_{iS})^2) \ln(\frac{M_X}{\mu})$$

$$+ \frac{1}{2}(t^{\cdot\cdot}_{iV})^2 + 4(t^{\cdot\cdot}_{iF})^2 \ln(\frac{M_X}{M_F}) + \frac{1}{2}(t^{\cdot\cdot}_{iS})^2 \ln(\frac{M_X}{M_S})]$$

$$- \frac{1}{8\pi^2} \sum_j \beta'_{ij} \int_{\ln\mu}^{\ln M_X} \hat{\alpha}_j(\mu')d \ln\mu' + L(\hat{\alpha}_1^2) \tag{4.12}$$

where by going to the minimal subtracted form $\hat{\alpha}_i$ the dependence on $R_i(\xi)$ is removed.

In the standard SU(5) model, with Higgs mesons in a 24 and a 5 dimensional representation, eqn.(6.12) which gives the minimal subtraction coupling constants for $SU(3)_C \times SU(2)_L \times \times U(1)$ reads;

$$\frac{1}{\hat{\alpha}_s(\mu)} = C + \frac{1}{6\pi}[A_3 \ln(\frac{M_X}{\mu}) + \frac{3}{2} \ln(\frac{M_X}{M_S}) + \frac{1}{2} \ln(\frac{M_X}{M_{S'}}) + \frac{3}{2}] + \tau_3 ,$$

$$\frac{1}{\hat{\alpha}_2(\mu)} = \frac{\sin^2\theta_W(\mu)}{\hat{\alpha}(\mu)} = C + \frac{1}{6\pi}[A_2 \ln(\frac{M_X}{\mu})+\ln(\frac{M_X}{M_3})+ 1]+T_2 \ ,$$

$$\frac{1}{\hat{\alpha}_1(\mu)} = \frac{3\cos^2\theta_W(\mu)}{5\hat{\alpha}(\mu)} = C + \frac{1}{6\pi}[A_1 \ln(\frac{M_X}{\mu})+ \frac{1}{5} \ln(\frac{M_X}{M_S'})]+T_1, \quad (4.13)$$

where M_X is the heavy vector mass and M_S and M_S' are the masses of the superheavy Higgs mesons of the 24 and 5 respectively, and

$$A_3 = 4N - 33 \ ,$$

$$A_2 = 4N - 22 + \frac{1}{2} \ ,$$

$$A_1 = 4N + \frac{3}{10} \ ,$$

$$T_i = - \frac{3}{4\pi} \sum_j \frac{\beta_{ij}}{A_j} \ln(1 + \frac{A_j}{6\pi C} \ln(\frac{M_X}{\mu})) \ , \quad (4.14)$$

where N is the number of fermion generations and

$$\beta_{ij} = \begin{pmatrix} 0 & 0 & 0 \\ 0 & 136/3 & 0 \\ 0 & 0 & 102 \end{pmatrix} -N_G \begin{pmatrix} 19/15 & 3/5 & 44/15 \\ 1/5 & 49/3 & 4 \\ 11/30 & 3/2 & 76/3 \end{pmatrix} .$$

$$(4.15)$$

Using these equations we may obtain $\alpha_s^{-1}(80)$ and $\sin^2\hat{\theta}_W(80)$ as functions of M_X and $\alpha^{-1}(80)$.

$$\alpha_s^{-1}(80) = 54.68 - 3.18 \log_{10}(M_X) + \frac{3}{8}(\alpha^{-1}(80) - 127.4)$$

and

$$\sin^2\hat{\theta}_W(80) = 0.399 - 0.013 \log_{10}(M_X) \ . \quad (4.16)$$

This gives for 50 MeV $\leq \Lambda_{\overline{MS}} \leq$ 500 MeV

$$M_X = 3.6 \{ {}^{+3.4}_{-3.2} \times 10^{14} \text{ GeV for } M_t = 20 \text{ GeV}$$

$$\text{and } \sin^2\theta_W(80) = 0.206 \, {}^{+0.016}_{-0.004} \, . \tag{4.17}$$

The fermion mass predictions of eq.(4.1) must also be radiatively corrected when applying them at a scale μ. For the b quark and the τ lepton this gives [13,17]

$$[\frac{m_b(\mu^2)}{m_X(\mu^2)}] = [\frac{\alpha_3(\mu^2)}{\alpha_3(M_X^2)}]^{\frac{4}{11-2/3n_f}} \, [1 + O(\frac{\alpha_s(\mu^2)}{\pi}) + ..] \, . \tag{4.18}$$

Evaluating this at the e^+e^- threshold momentum for producing γ gives (using m_t as input)

$$m_b = 5 - 5.5 \text{ GeV} \quad , \tag{4.19}$$

a result which is in agreement with experiment. However the results for m_s and m_d are not so good

$$m_s = 500 \text{ MeV} ,$$

$$m_d/m_s = \frac{1}{200} \, . \tag{4.20}$$

The value of m_s commonly quoted is $\simeq 150$ MeV [18]. However it is extracted from hadronic mass differences where the contribuition is related to matrix elements of $m_s\bar{\psi}_s\psi_s$. The conventional value for m_s assuming the matrix element of $\bar{\psi}_s\psi_s$ normalised at μ is unity, but this may be incorrect. Bag model estimates [19] give $\frac{1}{2}$ for this which would give a value nearer 300 MeV for m_s. Moreover the relation of this mass to the current quark masses predicted by SU(5) is unclear. Nonetheless the value for m_s does not look a good prediction and certainly the prediction for the ratio m_d/m_s in eq.(4.20) is bad when compared to the current algebra prediction [19]

$$\frac{m_d}{m_s} = \frac{1}{20} \, . \tag{4.21}$$

It is not clear how bad this result is for SU(5).
It has been pointed out [20] that small additional contributions to fermion masses coming from structure above M_X
may reasonably be expected at a level of a few MeV. This
would significantly affect the prediction for m_d, but not
spoil the predictions for the heavier quarks. It is possible
to play around with more complicated Higgs structure [21] to
obtain better predictions for m_d (and m_s), but we postpone
a discussion of this until we consider extensions of GUTs
beyond SU(5).

4.2 Nucleon Decay in SU(5) [22]

One of the features we noted above was the fact that the
X and Y gauge bosons did not conserve B and L separately but
only the combination (B-L). GUTs seek to combine quarks and
leptons in a single multiplet and so it is a general property
of them that there will be gauge interactions couplings leptons and quarks. This does not mean that baryon number is
violated. For example in eq.(3.17) the terms coupling X and
Y to the $\bar{5}$ all have $B = +\frac{1}{3}, L = +1$, so if this was the only
fermion representation one could ascribe these quantum numbers
to the $\bar{X}$ and $\bar{Y}$ fields and B and L would separately be conserved. It is the combination of this plus the coupling to
the 10, with quarks and antiquarks in the same representation,
that means B and L are separately violated for the terms
involving $u_L^c \gamma^\mu u_L$ have $B = \frac{2}{3}$, $L = 0$ and no choice of B and L
for X and Y will conserve B or L.

In the "age of the gauge", where all (continuous)
symmetries are expected to be gauge symmetries, one expects
the only conserved numbers to correpond to massless gauge
fields. Apart from the photon,gluons and graviton we know
of no massless fields so it seems reasonable to expect there
are no new absolutely conserved quantities such as B or L.
In SU(5), B and L are violated, but at a very slow rate because
of the large scale of breaking. What about the residual (B-L)

symmetry? In SU(5) this is an accidental <u>global</u> symmetry
(it is easy to check that it is preserved by the Yukawa
couplings of Fig.3). However further Yukawa couplings involving new Higgs representation will violate it at a rate depending on the mass of these new scalar states. Alternatively,
following our principle that all symmetries must be gauge
symmetries one may enlarge the gauge group to include the
(B-L) generator. Then (B-L) will be violated at a rate depending on the mass of new gauge boson. We will return to a
discussion of this point later.

In the minimal SU(5) scheme the X and Y interactions
of Fig.3 give rise to the interactions of Fig.7.

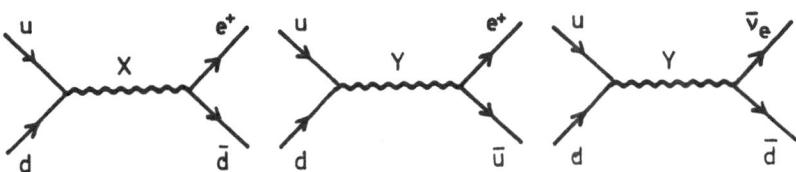

<u>Fig.7:</u> Graphs contributing to nucleon decay in SU(5).

At low energies ($<<M_X$) these give an effective 4-Fermi interactions which (neglecting mixing for the moment) is

$$\frac{1}{4} \, L = \frac{g^2}{8M_X^2} \, [(\varepsilon_{ijk} \bar{u}^C_{kL} \, \gamma_\mu \, u_{j\lambda}) (2\bar{e}^+_L \gamma^\mu d_{iL} + \bar{e}^+_R \gamma^\mu d_{iR})$$

$$- (\varepsilon_{ijk} \bar{u}^C_{kL} \gamma_\mu d_{jL}) (\bar{\nu}^C_e \gamma^\mu d_{iR}) + h.c.] \quad . \tag{4.22}$$

As discussed in Section 3.3, the mixing angles are those of the
Kobayashi Maskawa model with additional phases. This is important for there is no freedom to inhibit proton decay by
choosing mixing angles such that it decays only into heavy

generations. Keeping only the large Cabibbo mixing between
the first two generations eq.(4.22) becomes [14]

$$\frac{1}{4}L = e^{i\phi} \frac{g^2}{8M_X^2} [(\varepsilon_{ijk} \bar{u}_{kL}^c \gamma_\mu u_{jL}) \{ [(1+\cos^2\theta_c) \bar{e}_L^+ + \sin\theta_c \cos\theta_c \bar{\mu}_c^+]\gamma^\mu d_{iL}$$

$$+ [(1+\sin^2\theta_c) \bar{\mu}_L^+ + \sin\theta_c \cos\theta_c \bar{e}_L^+]\gamma^\mu s_{iL} + \bar{e}_R^+ \gamma^\mu d_{iR} + \bar{\mu}_R^+ \gamma^\mu s_{iR} \}$$

$$- [\varepsilon_{ijk} \bar{u}_{kL}^c \gamma_\mu (d_{jL}\cos\theta_c + s_{jL}\sin\theta_c)][\bar{\nu}_{eR}^c \gamma^\mu d_{iR} + \bar{\nu}_{\mu R}^c \gamma^\mu s_{iR}]]+ \text{h.c.}$$

$$(4.23)$$

This gives quantitative predictions for relative decay rates

$$\frac{\Gamma(N\to\mu^+ + \text{non-strange})}{\Gamma(N\to e^+ + \text{non-strange})} = \frac{\sin^2\theta_c \cos^2\theta_c}{(1+\cos^2\theta_c)^2+1}$$

$$\frac{\Gamma(N\to e^+ + \text{strange})}{\Gamma(N\to\mu^+ + \text{strange})} = \frac{\sin^2\theta_c \cos^2\theta_c}{(1+\sin^2\theta_c)^2+1} \qquad (4.24)$$

These predictions are specific to the minimal SU(5) scheme. If,
for example, there were an antisymmetric contribution to M^u
in eq.(3.34) they would not follow. They also rest on the
family groupings assumed in constructing the SU(5) model
which has only weak support from the mass relations derived
above. Measurement of these ratios would give detailed information about the structure of the underlying GUT.

4.3 Proton Lifetime

We now can try to estimate the proton lifetime following
from the Lagrangian of eq.(4.22). Gluon corrections to the
Born diagram give rise to terms involving $[\alpha_3 \log(\frac{M_X}{M_P})]^n$. These
potentially large terms may conveniently be summed by the
usual operator renormalisation group techniques [23]. The

terms of eq.(4.22) give rise to the combination $2\ O_1 + O_2$ where the operators $O_{1,2}$ are

$$O_1 = (\varepsilon_{ijk}\bar{u}^C_{RL}\gamma_\mu u_{jL})(\bar{e}^+_L\gamma^\mu d_{iL}) \quad,$$

$$O_2 = (\varepsilon_{ijk}\bar{u}^C_{kL}\gamma_\mu)(u_{jL}\bar{e}^+_R + d_{jL}\bar{\nu}^C_{eR})\gamma^\mu d_{iR} \quad. \tag{4.25}$$

The anomalous dimensions in leading order of O_1 and O_2 coming from gluon exchange give rise to the enhancement factors

$$A_3 = [\frac{\alpha_3(1\ \text{GeV})}{\alpha_3(M_X)}]^{\frac{2}{11-4/3N_g}} \quad. \tag{4.26}$$

Similarly one may compute [23] the enhancement due to $W^\pm$, Z and γ give

$$A_{21} = [\frac{\alpha_2(M_W)}{\alpha_2(M_X)}]^{\frac{27}{86-4N_g}} \times \begin{cases} [\frac{\alpha_1(M_W)}{\alpha_1(M_X)}]^{-\frac{69}{6+20N_g}} & \text{for } O_1 \\ \\ [\frac{\alpha_1(M_W)}{\alpha_1(M_X)}]^{-\frac{33}{6+20N_g}} & \text{for } O_2 \end{cases} \quad. \tag{4.27}$$

These gauge boson corrections sum the large logs coming from perturbative corrections to the fundamental processes of Fig.7. In addition one must estimate non-perturbative effects in going from the quark fields of eq.(4.25) to the proton i.e. we must compute the hadronic matrix elements of the operators O_1 and O_2. There are two contributions that have been estimated in several ways [24] to be of comparable magnitude (cf. Fig.8). The first involve a spectator quark and may be estimated using the bag model of SU(6). The second involves the meson emission first followed by the 3 quark overlap probability at the origin. It may be estimated by

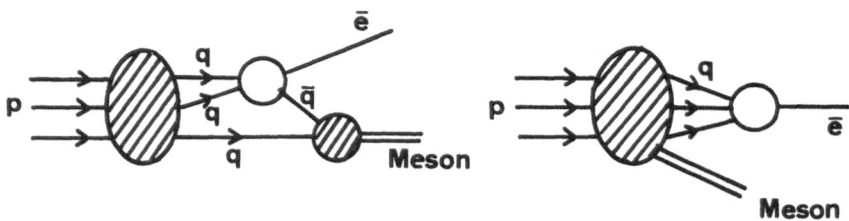

Fig.8: Operator matrix elements.

current algebra techniques. A variety of estimate have been
made [24] giving the range

$$t_{p,n} = (0.1 \text{ to } 5) \times 10^{30} \text{ yrs } \left(\frac{M_X}{5 \times 10^{14} \text{ GeV}}\right)^4 \qquad (4.28)$$

where a factor of $3\frac{1}{2}$ to 4 has been included from eqs.(4.27).

M_X as we discussed in the previous section depends
almost linearly on $\Lambda_{\overline{MS}}$ giving for a range (150 to 500) MeV
for $\Lambda_{\overline{MS}}$

$$m_X = (1 \text{ to } 6) \times 10^{14} \text{ GeV} \qquad (4.29)$$

and

$$t_{p,n} = \frac{1}{5} 10^{27} \text{ to } 10^{31} \text{ yrs } . \qquad (4.30)$$

The estimates of operator matrix elements also predict
the decay modes expected and these are given in Table 1 for
three representative calculations. The favoured mode is
$p \rightarrow e^+ \pi^0$ and currently many experiments are searching for this
signal [25]. A recently published result [25]gives a lower
limit for the lifetime of the channel of

Table 1 - Nucleon decay branching ratios in minimal SU(5)

	Decay Mode	Non-relativistic Model	Prefered "Recoil" Model	Relativistic Model
P	$e^+\omega$	21	25	26
	$e^+\rho^0$	2	7	11
	$e^+\pi^0$	36	40	38
	$e^+\eta$	7	2	0
	$\bar{\nu}\rho^+$	1	3	4
	$\bar{\nu}\pi^+$	14	16	15
	μ^+K^0	18	8	5
	$\nu_\mu K^+$	0	0	1
n	$\bar{\nu}\omega$	5	5	5
	$\bar{\nu}\rho^0$	1	1	2
	$\bar{\nu}\pi^0$	8	7	7
	$\bar{\nu}\eta$	2	0	0
	$e^+\rho^-$	6	12	19
	$e^+\pi^-$	79	72	68
	$\bar{\nu}_\mu K^0$	1	3	1

From Kane and Karl, ref.24.

$$\tau_{p \to e^+ \pi^o} \geq 6 \times 10^{31} \text{ yrs} \quad . \tag{4.31}$$

If this is true it is already outside the favoured range following from eq.(4.28) and Table 1.

4.4 Neutrino Masses

The Dirac terms introduced in eq.(2.5) and (3.34) are of the form $\bar{\nu}_L \nu_R$ +h.c.and the absence of a right handed ν_R state in both the standard model and in SU(5) means there is no such term for neutrinos. There is another possible Lorentz invariant mass term one can form using ν_L alone. This is the Majorana mass term $m\nu_L \nu_L$ (strictly it is $(m\nu_L^T C\nu_L +$ $+ m^* \nu_R^{cT} C\nu_R^c)$). This term is not possible in the minimal Weinberg-Salam model because it transforms as an $I = 1$, $I_3 = 1$ object and there are no $I = 1$ Higgs fields which could generate such a mass term on SSB. It also violates L number and L is conserved in the standard model. In SU(5) it is also forbidden because there are no $I = 1$ Higgs coupled to fermions and B-L is exactly conserved.

Thus in both the standard model and in minimal SU(5) there is no neutrino mass term either of the Dirac or Majorana type.

This is not an entirely convincing result for the absence of ν_R is put in by hand (nothing prevents us from adding an SU(5) singlet). Indeed in most generalisations of SU(5) such fields appear. Then SU(5) does not prevent a Dirac mass term $\bar{\nu}_L \nu_R$ arising from a coupling of the $\bar{5}$ fermions to the $\bar{5}$ of Higgs and the singlet ν_R field; and its natural scale would be of the order of a quark mass, m_q.

Moreover as discussed above it seems likely that (B-L) will at some level be broken so why we should expect higher order corrections to generate a term of the form $[H^o \nu_L][H^o \nu_L]$ where the two $I = \frac{1}{2}$ H fields have combined to supply the necessary $I = 1$ Higgs. One of the most elegant results of grand unification is that even if we include these terms we can understand [26] why neutrinos are light and different from other fermions. Let us see how this works.

If we allow both ν_L and ν_R fields the mass matrix involving both Dirac and Majorana masses is of the form [26]

$$(\nu_L, \ \bar{\nu}_R) \begin{pmatrix} m_1 & m_2 \\ m_2 & m_3 \end{pmatrix} \begin{pmatrix} \nu_L \\ \bar{\nu}_R \end{pmatrix} \ . \tag{4.32}$$

As we have already discussed it is reasonable to choose $m_2 = O(m_q)$ on purely dimensional grounds. m_1 coming for example from a $(H^0 \nu_L)^2$ term will be of order $\nu_0^2/M_{X'}$, where $M_{X'}$ is the scale associated with (B-L) violation ($M_{X'} > M_X$). One might worry that a larger contribution will arise if there is an $I = 1$ Higgs field in the complete theory. It has been argued however [27] that this field will in general acquire a vev in a manner analogous to the way the I-1 component of the Σ field as in eq.(3.29) acquired a vev ε. Then $\varepsilon = O(\nu_0^2/M_{X'})$ where $M_{X'}$ is the new scalar state in the same prediction for the order of magnitude of m_1 applies.

Finally what is the expected size of m_3? Since ν_R is an SU(5) singlet there is no symmetry reason forbidding a $\nu_R \nu_R$ mass term. It is likely therefore that this term will be of magnitude $O(M_X)$. Putting all this together we need to diagonalise a mass matrix of the form

$$(\nu_L, \ \bar{\nu}_R) \begin{pmatrix} \sim \dfrac{m_q^2}{M_X} & m_q \\ m_q & M_X \end{pmatrix} \begin{pmatrix} \nu_{L'} \\ \bar{\nu}_R \end{pmatrix} \ . \tag{4.33}$$

The mass eigenstates are $\nu_L + \varepsilon \nu_R$ and $\nu_R + \varepsilon \nu_L$ where $\varepsilon = m_q^2/M_X^2$ with masses $\sim \dfrac{m^2}{M_X}$ and $\sim M_X$ respectively. Due to the enormous mass M_X particular to GUTs the neutrino mass is expected to be very small $\sim O(10^{-6} - 10^{-3}$ eV). Somewhat larger neutrino masses are possible in particular models, but we will postpone a discussion of these until the next lecture.

4.5 Θ_{QCD} and Axionatics

It has been pointed out that non-perturbative phenomena in QCD require the addition of a new parameter Θ_{QCD} [18] to characterise the way topologically distinct contributions of gluon fields must be added in the functional integral

$$\int D[G_\mu] \rightarrow \sum_{n=-\infty}^{\infty} e^{in\Theta_{QCD}} \int D[G_\mu] \tag{4.34}$$

where n is the Pontryagin index of the gauge fields. This has the effect of introducing to L_{QCD} the term

$$\frac{\Theta_{QCD}}{32\pi^2} G_{\mu\nu}^a \tilde{G}^{a\mu\nu} \tag{4.35}$$

where G and $\tilde{G}$ are defined in eqs. (2.2) and (2.17). This para- meter violates parity (note the ε symbol in the defn. of $\tilde{G}$) and CP (The colour singlet product of two gauge fields is C even).

A term such as in eq.(4.35) also arises from the ano- maly in the axial current

$$A_\mu = \sum_{q=1}^{N} \bar{q} \gamma_5 \gamma_\mu q \tag{4.36}$$

which gives

$$\partial^\mu A_\mu = i\sum_q m_q \bar{q} \gamma_5 q + \frac{g^2}{16\pi^2} G_{\mu\nu}^a \tilde{G}^{a\mu\nu} \quad .$$

Clearly an axial transformation through an angle σ will rotate

$$\Theta_{QCD} \rightarrow \Theta_{QCD} - 2\sigma N \tag{4.37}$$

but the CP violating effect will reappear in the quark mass

matrix

$$-\sum m_q \; \bar{q} \; \gamma_5 \; q \; \rightarrow \; -\sum m_q \; \bar{q} \; \gamma_5 \; q \; + \; i\Theta_{QCD}\left(\frac{m_u m_d m_s}{m_u m_d + m_d m_s + m_u d_s}\right)$$

$$\times \; (\bar{u} \; \gamma_5 \; u \; + \; \bar{d} \; \gamma_5 \; d \; + \; \bar{s} \; \gamma_5 \; s) \; . \tag{4.38}$$

Θ_{QCD} is severely limited because a flavour conserving CP violating term of this form contributes to the neutron dipole electric moment d_n via the amplitude

$$<n|i\int d^4x \sqcup J_\mu^{em}(0)|n> \quad .$$

Using eq.(4.35) for the CP violating part of L one may estimate d_n. This gives [28]

$$(d_n/e) \; \approx \; 3 \times 10^{-16}\Theta_{QCD} \; cm \tag{4.39}$$

to be compared with the experimental bound

$$|(d_n/e)| \; \leq \; 6 \times 10^{-25} \; cm \tag{4.40}$$

giving the bound

$$|\Theta_{QCD}| \lesssim 2 \times 10^{-9} \quad . \tag{4.41}$$

How can this come about? Ellis et al [29] have argued that if Θ_{QCD} is zero at some high energy scale (M_{Planck}?) for a reason we will learn about only when we construct the final theory then radiative corrections will make Θ_{QCD} nonzero at low energies. However their estimate of these corrections is very small, $\delta\Theta_{QCD} \lesssim O(10^{-32})$, because they appear only in very high order of perturbation theory.

However one may not be satisfied with an explanation that appeals to our ignorance and another route has been suggested by Peccei and Quinn [9,30]. Note first that if $m_q=0$ then the effect of Θ_{QCD} vanishes in eq.(4.38). This is because there is then an exact axial symmetry rotating the q_i fields. We know that $m_{qi} \neq 0$ so this route is denied us but it is possible an exact axial symmetry exists when the Higgs sector generating masses is considered.

$$L_{Yuk} = \sum_{j,k} c_{jk}(u_j',d_j')_L \binom{\phi_2^o}{\phi_2^-} u_{k_R}' + \tilde{c}_{jk}(u_j',d_j')_L \binom{\phi_1^+}{\phi_1^o} d_{k_R}' \quad . \quad (4.42)$$

Performing rotations σ_L, σ_R, σ_R^D, σ_1 and σ_2 on the left handed fields, the right handed up fields, the right handed down fields ϕ_1 and ϕ_2 respectively we find L_{Yuk} is invariant if

$$-\sigma_L + \sigma_1 + \sigma_R^D = 0 \quad ,$$

$$-\sigma_L + \sigma_2 + \sigma_R^U = 0 \quad . \quad (4.43)$$

In the standard model $\phi_2 = \phi_1^+$ and $\sigma_2 = -\sigma_1$. Solving eq.(5.19) gives

$$\sigma_R^U + \sigma_R^D = 2\sigma_L \quad . \quad (4.44)$$

So there is no net axial rotation and Θ_{QCD} may not be rotated away. However if a new double Higgs field ϕ_2 is added the constraint eq.(4.44) does not follow and there is a (Peccei-Quinn) axial U(1) which rotates Θ_{QCD} to zero.

Of course there are now two residual charged and two neutral Higgs bosons after spontaneous symmetry breakdown. One is the pseudo Goldstone boson [31] of the $U(1)_{PQ}$. It acquires a mass because the $U(1)_{PQ}$ is broken by non-perturbative QCD effects which generate nonzero $\langle 0|\bar{q}_L q_R|0\rangle$ and break chiral symmetry. Using current algebra techniques one finds

$$m_a = \frac{N_G m_\pi f_\pi}{(m_u + m_d)^{1/2}} \left[\frac{m_u m_d m_s}{m_u m_d + m_d m_s + m_s m_u}\right] \frac{2^{1/4} G_F^{1/2}}{\sin 2\alpha} \quad (4.45)$$

where

$$\tan\alpha = \frac{<0|\phi_1|0>}{<0|\phi_2|0>} \qquad . \qquad (4.46)$$

The axion couplings are

$$L = i(2^{1/4}G_F^{1/2})a[m_u\bar{u}\gamma_5 u \tan\alpha + m_d\bar{d}\gamma_5 d \cot\alpha$$

$$+ m_e \bar{e}\gamma_5 e[\tan\alpha \text{ or } \cot\alpha]] \qquad . \qquad (4.47)$$

It seems that such an object does not exist, for experiments sensitive to it have not found it [30]. However the picture dramatically changes in GUT's for in such theories fields with vev's of $O(M_X)$ exist. If $U(1)_{PQ}$ is broken by such a field eq.(4.45) is modified by a factor (V_o/V) and so is eq.(4.47) for the couplings; making the axion "invisible" [32]. The simplest modification simply introduces two 5's H_1 and H_2 and a complex 24. With fermion couplings

$$f_{\bar{5}}\ f_{10}\ \bar{H}_1 \ , \qquad f_{10}\ f_{10}\ H_2 \ ,$$

and Higgs self couplings between ϕ and H_i

$$\bar{H}_1\ \phi^2\ H_2 \ , \qquad \bar{H}_1\ H_2\ \text{Tr}(\phi^2) \ ,$$

the axion mass becomes

$$m_a \sim \frac{f_\pi m_\pi}{10^{15}\ \text{GeV}} \sim 10^{-8}\ \text{eV} \ ,$$

$$g_{a\bar{f}f} \sim \frac{m_f}{10^{15}\ \text{GeV}} \sim 10^{-15} \qquad (4.48)$$

invisible to all laboratory experiments.

However there is another twist to the story, for Sikivie [33] has pointed out cosmological problems with the invisible axion. Due to the $U(1)_{PQ}$ the vacuum has a continuous degeneracy split only by the nonperturbative effects which give the axion its mass. These however leave a Z_N discrete subgroup of $U(1)_{PQ}$ invariant (cf. eq.(4.37) with $\sigma = \pi/N$) giving degenerate vacua separated by domain walls, which dominate the energy of the universe. It is possible to complicate the model to overcome this problem [34] but one final problem remains [35]. This is that the rate the axion field acquires its vev to minimise the PQ symmetry breaking is too slow and the stored energy in the axion field again dominates the energy of the universe.

4.6 SU(5) - A Critique

In section 1 we motivated the need for GUTs by listing many unsatisfactory features of the standard model. Let us take a look at how well SU(5) answers these critisisms.

(1) By fitting a family into a $\bar{5}$ + 10 the multiplet structure has simplified and the LH structure of quarks and leptons is better. However there is no explanation of the family structure and there is no connection between the vector, fermion and scalar representations. Although anomalies cancel between the $\bar{5}$ and 10 it is unclear why this happens.

(2) Charge is quantised and the third integral nature of quark charges is explained - a remarkable result. The gauge couplings are all related and the prediction for $\sin^2\theta_W$ is in good agreement with experiment. The value for M_X is such that proton decay should be seen soon.

(3) The situation with Yukawa couplings is not so good. There are three mass predictions which have varying success. There is no prediction for the Kobayashi Maskawa angles and there are (N_g-1) additional phases which contribute in proton decay. Similarly there is no im-

provement in the scalar potential where more couplings are needed in SU(5) than in the standard model.

(4) The appearance of a large mass scale is a mixed blessing. It explains why the proton should decay slowly and why neutrinos should have a low mass, but there is no explanation of why this mass should be so much greater than the weak interaction breaking scale. Indeed to accomodate these two scales it is necessary to fine tune parameters in the scale potential to one part in 10^{13}. It does not include gravity even though M_X is approaching the Planck mass.

It is clear that SU(5) is not the final theory. However many of its properties are so pretty that many of us would be very unhappy to give them up. The GUT addict sees SU(5) as the first approximation and hopes that the missing elements will be supplied by more complete theory. What could this theory be?

5.1 Beyond SU(5)

SU(5) is the unique rank 4 group which can grand-unify SU(3)×SU(2)×U(1). However, as we have emphasized, it leaves many questions unanswered. Can we improve on it? The first, obvious, possibility is to try larger groups which may contain in a single irreducible representation the states of a given spin. Here we discuss briefly some of the attempts to generalise SU(5).

5.2 SO(10) [36]

The only group of rank 5 which contains SU(3)×SU(2)×U(1) and accomodates the fermions is SO(10) whose complex 16-dimensional spinorial representation has the SU(3)×SU(2) decomposition

$$\underline{16} = (3,2) + 2(\bar{3},1) + (1,2) + 2(1,1) \quad . \tag{5.1}$$

This accomodates a generation (eq.(2.4)) supplemented by a neutrino state ν_R or $\bar{\nu}_L$. The SU(5) decomposition is

$$16 = 10 + \bar{5} + 1 \; . \tag{5.2}$$

Orthogonal groups are anomaly-free and so embedding SU(5) in SO(10) neatly explains why the SU(5) anomalies arising from the $\bar{5}$ and 10 miraculously cancelled. Unfortunately there is considerable freedom introduced by enlarging our group to SO(10) for there are many ways of breaking it. For example

$$SO(10) \begin{array}{l} \nearrow SU(5) \longrightarrow \\ \searrow SU(4) \times SU(2) \times U(1) \nearrow \end{array} SU(3) \times SU(2) \times U(1) \rightarrow SU(3) \times U(1) \; . \tag{5.3}$$

In SU(5) the unification scale M_X and $\sin^2\theta_W$ are unique because there is no larger subgroup of SU(5) that contains SU(3)×SU(2)× ×U(1) but clearly this is not true for SO(10). Since SU(5) is a subgroup of SO(10), charge quantisation works the same way and $\sin^2\theta_W$ in the symmetry limit is as derived in eq.(3.12). However the renormalized prediction varies according to the method of breaking SO(10) (some involve several different scales) and it is easy to change the proton decay rate predic- tion by a factor 10^3. The minimal version of SO(10) has a breaking pattern

$$SO(10) \rightarrow SU(5) \rightarrow SU(3) \times SU(2) \times U(1) \rightarrow SU(3) \times U(1)$$

$\underline{16}$ of Higgs   45 of Higgs   10 of Higgs   (5.4)

The $10 = 5 + \bar{5}$ of SU(5) and gives the mass relations of eq.(3.36) plus new relations for the Dirac masses

$$m_u = m_{\nu_e} \; , \quad m_c = m_{\nu_\mu} \; , \quad m_t = m_{\nu_\tau} \; . \tag{5.5}$$

Majorana (diagonal) mass terms for ν's can only be generated via $\underline{126}$ of Higgs absent in the minimal scheme of eq.(5.3). If it is included the SU(5) singlet component has a vev of order $>10^{15}$ GeV but the SU(5) 15 component is naturally of order v_o^2/v. It thus leads to the neutrino mass matrix of eq.(4.33).

A more interesting case was originally discussed by Witten [37]. With only the minimal Higgs content of eq.(5.3) higher order corrections of the form in Fig.(9) generate an effective 126 of order (α/π^2), giving ν_R a Majorana mass

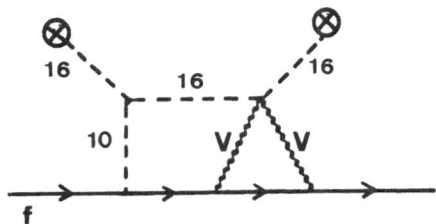

Fig. 9: Diagram generating Majorana masses for ν_R in the minimal SO(10) model

Because this mass term m_3 enters in the denominator for the light neutrino mass, in this model neutrinos are expected to be heavier and proportional to the associated up quark in their generation. Thus $m_{\nu t}$ could be as large as O(10) eV.

It is of course possible to enlarge the Higgs sector [36] allowing for more general mass formulas than those of eqs.(3.36) and (5.4). In SO(10) a relatively simple generalisation has been devised leading to the (phenomenologically better) relations

$$m_d = 3m_e, \qquad m_s = 3m_\mu, \qquad m_b = m_\tau . \tag{5.6}$$

However the moment the minimal Higgs structure is relaxed one realises there is enormous freedom to get practically any mass relation one wants. We will discuss a more promising method for generating the fermion mass sprectrum in the next section.

5.3 Family Groups [38]

Although SO(10) simplifies the fermion representation content it still requires three Xerox copies. Can larger groups accomodate all these in a single representation? The group should contain complex representations to allow us to distinguish between L and R, but should be real with respect to $SU(3) \times U(1)_{em}$. It should also be anomaly free, E_6 has a complex representation but must be Xeroxed. $E_7 + E_8$ have only real representations; as do the symplectic groups. This leaves orthogonal and unitary groups.

Orthogonal groups are anomaly free, SO(4n+2) for $n \geq 3$ has the SO(10) decomposition for its complex spinorial representation

$$2^{2n} = 2^{2n-5} \ \underline{16} + 2^{2n-5} \ \underline{\overline{16}} \quad . \tag{5.7}$$

Although clearly real with respect to SO(10) it may have a complex representation with respect to some other subgroup. For example SO(18) has a spinorial representation which decomposes with respect to $SU(4) \times SU(3) \times SU(2)$ as

$$\underline{256} = [\underline{4} + \underline{\overline{4}}] \ [(3,2) + 2(\overline{3},1) + (1,2) + 2(1,1)]$$

$$+ [\underline{6} + 2 \times \underline{1}] \ [(\overline{3},2) + 2(3,1) + (1,2) + 2(1,1)] \ . \tag{5.8}$$

It has been suggested the SU(4) may be used as technicolour group to make a technicolour GUT. SO(18) can give three normal families with a technicolour SO(5). However there are so many fermions that gauge couplings get large before unification and the success of $\sin^2\theta_W$ prediction would be purely accidental.

Unitary groups have also been tried for family groups. The [m]-fold totally antisymmetric representation of SU(n) contains just the SU(5) representations 1, 5, $\overline{10}$, 10 and $\overline{5}$. One may use these representations to build up the observed families. For example the minimal such model with just three generators is SU(7) with the antisymmetric representations

$$1[1] + 2[2] + 8[6] \qquad (133 \text{ states}). \qquad (5.9)$$

Another example with no repeated representations occurs for SU(11)

$$[4] + [8] + [9] + [10] \qquad (561 \text{ states}). \qquad (5.10)$$

However no-one yet understands why these representations should be used; there is clearly something missing.

Another possibility that has been explored is a discrete unification in which there is a family symmetry with the same structure as the underlying group giving for example SU(5)×SU(5) with a discrete symmetry connecting the groups.

In all these approaches in which the family structure is gauged one has to beware of flavour changing neutral currents. The GIM mechanism works when gauge interactions act the same way on each family. The additional gauge symmetries associated with a family group will produce flavour changing neutral currents and to avoid conflict with experiment the family symmetries must be broken at a scale higher then 300 GeV. For this reason it is extremely important to look for exotic processes such as $K_L \to \mu^- e^+$ signalling a breakdown of GIM.

5.4 Radiative Fermion Masses [39]

Associated with the family problem is the question of
fermion masses. Why are families ordered in mass? SU(5)
gives no explanation, the masses being related to Yukawa-
couplings which are input parameters of the theory. An
appealing possibility is that fermions may be massless at
tree level, acquiring their mass only in radiative order.
Then it could be different families acquire their mass
in different order so that family masses will be ordered in
the ratio $1:\alpha:\alpha^2$. There are two ways in which fermions may
have calculable masses (1) the Yukawa coupling is present
at tree level but the Higgs does not get a vev at tree
level; (2) the Yukawa-coupling is not present at tree level
but no chiral symmetries prevent their acquiring a mass by
non-renormalisable terms.

(1) may happen when for symmetry reason L contains no terms
of the form ϕh, ϕh^2, ϕh^3 but nonrenormalisable terms like
ϕh^4 can occur radiatively. In this case no ϕ tadpole is
generated at tree level. Another possiblity is that a
symmetry (supersymmetry?) guarantees $<\phi> = 0$.

If all fermions are unified in one multiplet, then only
mechanism (1) works. In mechanism (2) there are no $ff\phi$ terms
in L by symmetry but the fermion representation is reducible
so that chiral symmetries do not forbid the generation of a
mass term and non-renormalisable terms such as ffh^2 will
generate calculable fermion masses. As an example consider
the three families falling into a doublet (e and μ) and singlet (τ) of an $SU(2)^F$ family group. The Yukawa terms in the
Lagrangian are

$$(16,2)_f(16,1)_f(16,2)_H + (16,1)_f(16,1)_f(16,1)_H$$

where we have exhibited (SO(10), $SU(2)^F$) transformation properties. The absence of a Higgs transforming at $(16,3)_H$ ensures the lightest family is massless at tree level, and its

mass arises only in radiative order and is calculable. Various
schemes have been constructed [39] which give masses to
the third generation at tree level, to the second generation
at one loop order and to the first generation at two loop
level. They have masses in the ratio $1:\alpha:\alpha^2$ as desired.
However these models still look contrived; one feels something is missing.

5.5 The Hierarchy Problem

Although the extensions of SU(5) proposed above offer
some hope for understanding the low energy parameters of
the standard model and the family structure they are still
fundamentally unsatisfactory candidates for the GUT because
they still treat fermions, vectors and scalars as independent
quantities and do not explain the origin of the parameters
of the model principally connected with the scalar sector.
This does not mean that the GUT will not be good as appro-
simate models valid until energy scalar M_X. However there
is a flaw in this picture which suggests that the models
break down at energy scales $\simeq 1$ TeV and cast doubt on the
whole philosophy of grand unification. This flaw is known
as the hierarchy problem. In the discussion of the breaking
of SU(5) we met this problem when trying to keep the doublets
of Higgs scalars $\bar{H}$ necessary to break SU(2)×U(1) light
($O(M_W)$) while breaking SU(5) at large ($O(M_X)$) scales. Without
an unnatural cancellation between parameters (accurate to
1 part in 10^{13}) the natural mass scale for H is M_X. That this
problem persists in all GUTs may be seen from the following
argument.

Once we accept that ultimately the parameters in the
standard model will be related (as in a GUT) we can treat
the radiative corrections to them as meaningful and not just
to be absorbed in a counter term. For example the graphs in
Fig.10 contribute to the mass M of the Higgs boson H. This
(running) mass depends on the scale μ at which it is measured

and calculation of the graphs of Fig.10 gives the result

$$M^2(\mu) = M^2(M_X) + \sum_i c_i M_X^2 \qquad\qquad (5.11)$$

where c_i and α_i are coefficients and couplings depending on M_X. For the necessary result [3] that $M^2(1\ \text{TeV}) \lesssim 1\ \text{TeV}$ we see that this may only be achieved by an unnatural cancellation of the large terms on the RHS of eq.(5.11). It is unnatural in the sense that the parameters of the microscopic scale M_X combine to give a low mass ($\approx$ massless) scalar only at the macroscopic scale μ,i.e.simplicity is only apparent at the macroscopic scale [40].

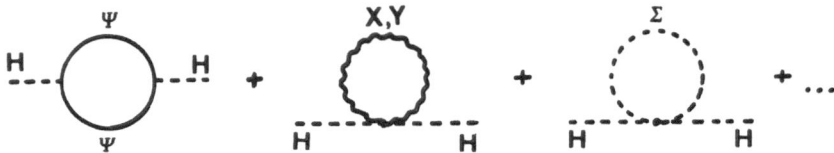

Fig.10: Graphs contributing to the Higgs scalar mass H. Σ and X are the additional Higgs and gauge bosons of the GUT and ψ are the fermions coupling to H.

This means that the standard model will break down at a scale $\approx$1 TeV, a scale which will be probed by the next generation of experiments! What could this new physics be?

There are three obvious possibilities. One is that the perturbative analysis leading to eq.(5.11) fails due to an interaction becoming strong. If this happens only in the scalar sector the GUT predictions may still be valid for gauge and Yukawa couplings will not be greatly affected. In

this case the departure from the standard model will be relatively difficult to see until the scalar mode is measured.
For example $e^+e^- \to W_L^+ W_L^-$ will have strong interactions in the
final state as longitudinal W^S are generated by Higgs scalars
(clearly this is not a first generation experiment!).

A second possibility is that there are no elementary
scalars, as in technicolour models [41] or in other composite
models. In this case the calculation leading to eq.(5.11)
fails for M_X scalar binding energy. Provided this energy
is $< \frac{1 \text{ TeV}}{\alpha}$ there will be no hierarchy problem.

Finally it may be that $(\sum c_i \alpha_i) = 0$. However it is not
enough to arrange this only in leading order perturbation
theory for then the next order will require $M_X < \frac{1 \text{ TeV}}{\alpha^2}$ and
so on. If we are to arrange this cancellation involving both
fermion and boson loops as in Fig.10 we need a symmetry and
the only known symmetry that can do this is supersymmetry [40].
Supersymmetry achieves this [43] by associating all scalars
with fermion partners so that the scalar mass is equal to
the fermion mass. Since fermion masses can be forbidden by
chiral symmetries the associated scalar mass too may be forbidden by a chiral symmetry. As a result in a supersymmetric
theory there are cancellations between fermion and boson
graphs of fig.(8) and eq.(5.11) is replaced by

$$M^2(\mu^2) = M^2(M_X^2) \left\{ \frac{\bar{g}^2(M_X^2)}{\bar{g}^2(\mu^2)} \right\} \tag{5.12}$$

so that $M^2(\mu^2)$ vanishes if $M^2(M_X^2)$ does. Thus there is no
strong constraint on M^2, and the supersymmetric model may be
valid up to the GUT scale.
Supersymmetry breaking in the gauge nonsinglet sector cannot
occur much above 1 TeV, otherwise this evasion of the hierarchy
problem, meaning that these theories will also be testable
in the immediate future.

5.6 Technicolour

If we consider the standard $SU(3)\times SU(2)\times U(1)$ model with no Higgs scalars we would naively think the W and Z bosons remain massless. This is not the case! Consider for example the inverse W boson propagator $P_{\mu\nu}(k)$. It has the form

$$P_{\mu\nu}(k) = \frac{g_{\mu\nu} - \dfrac{k_\mu k_\nu}{k^2}}{k^2\{1-g_W^2\dfrac{\pi(k^2)}{2}\}} \qquad (5.13)$$

where the term involving $\pi(k^2)$ arising from radiative correc-tions of the type shown in Fig.(11).

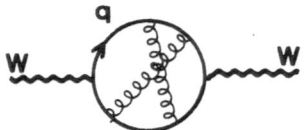

Fig.11: Typical graph contributing to $\pi(k^2)$. The curly lines are gluons.

Unless $\pi(k^2)$ is singular at $k^2 = 0$, $P_{\mu\nu}(k)$ will have a zero mass pole. However we know that, due to the strong interactions, $\pi(k^2)$ does have a zero mass pion pole. This results from the dynamical breaking of the chiral $SU(2)_L\times SU(2)_R$ symmetry and the pion in the resultant Goldstone boson. The contri-butions of this pole to $\pi(k^2)$ is

$$\pi(k^2) = \frac{f_\pi^2}{2}\,\frac{1}{k^2}$$

and

$$P_{\mu\nu}(k) = \frac{g_{\mu\nu} - \dfrac{k_\mu k_\nu}{k^2}}{k^2 - g_W^2 f^2/4} \quad , \tag{5.14}$$

with f_π = 100 MeV this gives a mass of 30 MeV to M_W.

Of course this mass is far too small. However a reasonable theory may be built with realistic masses for the W bosons by postulating [41] a new strong gauge interaction called technicolour and a type of quark, the techniquark, carrying both flavour and technicolour quantum numbers. The techniquarks and technicolour gluons replace the quarks and gluons in Fig.(11) and, provided the technicolour interaction becomes strong at a scale $\stackrel{\sim}{\sim}1$ TeV the mass generated for the W boson due to dynamical breakdown of the technichiral symmetry is of the right magnitude as can be seen in the translation table (2). An extremely attractive feature of this idea is that as there are no elementary scalars with their associated couplings everything can, in principle, be determined in terms of the gauge couplings of the theory.

Technicolour theories are rich in new phenomena, which should occur at a scale 1 TeV. The techniquarks will bind forming technihadrons with a mass 1 TeV. Moreover there are many pseudo-Goldstone bosons whose masses can be estimated in a specific model and some of which are 1 TeV. An example of the pseudo-Goldstone spectrum from an SU(N) technicolour theory is, with SU(2) doublets $\binom{U}{D}_{i=1...N}$ and $\binom{N}{E}_{i=1...N}$ transforming as quarks and leptons respectively und SU(3)$_c$, given in Table 3 [11,44]. In the table we see that there are light charged scalars $P^{\pm}$ whose mass is 8-14 GeV and light natural scalars P^0, P^3 whose mass is $\stackrel{<}{\sim}2.5$ GeV.

Charged scalars with mass $\stackrel{<}{\sim}13$ GeV would already have been seen at PETRA so this model is close to becoming disproved. There is another problem associated with technicolour theories which already rules out all models so far constructed.

Table 2 : Technicolour properties inferred from QCD

QCD	QT_cD
$\frac{g_c^2}{4\pi}$ (1 GeV) = 1	$\frac{g_{Tc}^2}{4\pi}$ (1 GeV) = 1
f_π = 100 MeV	$f_{\pi Tc}$ = 250 GeV
M_W = 30 MeV	M_W = 80 GeV
Hadrons $\sim$ 1 GeV	Technicolour $\sim$ 1 TeV
Pseudo Goldstone Bosons $\pi \sim$ 100 MeV	Pseudo Goldstone Bosons 2 GeV Upwards

The problem is that the conventional quarks and leptons remain massless as the technicolour condensate does not break the chiral symmetries associated with the light fermions. (In the standard model these symmetries are broken by Yukawa couplings with elementary Higgs scalars). To give masses to the light fermions it is necessary to introduce another gauge interactions, "extended technicolour", which couples light quarks to techniquarks, and generates light fermion masses via the graph of Fig.(12).

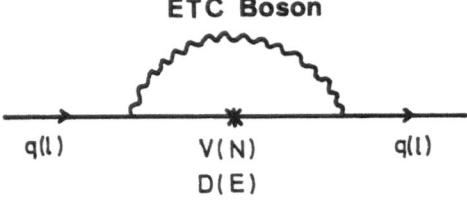

Fig.12: Graph generating light fermion (q,ℓ) masses in extended technicolour.

Table 3: Pseudo-Goldstone Boson States

	Colour	Charge	Mass(GeV)
P_8^0, P_8^3, $P_8^{\pm}$	8_C	$0,0,\pm1$	245
$P_{\bar{E}U}$, $P_{\bar{E}D}$, $P_{\bar{N}U}$, $P_{\bar{N}D}$	3_C	$\frac{5}{3},\frac{2}{3},\frac{2}{3},-\frac{1}{3}$	160
$P^{\pm}$	1_C	±1	5 to 8
P^0, P^3	1_C	0	0
$\pi^{0,\pm}$ (eaten up by z^0,$\bar{w}^{\pm}$)	1_C	$0,\pm1$	0

In order to generate the Cabibbo angle it is necessary for different generations of quarks to couple to the same generation of techniquarks so the SU(2) doublets $\binom{U}{D}$ and $\binom{N}{E}$ give masses to the quarks and leptons respectively via the graph of Fig.(12).

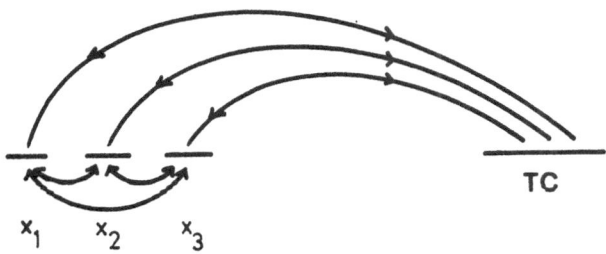

Fig.13: Pattern of ETC couplings. X_i are the light fermion generations and TC are the technifermions.

The (so far) insurmountable problem [45] follows from the fact that the extended technicolour couples quarks of different generations (cf.Fig.(13)) and generates large strangeness changing neutral currents. As usual the most dangerous is the $\Delta S=2$ $\Delta Q = 0$ processes, the amplitude being inversely proportional to the mass squared of the ETC boson (m^2_{ETC}). However this mass is constrained by the fact that light quark and lepton masses are also inversely proportional to m^2_{ETC}. In order to get the required light quark mass m_{ETC} must be so small as to generate an unacceptably large $\Delta S = 2$, $\Delta Q = 0$ amplitude [45].

To date no clearly acceptable way out of this problem has been proposed. Two solutions to the problem of flavour changing currents have been suggested. If asymptotic freedom fails, the estimates of Fig.10 have to be revised [46] and it is possible to accomodate a larger m_{ETC}, but what about the unseen pseudos? Another suggestion is to generate light quark masses via heavy scalar exchange [47] but again this requires a non-asymptotically free theory and looks contrived.

6. SUPERSYMMETRY

In our critique of SU(5) an outstanding problem was that it is not truly unified in the sense that scalars, fermions and vectors are not related and as a result there is a profusion of multiplets and couplings for which there is no good explanation. We have seen that technicolour, which simplifies the problem to one with only vectors and fermions, and no fundamental Yukawa couplings does not seem to work. We turn now to another approach which attempts to relate fermions, scalars and vectors by a new symmetry, supersymmetry.

In fact it has been shown [48] that it is the only

possible type of symmetry not yet used in addition to the
Lorentz group and internal symmetries based on Lie groups
(which commute with the Lorentz group). Supersymmetry relates
fermions and bosons and may provide a natural home for the elu-
sive scalar needed to spontaneously create gauge symmetries. It
relates matter and radiation, a massive supermultiplet containing spin components (the gauge fields) together with
spin $\frac{1}{2}$ fields and spin 0 Higgs scalar. It is presently the
only candidate for unifying particle physics and gravity (local supersymmetry generates gravitational interactions),
although the largest viable group with a nontrivial connection
between space-time and internal symmetries has a local gauge
symmetry SO(8) which is too small to accommodate the standard
model. As such it is probably relevant only as a theory for
constituent fields (at the Planck scale?).

Recently there has been renewed interest in a more
modest class of supersymmetric models, namely those based on
a direct product of the gauge group G with a global supersymmetry. The reason for this interest is that it may solve
the hierarchy problem discussed above. If it does, as we will
discuss in more detail, supersymmetry must be a good symmetry
at low energies O(1 TeV) (low, that is, relative to M_X or the
Planck scale!) and predicts the existence of new states below
the W mass and accessible to the next generation of accelerators.

6.1 Construction of a SUSY-GUT

The simplest supersymmetric model which can be constructed is a direct product of the internal symmetry gauge group
with a global (N = 1) supersymmetry [49,50,51,52,53]. The
basic building blocks are massless supersymmetry multiplets
[54] of the chiral or vector type as shown in Table (4).

A supersymmetric version of a GUT contains at least
twice the number of particles needed in the nonsupersymmetric

<u>Table 4:</u> Fundamental massless supersymmetric multiplets in
N = 1 supersymmetry

Chiral	$\binom{\psi}{\phi}$	2 components majorana fermion
		2 real scalar fields ($\equiv$ 0 complex)
Vector	$\binom{V_\mu}{\psi}$	2 component massless vector
		2 component majorana fermion

version. This happens because we cannot partner the quarks
and leptons with the gauge bosons because the former are
in an adjoint (real) representation while the latter are in
a fundamental (complex) representation. The only vectors in
a renormalisable theory are gauge bosons so we are forced to
assign quarks and leptons to a chiral supermultiplet with
scalar partners - the so called squarks and sleptons. The
gauge bosons are assigned to vector supermultiplets with fermion partners, the 'ino states. In addition, as we will see,
we cannot use these scalars to break the weak group so we
are forced to include separate Higgs chiral supermultiplets.
As a result we have failed in one of our immediate aims to
relate fermions, scalars and vectors. This will have to wait
until an energy scale (M_{Planck}?) at which the N>1 supersymmetries are relevant. However N = 1 supersymmetry still has the
potential to solve the hierarchy problem for, by associating
scalars with fermions, it is possible to forbid scalar masses
by forbidding(eq. by a chiral symmetry) the equivalent fermion
mass. What about the radiative corrections of Fig.(8)? Because
of supersymmetry quark and lepton Yukawa couplings are related to the couplings of their scalar partners in just such
a way to make fermion contributions (which have a minus sign
because of the fermion loop) cancel the scalar contribution.
Similarly gauge couplings are related to 'ino Yukawa couplings

in such a manner as to cancel the gauge boson contributions in Fig.(8) against fermion 'ino contributions.

Of course supersymmetry must be broken. We have seen no scalars degenerate with quarks or leptons, nor fermions degenerate with quarks or leptons, nor fermions degenerate with the photon or gluons. In this case the cancellations of the contributions of Fig.(8) are spoilt leaving a residual contribution to the light scalar masses

$$m^2_L = O(\Delta m^2_H \frac{\beta}{\pi}) \tag{6.1}$$

Thus the mass splitting Δm^2_H between supersymmetric partners must be limited

$$\Delta m^2_H \lesssim O(\frac{1 \text{ TeV}^2}{\beta})$$

where β is the coupling of the heavy multiplet to the light sector, e.g.the scalar doublet needed to break $SU(2) \times U(1)$.

If we consider the light gauge bosons (LG) and their gaugino partners, eq.(6.1) tells us that

$$\Delta m^2_{LG} \lesssim O(1 \text{ TeV}^2) \times \frac{\pi}{\alpha} \tag{6.2}$$

so their breaking scale may be somewhat larger than 1 TeV. If we consider chiral supermultiplets, their Yukawa couplings may be arbitrarily small, and their supersymmetry breaking mass splittings Δm^2 could be arbitrarily large. In the absence of an R-invariance forbidding gaugino masses, mass splittings between gauge non-singlet chiral superpartners would however feed through to the gauge supermultiplets by virtue of their gauge couplings, so that their supersymmetry breaking mass splittings are subject to the extra constraint

$$\Delta m^2_{GNS} \lesssim \Delta m^2_{LG} \times (\frac{\pi}{\alpha}) \lesssim O(1 \text{ TeV}) \times (\frac{\pi}{\alpha})^2 \tag{6.3}$$

However Yukawa shielding could easily enable gauge singlet chiral superfields to have arbitrarily large mass splittings perhaps even as large as the Planck mass M_p. In this case we would have to worry about the possible effects of gravitational couplings between the gauge singlet and non singlet sector.

To summarise the requirement that the hierarchy problem should be solved by supersymmetry puts strong bounds on the mass of gauge non singlet superpartners. This has enormously important phenomenological implications for it means these states should be found by the next generation of machines - supersymmetry theories are testable!

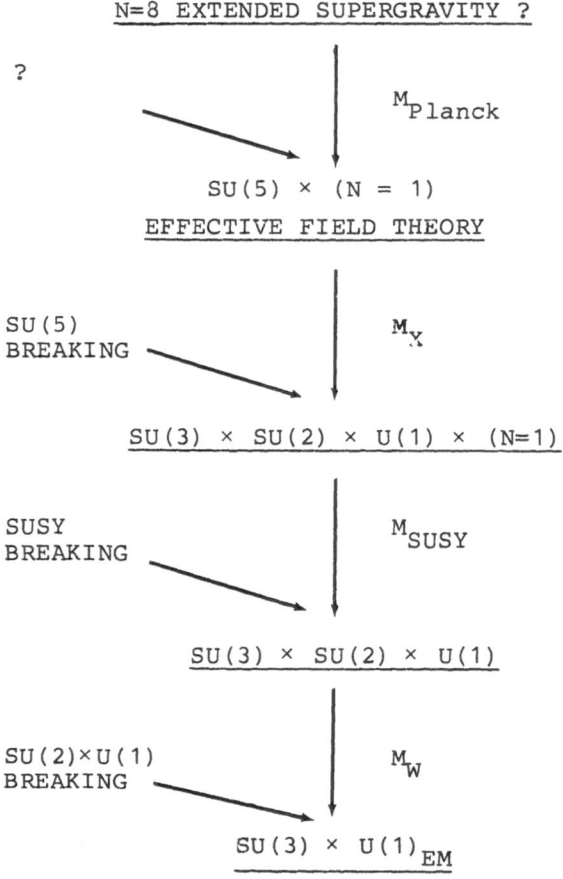

N=8 EXTENDED SUPERGRAVITY ?

Fig.14: Flow diagram for construction of your favourite SUSY MODI

We are now able to specify the necessary ingredients of a viable SUSY GUT. This is summarised in the flow chart of Fig.(14). At a scale $O(M_{Planck})$ we have an effective theory e.g. $SU(5) \times (N=1$ supersymmetry) (presumably originating in extended supergravity). It is necessary to introduce three stages of symmetry breaking with scalar M_X, M_{SUSY} and M_W as shown and it is at this stage that there is considerable freedom in choosing the model. In the next section we discuss the simplest SUSY-GUTs based on global supersymmetry and after that we consider the extension of these models to local super-symmetry.

6.2 An SU(5) SUSY GUT

The first step in our program is to choose our $N = 1$ supermultiplets. As discussed above the gauge fields must be assigned to an adjoint vector supermultiplet. Their fermion partners are known as the gaugino states - gluons have gluino partners, W's have Wino partners, the photon the photino etc. The quarks and leptons are assigned to chiral multiplets transforming as $N_g \times (\bar{5} + 10)$ under SU(5). Finally we have to choose the Higgs scalars. It is obviously tempting to try to use some of the new scalar states introduced as partners to the $\bar{5}$ of fermions, for example the sneutrinos. However the Higgs doublet scalars (H_D) must have Yukawa couplings to fermions to give them mass as in eq.(3.34). Consequently their colour triplet partners (H_T) in the $\bar{5}$ will mediate baryon number violating interactions. These will mediate proton de-cay and if this is not to occur we must have $M_{H_T} \gtrsim 10^{10}$ GeV. If we identify H_D with the sneutrino, selectron members of a $\bar{5}$ then H_T will be partners of the down antiquarks. Since, as discussed above, supermultiplet splitting in a gauge non sing-let representation may not be $O(1 \text{ TeV}^2) \frac{\pi}{\alpha}$ the H_T bosons will be much lighter then 10^{10} GeV and mediate proton decay for too fast. Consequently we are forced to choose new chiral super-multiplets to accommodate the Higgs scalars. Before we do this

we need to discuss the interactions of the states in our model. The gauge transformation properties of our states specify their interaction with gauge bosons via the gauge
principle in the usual way. In addition there are couplings
with gauginos related to the corresponding couplings of gauge
bosons by the replacement of the gauge boson by gaugino and
the replacement of one of the other states at the vertex by
its supersymmetric partner.

The Yukawa and scalar couplings of the model are also
related by supersymmetry. This is most simply expressed by describing these couplings in terms of a superpotential P. P
is a gauge invariant function of dimension ≤ 3 constructed
from the chiral superfields of the model (but not their
complex conjugates). Then the Yukawa and scalar couplings are
given in terms of P by

$$\text{Yukawa} = \sum_{i,j} \frac{\partial^2 P}{\partial \phi_i \partial \phi_j} \psi_i \psi_j \quad , \qquad (6.4)$$

$$\text{scalar} = \sum_i \left| \frac{\partial P}{\partial \phi_i} \right|^2 = \sum_i F_i^* F_i . \qquad (6.5)$$

Here ψ, ϕ refer to the (LH) fermion and scalar components of
the chiral supermultiplets respectively and the sums over
i,j run over all the chiral supermultiplets. The F_i are the
auxiliary fields.

We can now return to the question of how to choose the
Higgs supermultiplets. In order to break SU(5) it is necessary
to introduce an adjoint chiral supermultiplet. What about
SU(2)×U(1) breaking? The coupling of eq.(3.34) giving rise
to quark masses cannot be duplicated here using a single $\bar{5}$
of Higgs fields for it would need the complex conjugate state
transforming as a 5 and the rules for constructing P forbid
this. Consequently we are forced to choose two chiral supermultiplets $\bar{H}$ and H transforming as $\bar{5}$ and 5 respectively under
SU(5). The superpotential generalising eq.(3.34) giving quark

and lepton masses is

$$P_{5M} = M_{ij}^{(d)} \psi_i^\alpha \chi_{\alpha\beta} \bar{H}^\beta + M_{ij}^{(u)} \epsilon^{\alpha\beta\gamma\delta\rho} \chi_{i_{\alpha\beta}} \chi_{j_{\gamma\delta}} H_\rho \qquad . \qquad (6.6)$$

In addition we must ensure that $\bar{H}$ and H are split so that their triplet components have mass $\geq 10^{10}$ GeV and the doublet components are light $<O(M_W)$. The final multiplet choice of our $SU(5) \times [N=1]$ model is given in Table 5.

Table 5: Chiral supermultiplets used in model-building.

Role	Notation	SU(5) Representation Content
Matter	ψ_a^α, $\chi_{a\ \alpha\beta}$	$N_G \times (\underline{5} + \underline{10})$
Higgs	$H_{1\alpha}$, H_2^α Σ_α^β Z	$(\underline{5} + \underline{5})$ $\underline{24}$ $\underline{1}$
O'Raifeartaigh Sector	A,B,C	$3 \times (\underline{1})$
Couplings to Gauge Sector	$\phi_{1\alpha}$, ϕ_2^β	$(\underline{5} + \underline{5})$

6.3 SU(5) Breaking

Following our flow diagram of Fig.(12) we now discuss SU(5) breaking. Here there is considerable freedom which we will discuss later but first we consider a simple example in which terms $P_{5\Sigma}$ and P_{5T} are added to the superpotential

$$P_{5\Sigma} = \frac{\beta_3}{3} \text{Tr}(\Sigma^3) + \frac{\beta_2}{2}T \qquad + \beta_1 X M^2 \ , \qquad (6.7)$$

$$P_{5T} = \beta_4 X Y^2 + \beta_5 M Y^2 \ . \qquad (6.8)$$

From eq.(6.5) we see that $F_\Sigma^* F_\Sigma$ may be zero for multiple choices of the vev of Σ. In particular,

$$\langle \Sigma \rangle = 0 \ , \qquad (6.9)$$

$$\langle \Sigma \rangle = \frac{\beta_2}{3\beta_3} \begin{pmatrix} 1 & & & & 0 \\ & 1 & & & \\ & & 1 & & \\ & & & 1 & \\ 0 & & & & -4 \end{pmatrix} \langle X \rangle \ , \qquad (6.10)$$

$$\langle \Sigma \rangle = \frac{\beta_2}{\beta_3} \begin{pmatrix} 2 & & & & 0 \\ & 2 & & & \\ & & 2 & & \\ & & & -3 & \\ 0 & & & & -3 \end{pmatrix} \langle X \rangle \ , \qquad (6.11)$$

will leave $F_\Sigma^* F_\Sigma$ zero. These vev's leave SU(5) unbroken, break SU(5) to SU(4)×U(1) or to SU(3)×SU(2)×U(1), respectively. It has been pointed out [54] that there are further possibilities which may be obtained from the above by an SL(5,C) transformation which breaks the gauge symmetry further. We will consider these other possiblities later.

It is a characteristic of many superpotential models, following from the positive definiteness of the potential, that there should be multiple degenerate minima before super-symmetry breaking effects are included. Which of the minima will be selected? One might hope that radiative corrections will split the degeneracy and the absolute minimum will be the one required. However, due to the non-renormalisation of terms in the superpotential in the supersymmetry limit, this degeneracy will only be lifted by supersymmetry breaking effects. In the gauge sector supersymmetry breaking must be very small on the scale of M_{Planck}, typically of order 1 TeV, otherwise the hierarchy problem will reappear. Thus any splitting of the degenerate minima will be of order 1 TeV, whereas the barrier between the minima is of typical height $O(m_X)$. This means that even though the true minimum may be the desired $SU(3) \times SU(2) \times U(1)$ one, if the universe is initially in one of the other minima the probability for tunelling into the true one, proportional to $\exp[-O(m_X/1 \text{ TeV})^4]$, will be quite negligible [55,56]. In this case, it is the temperature-dependent terms in the effective potential that determines the minima chosen [55]. These terms, irrelevant at present temperatures, were crucial in the evolution of the Universe after the Big Bang and split the degeneracy of the zero temperature potential. To a good approximation the temperature-dependent terms given by [57]

$$V_1(t) = \frac{-\pi^2}{90} (N_B + \frac{7}{8} N_F) + \sum_{ij} \left| \frac{\partial^2 P}{\partial \phi_i \partial \phi_j} \right|^2 \frac{T^2}{8} + \frac{T^2}{2} g^2 \sum_i C(\phi_i) |\phi_i|^2$$

(6.12)

where $N_{B,F}$ are the numbers of light (mass $<<T$) degrees of freedom, the sum i,j runs over these light states and the $C(\phi_i)$ are the quadratic Casimirs of the multiplets.

At very high temperature, the T^4 terms dominate and the $SU(5)$ invariant solution is preferred. At lower temperatures the T^2 terms become important. Consider the terms involving X and Y coming from $P_{5\Sigma}$ and P_{5T} in eq.(6.7) and (6.8)

$$V_a T^2 = \frac{1}{8} \{ 8|\beta_4|^2 |Y|^2 + 4|\beta_4 X + \beta_5 M|^2 + 24|\beta_2|^2 |X|^2 \} T^2 .$$

(6.13)

This clearly shows $\langle Y \rangle = 0$ and $\langle X \rangle \neq 0$. In addition there are terms involving the Σ field

$$V_b T^2 = \frac{1}{8} \{ (4|\beta_3|^2 + 2|\gamma_{24}|^2 + 2|\beta_2|^2 + 10g^2) \, \mathrm{Tr}(\Sigma^2) \} T^2 \; . \tag{6.14}$$

Let us consider the potential in the neighbourhood of one of the minima of eqs.(6.9)-(6.11). There finite temperature potential has the form

$$V_M(T) = 4|\beta_4 X + \beta_5 M|^2 |Y|^2 + |\frac{\beta_2}{2} \mathrm{Tr}\,\Sigma^2 + \beta_1 M^2 + \beta_4 Y^2|^2$$

$$+ (V_a + V_b) T^2 + O(NT^4) + \text{terms vanishing at the} \atop \text{minimum} \; . \tag{6.15}$$

For temperatures $T_C^2 \lesssim 1/N(V_a + V_b)$, the T^2 terms will be the most important in splitting the zero temperature degeneracy. If β_1 and β_2 are small relative to the other terms, the favoured minimum is $\langle\Sigma\rangle = \langle X\rangle = \langle Y\rangle = 0$ and SU(5) remains unbroken. If β_1, β_4 and β_5 are large, the minimum of eq.(6.15) is at

$$\langle Y^2 \rangle \simeq 0 \; , \tag{6.16}$$

$$\langle \beta_4 X + M\beta_5 \rangle \simeq 0 \; , \tag{6.17}$$

$$\langle \mathrm{Tr}(\Sigma^2) \rangle = \frac{-2\beta_1 M^2}{\beta_2} \; .$$

If $\quad \dfrac{2\beta_1}{\beta_2} \gtrsim \dfrac{30\beta_2^2}{\beta_3^2} \dfrac{\beta_3^2}{\beta_4^2}$

this minimum lies closer to the SU(3)×SU(2)×U(1) minimum of eq.(6.11) and the temperature-dependent terms will drive the vacuum towards this minimum[+]). As the temperature falls, this

---

[+]) The above analysis does not use the full form [57] of $V_M(T)$ valid for T in the neighbourhood of mass thresholds. We have verified that our scenario works if a more complete form of $V_M(T)$ is used. In particular the T_C^4 terms cannot trap the universe in a false minimum.

minimum will be the only one significantly occupied [53]. The phase transition occurs at $T = T_C$.

6.4 Multiplet Splitting

Having chosen the $SU(3) \times SU(2) \times U(1)$ minimum how do we arrange to give Higgs triplets a large mass? The original suggestion of Dimopoulos and Georgi [50] achieved this by introducing the superpotential

$$P_{5H} = \bar{H} (m'1 + \gamma_{24} \Sigma) H \quad . \tag{6.18}$$

This gives by eq.(6.4) terms in the potential

$$V = |\bar{H}(m'1 + \gamma_{24} \Sigma)|^2 + |(m'1 + \gamma_{24} \Sigma H|^2 \quad . \tag{6.19}$$

This gives scalar masses

$$m_{H_D}^2 = m' - m \quad ,$$

$$m_{H_T}^2 = m' + \frac{2}{3}m \quad ,$$

where $m = -\gamma_{24} <\Sigma_{55}>$. $\hspace{3cm}$ (6.20)

If $m' = m$ then the doublets remain massless while the triplets acquired a mass of order M_X. This certainly achieves the aim of splitting the Higgs multiplets although there is no symmetry giving the result $m = m'$.
Do radiative corrections spoil the result? It is a remarkable property of supersymmetry theories that the parameters in the superpotential are not renormalised. Thus when supersymmetry is exact (as it is at the scale M_X), there will only be wave function renormalisation affecting the scalar masses. Consequently m_{H_D} remains zero beyond tree level if $m = m'$ up to supersymmetry violation effects which we will discuss later.

The method of Georgi and Dimopoulos illustrates the first

method for splitting the H and $\bar{H}$ multiplets, namely fixing
it by hand at a scale M_X and relying on the non-renormalisation
theorems to maintain the splitting at low scales. A second
way has been suggested [51,58] which has the advantage of explaining why m' = m initially. It relies on energetics to do
this the condition m= m' being necessary to minimise the potential energy. This is achieved by replacing eq.(6.18) by

$$P_{5H} = \bar{H}(\gamma_1 Z 1 + \gamma_{24}\Sigma)H \quad , \tag{6.21}$$

where Z is an SU(5) singlet superfield. If this is the only
time Z is mentioned in the superpotential the potential will
only involve Z through the terms

$$V = |\bar{H}(\gamma_1 Z 1 + \gamma_{24}\Sigma)|^2 + |(\gamma_1 Z 1 + \gamma_{24}\Sigma)H|^2 . \tag{6.22}$$

Now if SU(2) is broken through $\bar{H}_5$ and H_5 acquiring vevs. (How
this happens we will come to later) minimisation of V with
respect to the vev of Z gives

$$\gamma_1 <Z> - m = 0 \quad . \tag{6.23}$$

Clearly $\gamma_1 <Z>$ plays the same role as m' in eq.(6.20) so
the above equation requires m' = m as is needed to split the
multiplets. For obvious reasons this method has become known
as the "sliding singlet" technique.

6.5 Supersymmetry Breaking

The anticommutation relation for the supersymmetry generators (for the full algebra see the lectures of Professor
Nicolai)

$$\{Q_\alpha, \bar{Q}_\beta\} = i\gamma_{\alpha\beta}^M P_\mu \tag{6.24}$$

immediately leads to the result for Hamiltonian

$$H = P_o = \frac{1}{2}\sum Q_\alpha Q_\alpha \qquad (6.25)$$

Thus the energy density in globally supersymmetric theories is positive semi-definite. If supersymmetry is not spontaneously broken

$$Q_\alpha |0\rangle = 0 \qquad (6.26)$$

and consequently H = O. If supersymmetry is spontaneously broken

$$Q_\alpha |0\rangle \neq 0 \qquad (6.27)$$

and H > O. Thus supersymmetry is spontaneously broken if and only if the energy density is positive definite. To see how this may happen consider the form for the scalar potential energy

$$V(\phi) = \sum_i |\frac{\partial P}{\partial \phi_i}|^2 + \sum_{a,L} |g_a|^2 |\phi_i^+ \lambda^a \phi_i|^2 \qquad (6.28)$$

The first term is the F term from the superpotential P. The second term is the D term contribution to V coming from the gauge interaction. If V is to acquire a positive value breaking supersymmetry, then either the F term [59] or the D term [60] must be non zero. The latter may only happen by adding a constant term

$$|\phi_i^+ \lambda^a \phi_i|^2 \rightarrow |\phi_i^+ \lambda^a \phi_i + \xi|^2 \qquad (6.29)$$

This will explicitly violate gauge invariance for all but a U(1) group. Since we wish to construct a GUT based on a semisimple group this route is not available to us.

The alternative is to arrange for the F term to acquire a nonzero vev. This is achieved through a term in the superpotential of the form

$$P_{OR} = \lambda_1 ABM + \lambda_2 (A^3 - M^2)C \qquad (6.30)$$

where A,B and C are further singlet chiral supermultiplets.
Then

$$V = |\lambda_1 \, AM|^2 + |\lambda_2 (A^2 - M^2)|^2 + |\lambda_1 \, BM + 2\lambda_2 \, AC|^2 . \qquad (6.31)$$

Clearly for no value of A is $V = 0$ and so supersymmetry is
violated.

There are two possibilities which have been tried. One
is to write a form of the O'Raifeartaigh potential in the
gauge non-singlet sector so that at tree level supersymmetry
breaking splits the gauge non-singlet supermultiplets. However it has been shown [49,61] that the spectrum at tree
level is unacceptable as it splits the real and imaginary
scalar components of a chiral supermultiplet, one lying above
the fermion mass and one below. In this case one has to rely
on radiative corrections to give an acceptable mass spectrum.
The second possibility is to keep supersymmetry breaking in
a gauge singlet sector but to couple it to nonsinglet fields
via Yukawa couplings. Supersymmetry breaking in the nonsinglet
sector occurs only in radiative order. Let us discuss this
second possibility first.

To couple the singlet sector to the nonsinglet sector it
is necessary either to introduce a Yukawa coupling of A to a
gauge nonsinglet field or to rely on gravitational coupling.
We will discuss the latter possibility later when considering
local supersymmetry. Here we introduce two new chiral super-
multipilets ϕ_1 and ϕ_2 with couplings

$$\lambda_3 \, \phi_1^a \, \phi_{2a} \, A + \lambda_4 \, AM^2 . \qquad (6.32)$$

The purpose of the last term is to break a symmetry,
known as R invariance, which would, if exact, forbid 'ino
masses. A theory is R invariant if the superpotential P transform as

$$P(\phi_i) \rightarrow P(e^{i\theta_i}\phi_i) = e^{i\alpha}P(\phi_i) \qquad (6.33)$$

where under R transformation each chiral superfield ϕ_i multi-plied by a phase θ_i and α is an arbitrary phase. Without the term in eq.(6.32) proportional to λ_4 the superpotential is in-variant under the transformation

$$A \rightarrow A \ ,$$

$$C,B \rightarrow e^{i\alpha}C,B \quad ,$$

$$\phi_1 \rightarrow e^{i(\alpha+\beta)}\phi_1 \ ,$$

$$\phi_2 \rightarrow e^{-i\beta}\phi_2 \quad . \tag{6.34}$$

The term in eq.(6.32) requires an A transformation $A \rightarrow e^{i\alpha}A$ in conflict with eq.(6.34) so R invariance is broken and 'inos may acquire masses.

The superpotential we have chosen does not include all possible terms consistent with the symmetries of the theory. For example we do not include a term A^3 even though, in the absence of an R symmetry, it is allowed. We rely on the non - renormalisation of the terms in the superpotential to ensure that such terms do not arise if omitted initially. This "super-natural" condition is possibly an unpleasant feature for we have no explanation for the original omission of such terms. Models based on global supersymmetry which do include all possible terms have been constructed but are rather compli-cated [62]. We will return to a discussion of the role of R invariance in local supersymmetry.

Let us give a more careful analysis of the supersymmetry breaking coming from eqs.(6.30) and (6.32). Since these are the only ones to mention A,B,C,ϕ_1 and ϕ_2 we may minimize them independently of the other terms. We find [53]

$$<ReA>^2 = \{\frac{2|\lambda_2|^2-|\lambda_1|^2}{2|\lambda_2|^2}\} \ N^2 \quad , \tag{6.35}$$

$$<\text{ImA}> = 0 \quad , \tag{6.36}$$

$$\lambda_1 <B>M + 2\lambda_2 <A><C> + \lambda_4 M^2 = 0 \quad , \tag{6.37}$$

$$<\phi_1> = <\phi_2> = 0 \quad . \tag{6.38}$$

At the tree level the combination of fields $\{-2\lambda_2 <A>B + \lambda_1 MC\}$ has an undetermined vev. One-loop radiative corrections give this combination a positive mass [63] and thus it has zero vev. Thus

$$2\lambda_2 <A><B> = \lambda_2 M<C> = \frac{-2\lambda_1 \lambda_2 \lambda_4 M^3 <A>}{(4\lambda_2^2 <A>^2 + \lambda_1^2 M^2)} \quad . \tag{6.39}$$

In the approximation λ_1, $\lambda_2 >> \lambda_3$, λ_4 the fermions have mass^2 0 and $\lambda_1^2 M^2 + 4\lambda_2^2 <A>^2$. These correspond to the goldstino, the massless Goldstone fermion which is generated when global supersymmetry is broken, and to a massive Dirac fermion. These states are given respectively by

$$\frac{1}{[4\lambda_2^2 <A>^2 + \lambda_1 M^2]^{1/2}} \ (2\lambda_2 <A>B - \lambda_1 MC) \tag{6.40}$$

and

$$\frac{1}{[4\lambda_2^2 <A>^2 + \lambda_1^2 M^2]^{1/2}} \ (\lambda_1 MB + 2\lambda_2 <A>C)_L + \bar{A}_R \quad . \tag{6.41}$$

If (as we expect) the global supersymmetry we are considering is a relic of a local supersymmetry at a fundamental level, the goldstino will be "eaten" by the gravitino, the spin 2/3 partner of the graviton. (Local supersymmetry includes Einstein's gravity). The goldstino acquires a mass [64]

$$M_{\text{gravitino}} \simeq \frac{M^2}{M_{\text{Planck}}} \quad . \tag{6.42}$$

In the scalar sector there are two real scalar states ($\equiv$ one complex state) of mass^2 $(\lambda_1^2 M^2 + 4\lambda_2^2 <A>^2)$ given by

$$\frac{1}{[4\lambda_2^2<A>^2+\lambda_1^2M^2]^{1/2}} \ [\lambda_1'MB + 2\lambda_2<A>C] \quad . \tag{6.43}$$

There are two massless scalars (which acquire mass in higher order) given by

$$\frac{1}{[4\lambda_2^2<A>^2+\lambda_1^2M^2]^{1/2}} \ [2\lambda_2<A>B - \lambda_1BM] \quad . \tag{6.44}$$

Finally, there are two states, Re(A) and Im(A) with masses

$$\lambda_1^2M^2 + 6\lambda_2^2<ReA>^2 - 2\lambda_2^2M^2$$

and

$$\lambda_1^2M^2 + 2\lambda_2^2<ReA>^2 + 2\lambda_2^2M^2 \tag{6.45}$$

respectively. We see from these masses that supersymmetry is broken at the tree level for the A chiral supermultiplet, the scalar fields being split on either side of their fermion partners and the masses satisfying the sum rule

$$2 \sum_i m_{F_i}^2 = \sum_i m_{B_i}^2 \tag{6.46}$$

where $m_{F(B)}^2$ refer to fermion and boson mass^2, respectively. The breaking of supersymmetry in the gauge nonsinglet sector proceeds only through radiative corrections. The graphs of Fig.(15) generate a gluino mass of order

$$m_{\tilde{g}} = \frac{\alpha_s}{3\pi} \ \frac{\lambda_3^2\lambda_4}{16\pi^2} \ \ln \ (\frac{m_{\psi A}^2}{m_{\phi A}^2}) \ A \tag{6.47}$$

where ψ and ϕ represent the fermion and scalar components of A.

However the superpartners of the gluinos, the gluons, are prevented by $S(3)_c$ gauge symmetry from acquiring a mass radiatively. Similarly the Winos and Binos may acquire radiative masses.

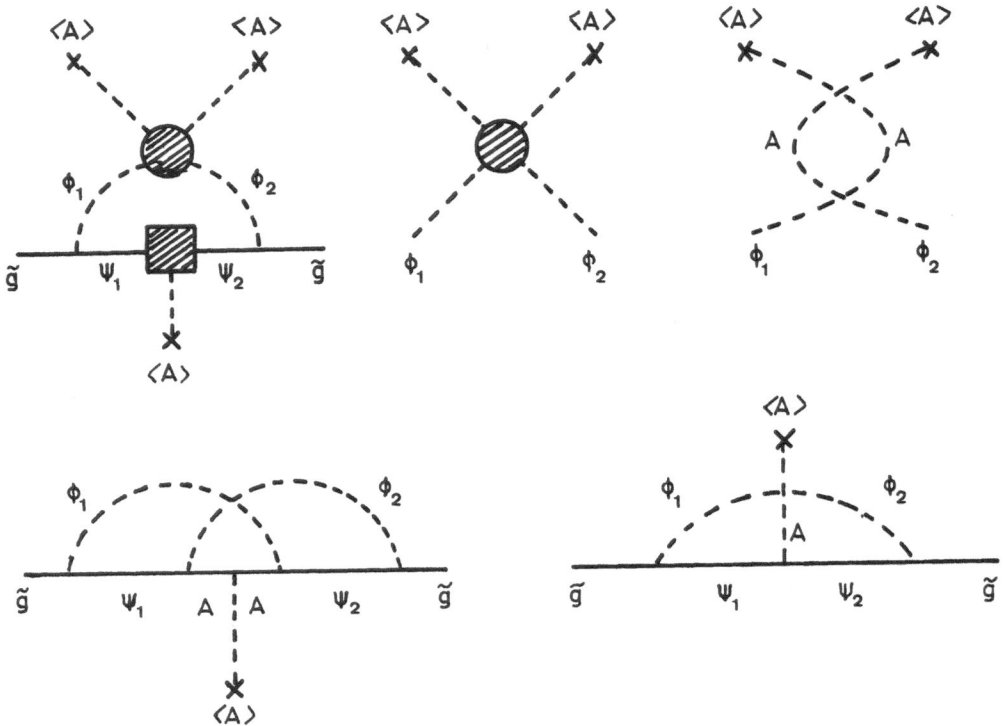

Fig.15: Graphs contributing to gluino in two loop order.

Once the 'inos acquire mass they can generate masses for the scalars via the graphs of Fig.(16a); because there is no longer perfect cancellation between fermion and boson contribution. This gives

$$m_{\tilde{q}}^2 = \frac{4\alpha_s}{3\pi} \ln(\frac{m_\phi^2}{m_{\tilde{g}}^2}) \, m_{\tilde{g}}^2 \quad . \tag{6.48}$$

and $m_{\tilde{\ell}}^2 = \delta m_{T,S}^2 = \frac{3\alpha_s}{4\pi} \ln(\frac{m_\phi^2}{m_{\tilde{W}}^2}) m_{\tilde{W}}^2 \, . \tag{6.49}$

While the contribution (6.49) to the light Higgs and slepton masses is positive, there are also negative contributions to the Higgses (but not the sleptons) from the squark loops in Fig.(16b) which are of order

$$\delta m_{T,S}^2 = -\frac{3}{8\pi^2}|h_{U,D}|^2 \ln\left(\frac{m_\phi^2}{m_{\tilde{q}}^2}\right) m_{\tilde{q}}^2 \qquad (6.50)$$

where $h_{U,D}$ are the quark-Higgs Yukawa couplings to up and down quarks respectively.

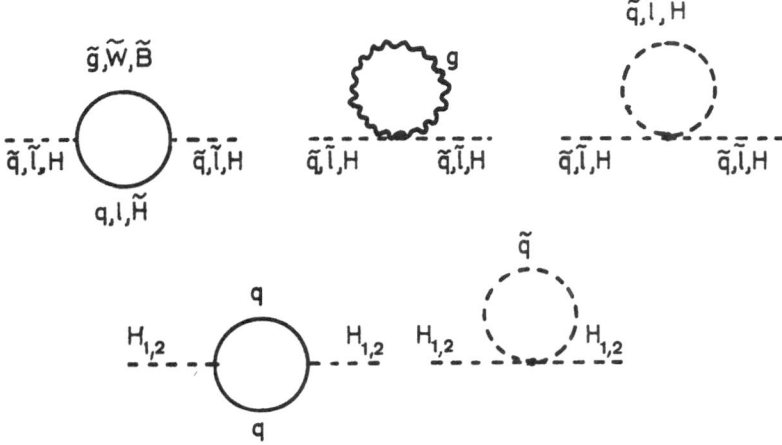

Fig.16: Graphs contributing to scalar masses.

It is easy to check that the negative contribution of eq.(6.50) overwhelms the positive piece of (6.49) and triggers

spontaneous symmetry breaking of SU(2)×U(1) by the Higgses
(and not the sleptons) if we choose

$$h_{U_i}^2 \geq 0 \left(\frac{1}{2 \ln \frac{m_\phi^2}{m_{\tilde{W}}^2}} \right) \quad i.e. \ m_t \gtrsim 25 \ \text{GeV} \quad . \tag{6.51}$$

One of the nice features of this model is that the fact SU(2)
is broken but SU(3) remains exact follows automatically from
the different magnitudes of the couplings α_2 and α_3. The
negative contribution of (6.50) is always smaller then the
positive contribution of eq.(6.48) for the triplet of Higgs
scalars proportional to α_3 but may be larger then that for
the Higgs doublets eq.(6.49) since this is proportional to
α_2.

To summarise, scalar partners of quarks and leptons
and fermionic partners of gauge bosons are heavier than their
conventional partners and thus need not have been seen yet.
In addition the supersymmetry breaking scale is related via
radiative corrections to the SU(2)×U(1) breaking scale: there
is no need to input by hand a second scale. It is also easy
to introduce a term that automatically splits the Higgs multiplets, as is required if proton decay is not to proceed too
quickly. The final mass spectrum of the model is summarised
in Table 6 with a more visible version in Fig.(17) [53]. It
is remarkable that radiative corrections have automatically
generated a reasonable spectrum.

Table 6: Broken Supersymmetric Spectroscopy

Fermions	Mass	Characteristics
$\tilde{W}^{\pm} + \ldots$	$m_{\tilde{W}} + \ldots$	Mainly I = 1, vector-like neutral weak couplings
$\tilde{H}^{\pm} + \ldots$	$\dfrac{g_2^2 v_1 v_2}{m_{\tilde{W}}}$	Mainly I = $\frac{1}{2}$, vector-like neutral weak couplings decaying into $\bar{H}^0 + (\ell\nu$ or $\bar{q}q)$
$\tilde{W}^3 + \ldots$	$m_{\tilde{W}} + \ldots$	Mainly I = 1 wino
$\tilde{B}^0 + \ldots$	$\dfrac{5}{3}(\dfrac{\alpha_1}{\alpha_2})m_{\tilde{W}} + \ldots$	Mainly I = 0 bino
$\tilde{A}^0{}' \equiv \dfrac{v_1 \tilde{H}_1^0 - v_2 \tilde{H}_2^0}{v}$	$\dfrac{g_2^2 v^2}{5 m_{\tilde{W}}} + \ldots$	Mainly I $\equiv \frac{1}{2}$ shiggs decaying into $\tilde{H}^{\pm} + (\ell\nu$ or $\bar{q}q)$
$\tilde{S}^0 \equiv \dfrac{v_2 \tilde{H}_1^0 + v_1 \tilde{H}_2^0}{v}$ OR $\tilde{S}^0_{\pm} \equiv \dfrac{1}{\sqrt{2}}(\tilde{S}^0 \pm \tilde{Y}^0)$	$= 0$ $\dfrac{\lambda v}{2} \pm \ldots$	Mainly I = $\frac{1}{2}$ shiggs: and of all supersymmetric decay chains Mixture of I = $\frac{1}{2}$ and I=0, lighter one stable, heavier one quasistable decaying into lighter one $+(\ell^+\ell^-$ or $\bar{q}q)$
Bosons	**(Mass)^2**	**Characteristics**
$\dfrac{v_1 H_2^0 + v_2 H_1^0}{v}$	$4\lambda^2 v_2^2 + \ldots$	I = $\frac{1}{2}$, mainly coupled to charge $-1/3$ quarks
$\dfrac{v_1 H_1^0 - v_2 H_2^0}{v}$	$\dfrac{g_2^2 + g_1^2}{2} v^2 + \ldots$	I = $\frac{1}{2}$, mainly coupled to charge $+2/3$ quarks
$a \equiv \dfrac{\eta_1 v_2 - \eta_2 v_1}{v}$	few GeV?	Pseudoscalar coupling, larger for charge $-2/3$ quarks
$\dfrac{v_1 H_2^{\pm} - v_2 H_1^{\pm}}{v}$	$\dfrac{g_2^2}{2} v^2 + \ldots$	Conventional charged Higgs boson
Y?	$0(\lambda^2 v^2)?$	I = 0 scalar and pseudoscalalar, very weakly coupled to light quarks.

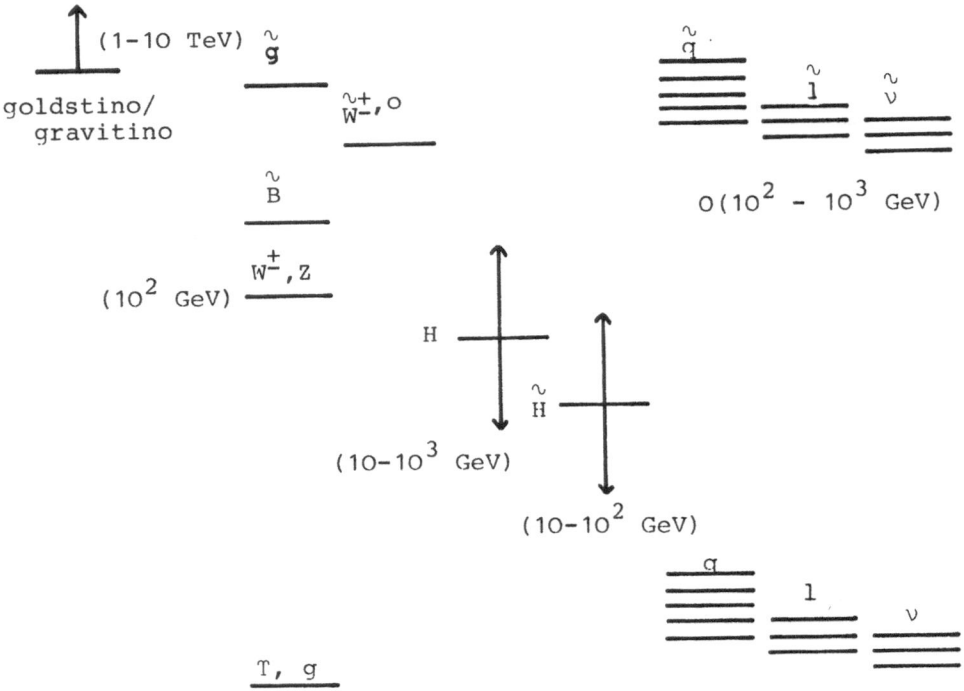

Fig.17: Low Mass spectrum for SU(5) SUSY-GUT.

6.6 Geometric Hierarchy - An Alternative Scheme

In the previous section we discussed in detail a supersymmetric scheme based on supersymmetry breaking in the gauge singlet sector. Here $M_W = (\lambda)^n M_{SUSY}$ where λ is generic for the Yukawa couplings of the theory. An alternative scheme the "geometric hierarchy" scheme [65] has been proposed in which $M_W = M_{SUSY}^2/M_X$. In this supersymmetry is broken in a gauge nonsinglet sector via the superpotential

$$P = \lambda_1 \, \text{Tr}(Y\Sigma^2) + \lambda_2 X(\text{Tr}\Sigma^2 - M^2) \tag{6.52}$$

where Y is an adjoint chiral supermultiplet and X is singlet. This has the advantage of a unique vacuum SU(3)×SU(2)×U(1). Supersymmetry is broken with a scale M^2. SU(5) is broken with a scale $\alpha<X>$. At tree level $<X>$ is not determined.

The supersymmetry is communicated to the light sector via massive states. As a result the quark masses, which arise from terms of the form $\int d^4\theta\psi^+\psi Tr(\Sigma^+\Sigma) = \psi^+\psi F_\Sigma^+ F_\Sigma$ has on dimensional grounds a magnitude

$$m_{\phi\tilde{q}}^2 = O(\frac{<F_\Sigma>^2}{M_\Sigma^2}) = O(\frac{M_{SUSY}^4}{M_X^2})$$

6.7 Stability of the Hierarchy

Of course it is important to check that mass relations valid at tree level or low order in perturbation theory persist in higher order. The most severe constraint comes from the sliding singlet sector for, in order for it to work, it is necessary that radiative terms involving the singlet Y field should not introduce a contribution to V larger than $O(M_W^4)$ for the latter is the maximum value for the contribution to V from eq.(6.22) if the Higgs doublet masses are to be $\lesssim M_W$. However this proves to be a very strong constraint. For example radiative corrections give rise to a term

$$m_R^2(\gamma_1 Y + \gamma_{24}\Sigma_1^1)^2 \qquad . \tag{6.53}$$

Because $<\gamma_1 Y + \lambda_{24}\Sigma_1^1> = O(M_X)$ the resultant constraint on m_R^2 is

$$m_R^2 \lesssim M_W^4 / M_X^2 \qquad . \tag{6.54}$$

For the scheme of section (6.2) this indeed is possible for the contribution to m_R^2 is via a heavy loop whereas the contribution to the Higgs mass was via a ϕ loop which may be light. Using the dimensional arguments given above we see m_R^2 is $O((M_\phi^2/M_X^2) \cdot m_H^2) = O(M_W^4/M_X^2)$ as required if $M_\phi^2 \simeq M_W^2$ (a detailed calculation relaxes this condition slightly)[53]. However for the model of the geometric hierarchy type the radiative corrections generate a mass m_R^2 without the (M_ϕ^2/M_X^2) suppression, in conflict with eq.(6.54) [66].

Consequently if models of the geometric hierarchy are to work an alternative method for splitting the Higgs multiplets is needed.

6.8 The Missing Doublet Scheme. [67]

This idea relies on choosing a new Higgs representation whose SU(3)×SU(2)×U(1) content does not include a SU(2) doublet, SU(3) singlet. As a result its Yukawa couplings to H and $\bar{H}$ cannot give their doublet components a mass but can give one to their triplet components. For example the (3,2,1) content of a 50 is

$$\underline{50} = (8,2) + (6,3) + (\bar{6},1) + (3,2) + (\bar{3},1) + (1,1) . \quad (6.55)$$

It has, as desired, no (1,2) component. We couple two 50s θ and $\bar{\theta}$ to avoid anomalies to H and $\bar{H}$ via the superpotential

$$P_{MD} = b\theta\Sigma H + b' \bar{\theta} \Sigma\bar{H} + \tilde{M} \bar{\theta}\theta \quad (6.56)$$

where Σ is here a 75, needed instead of a 24 to construct the θ,H mixing terms. $P_{5\Sigma}$ together with P_{MD} breaks SU(5) uniquely to SU(3)×SU(2)×U(1) giving

$$P_{MD} = \theta_3(bM/a)H_3 + \bar{\theta}_3(cM/a) + \overset{\sim\sim}{M}\theta_3\theta_3 \quad (6.57)$$

where the subscripts refer to the SU(3) transformation properties. Clearly the triplets acquire a mass $\geq (\tilde{m}^2/M)$ while the doublets remain massless.

6.9 Locally Supersymmetry Models

Recently it has been realised that significant contributions

to scalar masses may come from gravitational corrections.
That this should be so may be seen qualitatively from the
contribution expected from graphs of the type shown in
Fig.(16). The graviton, gravitino supermultiplet couples the
gauge singlet to gauge non-singlet sectors so irrespective of the
sector in which the supersymmetry breaking occurs we expect
a contribution to every scalar mass of order M_{SUSY}^4/M_{Planck}^2.
If M_{SUSY} is $\geq O(10^{10}$ GeV) then this contribution will be
$\gtrsim O(M_W)$ and may not be ignored.

In fact the gravitational corrections occur even at
tree level. The coupling of a (N=1) supersymmetry theory to
gravity has been worked out[68]. For example if one demands
the standard kinetic energy terms (since we are necessarily
dealing with a non-renormalisable theory there is no fundamental reason why we should not include higher derivatives)
then the scalar potential is modified from that given in
eq.(6.5) to

$$V = e^{\sum_i |\phi_i|^2 \kappa} \{|\sum_i P_{gi} + \kappa\phi_i^*P|^2 - 3|P|^2\kappa\} \tag{6.58}$$

where $\kappa = \dfrac{\gamma\pi}{M_{Planck}^2} \equiv \dfrac{1}{M^2}$ and the sum runs over all chiral
supermultiplets. When $\kappa = 0$ this clearly reduces to the stan-
dard supersymmetric form, eq.(6.5). The appearance of a negative term allows us to adjust V to be zero at the minimum
giving a zero cosmological constant. If this is done there
is a relation between $<P>$ and the supersymmetry breaking scale
M_{SUSY}. The gravitino mass is related to $<P>$ and with vanishing cosmological constant [64]

$$m_{3/2} = M_{SUSY}^2/M_{Planck} \quad .$$

Now from eq.(6.58) we see that there are contributions of
the form $\kappa^2|\phi_L|^2|P|^2 = m_{3/2}^2|\phi_L|^2$. Thus we see that at tree
level the gravitational corrections typically generate sca-
lar masses of order $m_{3/2}$. The hierarchy constraint that this
should be $\lesssim M_W$ implies $M_{SUSY} \lesssim 10^{10}$GeV. Though this tree level

result may be avoided [69] (eq. by arranging ϕ to be a pseudo Goldstone boson) it recurs beyond tree level so that in any case we expect $M_{SUSY} < 10^{13}$ GeV[76]. The fact that scalar masses arise easily through gravitational corrections had led to intense activity to try to simplify the supersymmetry breaking sector and eliminate the need for a special Yukawa interaction coupling supersymmetry breaking from the gauge singlet to the gauge nonsinglet sector.

6.10 SUGRA-GUTs - Tree Level Weak Breaking

We have seen that with $M_{SUSY} \sim 10^{10}$ GeV, gravitational correc- tions generate scalar masses at tree level of order $m_{3/2} = $ $=M_{SUSY}^2/M_{Planck} \sim M_W$. It is natural to ask if these can be used to trigger $SU(2) \times U(1)$ breaking and build a GUT with no hierarchy problem[71,72]. The simplest possibility is to choose a superpotential which separates the (3,2,1) singlet fields z_i responsible for supersymmetry breakdown and the y_a fields which include the quark, lepton and Higgs scalars.

$$P = h(z_L) + g(y_a) \quad . \tag{6.59}$$

The resulting scalar potential is [71]

$$V = \exp\{(|z_L|^2 + |y_a|^2)K\}[|h_{ji} + z_L^*PK|^2 + |g_{ja} + y_a^*PK|^2$$

$$- 3|P|^2K^2] + \tfrac{1}{2} D_\alpha D^\alpha \tag{6.60}$$

where summation on a, i and α is understood.

We assume V is minimised for

$$<z_L> = b_i M ,$$
$$<h_L> = a_L mM ,$$
$$<h> = mM^2 ,$$
$$<y_a> = 0 \quad . \tag{6.61}$$

Then the condition the cosmological constant vanishes at the minimum is

$$\sum_i |a_L + h_L|^2 = 3 \tag{6.62}$$

The effective low energy potential is found by keeping those terms nonvanishing in the limit $M \to \infty$

$$V = |\tilde{g}_a|^2 + m_{3/2}^2 |y_{ja}|^2 + m_{3/2}(A\tilde{g} + h.c.) + \frac{1}{2} D_\alpha D^\alpha \tag{6.63}$$

where

$$m_{3/2} = \exp(\frac{1}{2}|b_i|)m$$

$$\tilde{g} = \exp(\frac{1}{2}|b_L|)g$$

and

$$A = b_i^*(a_L + b_L) \tag{6.64}$$

by comparison with the formula

$$|\tilde{g}_a \pm m_{3/2}y_a^*|^2 = |\tilde{g}_a|^2 + m_{3/2}^2|y_a|^2 \pm 3m_{3/2}(\tilde{g} + \tilde{g}^*)$$

which follows since g is cubic in y_a, we see that V is positive definite for $|A| < 3$. For $|A| > 3$ there is a lower minimum with nonzero vev's for y_a so SU(2)×U(1) may in this case be broken. The simplest example of this is with

$$\tilde{g} = \lambda y H\bar{H} + \frac{1}{3} \sigma y^3 \tag{6.65}$$

where y is an SU(2)×U(1) singlet. For $|A| > 0$ there will be a minimum of V with nonzero vev amongst y, H and $\bar{H}$ and one may check that for

$$1 < \frac{\sigma}{\lambda} < \frac{1}{16}(A + \sqrt{A^2 - 8})^2 \tag{6.66}$$

this minimum has all three nonzero with $\langle H \rangle = \langle \bar{H} \rangle$. Provided

$m^2_{3/2} > |A|m_t$ the inclusion of scalar quark and lepton fields
does not affect this minimum for the common positive mass^2
contribution $m^2_{3/2}$ prevents the latter fields from acquiring
vev's.

What about our assumption of $|A| > 3$? The simple Polony [73] poten-
tial of the form

$$h(z) = mM(z + \beta M) \qquad\qquad (6.67)$$

breaks supersymmetry, but, for zero cosmological constant
has $|A| = 3 - \sqrt{3}$. More complicated potentials are needed
spoiling somewhat our hope of simplifying the Yukawa couplings needed in this type of model.

6.11 Radiative Corrections in SUGRA Models

More serious still is the fact that radiative corrections
tend to destabilise the hierarchies in this type of SUGRA
GUT[74]. In general the problem arises, when the superpotential P contains terms of the form

$$P = \lambda_1 B^2 L + \lambda_2 L^3 + M_x B^2 \qquad\qquad (6.68)$$

where B is a heavy field (mass $\sim M_x$) and L is a light field.
Following from eq.(6.58) we see in SUGRA models there are
additional supersymmetry breaking couplings of the form

$$m_{3/2}(\lambda_1 B^2 L + \lambda_2 L^3 + M_x B^2) \quad . \qquad\qquad (6.69)$$

These give rise to supergraphs of the form of Fig.18 giving
mass terms for the light sector of order $\lambda_1 \lambda_2 m_{3/2} M_x$, much
larger than the term of order $m^2_{3/2}$ discussed above.

For the model of section (6.9) the coupling $\lambda H \bar{H} Y$ contains couplings of the light field Y to the heavy colour triplets of H and $\bar{H}$ and also to the light doublet fields of H
and $\bar{H}$. Thus it gives rise to a contribution as in Fig.18 and
a mass term for $H\bar{H}$ of magnitude $O(\lambda^2 m_{3/2} M_x)$. If this is to be

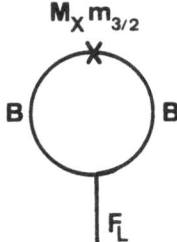

Fig. 18

less than $O(M_W)$ then $M_{SUSY} \lesssim 10^{15}$ GeV. In this case the gravitational corrections are too small to split the super-multiplets and to generate $SU(2) \times U(1)$ breaking. The models are approximately globally supersymmetric and must be con-structed along the lines of section (6.2).

It is possible to avoid the contributions of the type shown in Fig.18 and to build a viable SUGRA-GUT realising the strategy of section (6.7), but this requires the model be considerably complicated. The simplest such model [72] is given by the superpotential

$$P = \lambda_1 \Sigma^3 + M\Sigma^2 + \lambda_2 \Theta\Sigma\bar\Theta + \lambda_3 \Theta\Sigma H + \lambda_4 \bar\Theta\Sigma\bar H$$

$$+ M'\Theta\bar\Theta + \lambda_2'\Theta'\Sigma\bar\Theta' + \lambda_3'\Theta'\Sigma H' + \lambda_4'\bar\Theta'\Sigma\bar H' + M''\Theta'\bar\Theta' \qquad (6.70)$$

where there are now two sets of $(50 + \overline{50} + 5 + \bar5)$ namely $(\Theta,\bar\Theta,H,\bar H)$ and $(\Theta',\bar\Theta',H',\bar H')$. It is clear that there are no contributions of the type shown in Fig.18 and indeed the authors have shown the hierarchy is stable against all radia-tive corrections. To me this model is disappointing for, if anything, it has a more complicated Yukawa sector than that of the globally supersymmetric model of section (6.2).

SUGRA-GUTs - Radiative Breaking

A possible improvement has been suggested [75] which, following closely the idea of the model we studied in section (6.2), seeks to break SU(2)×U(1) by radiative corrections. As we noted above the gravitational corrections give a common mass, $m_{3/2}$, to all sectors via the effective potential of eq.(6.58). However this supersymmetry breaking mass terms receive (divergent) radiative corrections which must be regulated. If all gauge and Yukawa interactions were neglected the gravitational interactions would preserve a U(n) symmetry amongst all the n scalar fields. Accordingly it is usually assumed that the equal masses received at tree level are to be interpreted as the mass including all pure gravitational radiative corrections at the Planck scale. (It is not clear to me that this is a reasonable assumption for Yukawa interactions involving the top quark are not necessarily small. Moreover if extended supergravity is responsible for the effective theory of eq.(6.48) it means gravitational interactions are generating Yukawa couplings quite different for different multiplets; so why not different masses too?) Then radiative corrections coming from the usual renormalisable interactions will split the scalar masses as the scale is lowered and in particular the Yukawa couplings, if sufficiently large, will drive the Higgs mass^2 negative at some scale which is related to M_{Planck} by exponential factors and can be very large. Explicit calculations show that this may happen for $m_t > 50$ GeV [75].

However in this models it is still necessary to split the H and $\bar{H}$ multiplets. The missing partner technique of section (6.7) will work but radiative corrections of the type discussed in the previous section require it be complicated to the form of eq.(6.70). Again the SUGRA-GUT looks more complicated than the SUSY GUT it was supposed to simplify. (The simpler sliding singlet method does not work because the condition eq.(6.54) is inconsistent with the choice of

$m_{3/2} \tilde{} M_W$ used to split supermultiplets in SUGRA-GUTs.) This method can be extended at the expense of enlarging the GUT [76].

7. PHENOMENOLOGY OF SUSY-GUTs

The spectrum of Fig.17 illustrates the expected mass pattern in a minimal SU(5) class of supersymmetric models in which radiative corrections are responsible for splitting the gauge non-singlet supermultiplets and in which R invariance is broken, allowing the gauginos to become heavy. Other schemes, based on tree level breaking, have been considered [77] although they are more difficult to grand unify. The major difference is they have a light goldstino and/or photino, giving rise to a different decay pattern for the new supersymmetric states. This affects the pattern of decay of supersymmetric states and consequently the phenomenological signals. For the models with a light photino these signals have been extensively studied [78]. Here we will mention some of the novel features of the more recent models with the spectra of Fig.(17). First, however, there are several predictions insensitive to the low energy structure of the theory. In particular the predictions for $\sin\theta$ and M are changed from the usual grand unified predictions due to the new supersymmetric states (masses 1 TeV) which contribute to the β function. This gives [79]

$$M_x = \left\{ {1 \atop 4} \right. \times 10^{16} \text{ GeV and } \sin^2\theta_W = \left\{ {.236 \atop .229} \right. \text{ for } \Lambda_{\overline{MS}} = \left\{ {100 \text{ MeV} \atop 300 \text{ MeV}} \right., \text{ with}$$

the minimal SU(5) supermultiplet structure of Table 3 assuming light states are $\tilde{} \leqslant 1$ TeV and heavy states have a mass $\simeq M_x$. M_x is about 20 times the non-supersymmetric value mainly because the adjoint of fermions introduced in the vector supermultiplet make the β functions smaller in magnitude,

236

slowing up the evolution of the couplings. It may be seen
that $\sin^2\theta_W$ is increased from the SU(5) value of .215 and
is in poorer agreement with the current best (radiatively
corrected) value for $\sin^2\theta_W$ of [86] .210 $\pm$ .005. This result
is somewhat sensitive to the masses assumed for the heavy
states; for example the colour triplets of H and $\bar{H}$ had a
mass $\sim 10^{10}$ GeV, then [81] $\sin^2\theta_W = \{ {.225 \atop .218}$.

With such a high value for M_x it might be assumed
that proton decay will be negligible. This is not the case
for new processes occur involving the new superpartners which,
in the minimal SU(5) theory, mediate proton decay [82,83] (see
Fig.19).

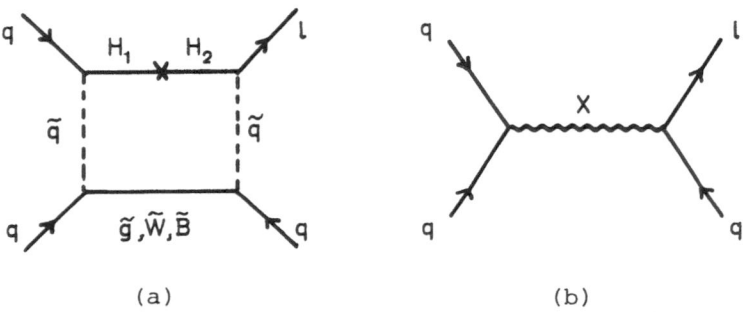

Fig.19: (a) Fermionic Higgs contributions to proton decay,
(b) Vector boson contribution to proton decay.

These give

$$\tau_p \propto M^2_{\tilde{H}_{1,2}} \quad M^2_{\tilde{W},\tilde{B},\tilde{g}} \times \text{Yukawa couplings.}$$

Because $M^2_{\tilde{W},\tilde{B},\tilde{g}} \lesssim 1$ TeV and $M^2_{H_{1,2}} \gtrsim M^2_x$ this rate is potentially
much less than the usual SU(5) of Fig.(19b) which gives
$\tau_p \propto M^4_x$. In fact, because of the Yukawa couplings, the rate turns

out to be $\simeq 10^{30}$ years. However, because the fermionic Higgs preferentially couple to heavy states the dominant decay mode is $P \to \bar{\nu} K^+$ in contrast to the $\pi^0 e^+$ mode in usual SU(5). Unfortunately this is not an unambiguous prediction of SUSY-GUTs. For example in some models the graph of Fig.(18a) does not exist and the dominant diagram may be through colour tri-plet Higgs exchange [81] with dominant decay modes $P \to \bar{\nu}_\mu K^+$, $\mu^+ K^0$ or even through the original SU(5) diagram of Fig.(19b).

Another sensitive testing ground for unified models is in rare K decays. The most sensitive process is $K^0 - \bar{K}^0$, contributing to the $K_L K_S$ mass difference. In addition to the usual graph of Fig.(19a), there are also scalar quark contri-butions coming from Fig.(20b).

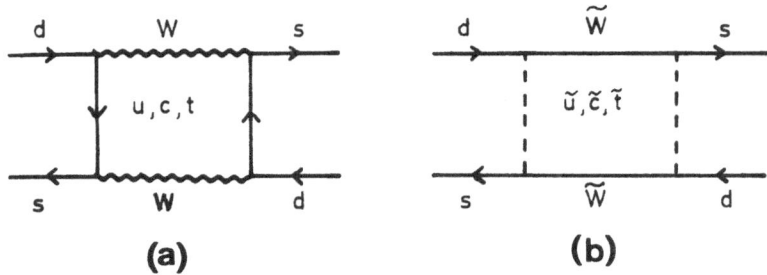

(a)   **(b)**

Fig.20: Graphs contributing to $\Delta S=2$ processes.

Evaluating these graphs gives the constraint on the scalar quark masses [83]

$$\frac{m_{\tilde{c}}^2 - m_{\tilde{u}}^2}{m_{\tilde{c}}^2} \lesssim 10^{-3} \; . \tag{7.1}$$

This is a very strong constraint on the mass difference of scalar quarks, but it is satisfied in models of the type

in section 6 in which scalar quarks acquire mass radiatively.
For them there is a large flavour independent term coming
from gauge couplings and a small flavour dependent term
coming from Yukawa couplings. Thus the scalar quark mass
differences are naturally of order of the quark mass differences.

We turn now to the specific properties and signals of
the supersymmetric model of section 6 [53,84]. There is a
conserved quantum number of the new supersymmetric states
introduced in Table 5 which means that these states are only
produced in pairs and that they cannot decay entirely into
conventional states. Thus all new symmetric states will
ultimately decay into the lightest such state; in the case
of the minimal model of section 6 it is mainly the fermionic
partner of the Higgs (mass $\gtrsim$ 20 GeV) (cf.Table 5). In models
with the "sliding singlet" Z, there are two such states $\tilde{S}_{\pm}$
and these may be produced via the Z vector boson and may give
rise to the characteristic signal of Fig.(21).

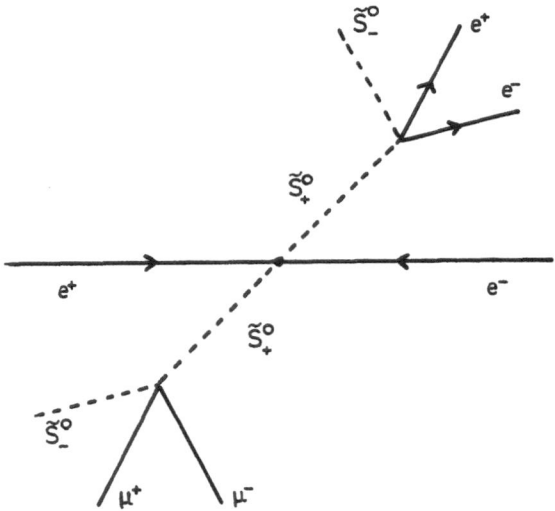

<u>Fig.21:</u> Characteristic production and decay pattern of the
lightest new fermion states $\tilde{S}_{\pm}^{0}$.

The other Higgs fermion states, the $\tilde{H}^{\pm}$, are also relatively light, will be produced in e^+e^- annhilation and will have characteristic neutral current couplings which will be able to distinguish them from a new family of heavy lepton or from a Wino (see Table 7).

Table 7: Neutral Current Couplings of Charged Fermions

Particle	g_V	g_A
e^-,μ^-,τ^-	$-\frac{1}{2}+2\sin^2\theta_W \approx 0$	$+\frac{1}{2}$
$\tilde{W}^- + \epsilon\tilde{H}^-$	$-2-\frac{1}{2}(\frac{v^2}{\alpha_2^2 M_o^2})g_2^2+2\sin^2\theta_W \approx -\frac{3}{2}$	$\frac{1}{2}(\frac{v_1^2-v_2^2}{\alpha_2^2 M_o^2})g_2^2 \approx 0$
$\tilde{H}^- - \epsilon\tilde{W}^-$	$-1-(\frac{v^2}{\alpha_2^2 M_o^2})g_2^2+2\sin^2\theta_W \approx -\frac{1}{2}$	$(\frac{v_1^2-v_1^2}{\alpha_2^2 M_o^2})g_2^2 \approx 0$

Another characteristic of supersymmetric models is that there are two multiplets of Higgs doublets together with Higgs singlets and consequently the Higgs scalar spectrum is quite rich (cf. Fig.(17) and Table 6).

Mass Scales in SUSY-GUTs

One of the reasons for studying SUSY-GUTs was the hope they would solve the hierarchy problem. In section 2 we found that they indeed allowed scalars to remain massless even in the presence of a superheavy scale M_x. However there remains the question of why the bosonic mass scales are what they are.

Ideally one would like to explain the relative sizes of M_{Planck}, M_X, M_{SUSY} and M_W.

M_{SUSY}?

In section 2 we discussed a mechanism by which a large value for the supersymmetry breaking scale M_{SUSY} would still give a small value for M_W, the difference in scales being given by a high power of the coupling. It is tempting to try to choose $M_{SUSY} \simeq M_{Planck} \simeq M_X$ as the single mass scale in the theory. However we saw that tree level gravitational terms give scalar masses $\propto m_{3/2}$. Thus the condition that the Higgs doublet should be less than 1 TeV translates to the condition that $\frac{M_{SUSY}^2}{M_{Planck}} \lesssim 1$ TeV i.e. $M_{SUSY} \lesssim 10^{10}$ GeV[69]. It is possible to avoid this bound by arranging that this mass squared term be negative, driving a Goldstone mode in which scalars remain massless at tree level because they are (pseudo) Goldstone bosons. However even in this case radiative corrections of the type shown in Fig.(22) will generate a scalar mass. Estimates of these graphs [70] give $M_{SUSY} \lesssim 10^{13}$ GeV.

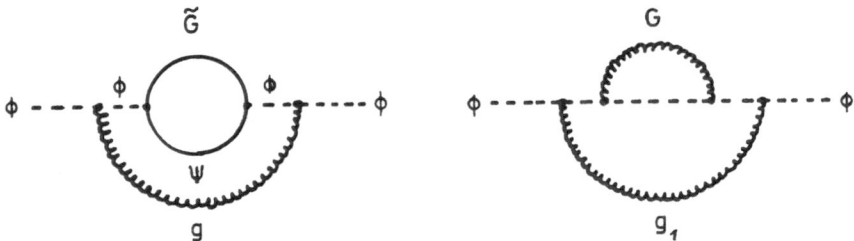

Fig.22: Graphs contributing to scalar masses. G and $G_{3/2}$ are the graviton and gravitino respectively.

Thus it appears that M_{SUSY} cannot be as high as M_{Planck}. Does this mean a new hierarchy problem has arisen? Probably not for there are several possible ways to get a small value for M_{SUSY}.

(i) It should not be forgotten that nonperturbative supersymmetry breakdown has not been ruled out for the models of interest with complex fermion representations. If this happens we would expect $M_{SUSY} \propto e^{-1/g^2} M_{Planck}$. However the lack of definite evidence for or against has caused people to look elsewhere for more tractable methods.

(ii) Since M_{SUSY} is related to a vacuum expectation value of a potential at its minimum it is not surprising if M_{SUSY} should be less than any of the mass scales M' in the potential. In fact it is easy to construct examples [85] with $M_{SUSY} \propto (\frac{M'}{M_{Planck}})^\eta$ giving an arbitrarily small value for M_{SUSY}.

M_W?

The simplest solution for M_W is $M_W \simeq M_{SUSY}$. However other possibilities have been discussed. The first [53], discussed in section 2, is that $M_W = f(\frac{\lambda}{4\pi}) M_{SUSY}$, and that a large number of loops combine to give a small ratio for M_W/M_{SUSY}. An alternative scheme, [65] the geometric hierarchy scheme, generates M_W through radiative corrections too, but in this case the equivalent graphs to Fig.(16) involve massive (gauge non singlet) states with mass $\simeq M_X$. Then it is easy to see on dimensional grounds

$$M_W = \frac{M_{SUSY}^2}{M_X} .$$

Unfortunately in this class of models the sliding singlet trick of section 2 fails due to radiative corrections and models constructed so far have to adjust parameters to arrange the necessary splitting in the Higgs multiplets.

M_X?

The simplest idea is to have $M_X = O(\alpha M_{Planck})$, or

$O(\frac{1}{\alpha} M_{SUSY})$. Both of these possibilities look feasible and reasonable. Another suggestion due to Witten [58] is that M_X (and also M_{Planck}!) may be related to $M_{SUSY} \simeq M_W$ by radiative corrections generating an inverse hierarchy

$$M_X = e^{c/g^2} M_{SUSY}$$

where c is a coefficient determined by calculation of (perturbative) radiative corrections. Unfortunately the original idea does not appear to work for with realistic values for c and g a hierarchy of only of order $e^{c/g^2} \simeq 10$ is natural [86]. However it may be possible to resurrect the idea of a low fundamental scale for M_{SUSY} in a different guise [87], although in these schemes there is no understanding of M_{Planck} relative to M_{SUSY}.

8. CONCLUSIONS

Global supersymmetry solves the hierarchy problem in a way which requires many new states. However it is expected these states should be heavy and would not yet have been observed, although not so heavy that they will elude detection soon if they exist. Viable models have been constructed and there are plausible explanations for the relative magnitude of the bosonic mass scales. The inclusion of gravity may render the models more elegant in the symmetry breaking sector and allows for the cancellation of the cosmological constant, although its cancellation is unnatural and understanding this problem remains a fundamental stumbling block. However it is an exciting fact that we may for the first time in particle physics have phenomenological reasons for including gravity and for moving towards a truly unified theory.

Of course the whole construction is pure theory at present; encouragingly these models are testable and experiment will decide.

REFERENCES

1. S.L. Glashow, Nucl. Phys. 22 (1961) 579; S. Weinberg, Phys. Rev. Lett. 19 (1967) 1264; A. Salam, Proc. 8th Nobel Symposium, Stockholm 1968, ed. N. Svartholm (Almquist and Wiksell, Stockholm, 1968) 367.

2. M. Kobayashi and T. Maskawa, Prog. Theor. Phys. 49 (1973) 652.

3. C.H. Llewellyn Smith, Phys. Lett. 46B (1973) 233; B.W. Lee, C. Quigg and H.B. Thacker, Phys. Rev. Lett. 38 (1977) 883; Phys. Rev. D16 (1977) 1519; M. Veltman, Acta. Phys. Pol. B8 (1977) 475; Phys. Lett. 70B (1977) 253.

4. S. Weinberg, Phys. Rev. D8 (1973) 1264; D.V. Nanopoulos, Nuovo Cimento Lett. 8 (1973) 873.

5. J.E. Kim, P. Langacker, M. Levine and H.H. Williams, Rev. Mod. Phys. 53 (1981) 211 and references therein.

6. W. Marciano and A. Sirlin, Phys. Rev. D22 (1980) 2695; A. Sirlin and W. Marciano, Nucl. Phys. B189 (1981) 442. C.H. Llewellyn Smith and T. Wheater, Phys. Lett. 105B (1981) 486.

7. M.A. Green and M. Veltman, Nucl. Phys. B169 (1979) 137, Err. B175 (1980) 547; F. Antonelli and L. Maiani, Nucl. Phys. B186 (1981) 269.

8. G. 't Hooft, Phys. Rev. Lett. 37 (1976) 8, Phys. Rev. D14 (1976) 3432; R. Jackiw and C. Rebbi, Phys. Rev. Lett. 37 (1976) 172; C.G. Callan, R.F. Dashen and D.J. Gross, Phys. Lett. 63B (1976) 334.

9. R.D. Peccei and H.R. Quinn, Phys. Rev. Lett. 38 (1977) 1440; Phys. Rev. D16 (1977) 1791.

10. H. Georgi and S.L. Glashow, Phys. Rev. Lett. 32 (1974) 438.

11. For review see: J. Ellis, Proceedings of the twenty-first Scottish Universities Summer School in Physics, ed. by K. Bowler and D. Sutherland, SuSSP (1980); J. Ellis, Proceedings of the Les Houches Summer School (1981) and CERN preprint TH 3174 (1982); C. Jarlskog, lectures given at Advanced Summer Institute 1978 on New Phenomena in Lepton

Hadron Physics, University of Karlsruhe, Germany (1978).

12. H. Georgi, H.R. Quinn and S. Weinberg, Phys. Rev. Lett. 33 (1974) 451.

13. A.J. Buras, J. Ellis, M.K. Gaillard and D.V. Nanopoulos, Nucl. Phys. B135 (1978) 66.

14. J. Ellis, M.K. Gaillard and D.V. Nanopoulos, Phys. Lett. 80B (1979) 360, E 82B (1979) 464; ibid 88B (1980) 320.

15. D.A. Ross, Nucl. Phys. B140 (1978)1,T. Goldman and D.A. Ross, Phys. Lett. 84B (1979) 208; Nucl. Phys. B171 (1980) 273, J. Ellis, M.K. Gaillard, D.V. Nanopoulos and S. Rudaz, Nucl. Phys. B176 (1980) 61.

16. C.H. Llewellyn Smith, G.G. Ross and J. Wheater, Nucl. Phys. B177 (1981) 263; S. Weinberg, Phys. Lett. 91B (1980) 51; P. Binétruy and T. Schücker, Nucl. Phys. B178 (1981) 293, 307; L. Hall, Nucl. Phys. B178 (1981) 75.

17. M.S. Chanowitz, J. Ellis and M.K.Gaillard, Nucl. Phys. B128 (1977) 506, D.A. Ross and D.V. Nanopoulos, Nucl. Phys. B157 (1979) 273.

18. S. Weinberg, I.J. Rabi Festschrift, Trans. N.Y. Acad. Sci. II, (1977) 38.

19. J.F. Donoghue, E. Golowich, W.A. Ponce and B.R. Holstein, Phys. Rev. D21 (1980) 186.

20. J. Ellis and M.K. Gaillard, Phys. Lett. 88B (1979) 315.

21. H. Georgi and C. Jarlskog, Phys. Lett. 86B (1979) 297.

22. For a general review see P. Langacker, Phys. Reports 72 (1981) 187.

23. J. Ellis et al. ref.15; F.A. Wilczek and A. Zee, Phys. Rev. Lett. 43 (1979) 1571; H.A. Weldon and A. Zee, Nucl. Phys. B173 (1980) 269; J. Ellis, M.K. Gaillard and D.V. Nanopoulos, Phys. Lett. 88B (1980) 320.

24. See references cited in ref. 22 and ref.11. Also G. Kane and G. Karl, Phys. Rev. D22 (1980) 2808; N. Isgur and M. Wise, Phys. Lett. 117B (1982) 179.

25. See the lectures by E. Fiorini, these proceedings.

26. M. Gell-Mann, P. Ramond and R. Slansky, unpublished (1979), R. Slansky, Talk at the Sanibel Symposium 1979,

T. Yanagida, Proc. Workshop on the Unified Theory and the Baryon number in the Universe (KEK, Japan, 1979).

27. R. Barbieri, D.V. Nanopoulos, G. Morchio and F. Strocchi, Phys. Lett. 90B (1980) 91; M. Magg and C. Wetterich, Phys. Lett. 94B (1980) 61.

28. V. Baluni, Phys. Rev. D19 (1979) 2227.

29. J. Ellis and M.K. Gaiilard, Nucl. Phys. B150 (1979) 141.

30. R. Peccei, Max-Planck Institute preprint MPI-PAE-TPh 81-45 (1981) has a good review of the axion.

31. S. Weinberg, Phys. Rev. Lett. 40 (1978) 223; F.A. Wilczek, Phys. Rev. Lett. 40 (1978) 279.

32. J. Kim, Phys. Rev. Lett. 43 (1979) 103; A. Zhitnitskii, Sov. J. Nucl. Phys. 31 (1980) 260; M. Dine, W. Fischler and M. Srednicki, Phys. Lett. 104B (1981) 409.

33. P. Sikivie, Phys. Rev. Lett. 48 (1982) 1156.

34. G. Lazarides and Q. Shafi, Phys. Lett. 115B (1982) 121; H. Georgi and M.B. Wise, Phys. Lett. 116B (1982) 126; S. M. Barr, D.B. Reiss and A. Zee, Phys. Lett. 116B (1982) 227.

35. J. Preskill, M.B. Wise and F.Wilczek, Phys. Lett. 120B (1982) 127; L. Abbott and F. Sikivie, Phys. Lett. 120B (1982) 133; M. Dine and W. Fischler, Phys. Lett. 120B (1982) 137.

36. H. Fritzsch and P. Minkowski, Ann. of Phys. 93 (1975) 193; H. Georgi, "Particles and Fields - 1974" ed. C.E. Carlson (AIP., NY 1975); H. Georgi and D.V. Nanopoulos, Nucl. Phys. B159 (1979) 16.

37. E. Witten, Phys. Lett. 91B (1980) 81.

38. F. Wilczek and A. Zee, Grand Unified Theories and Related Topics (World Scientific Pub., Singapure) (1981) 143; M. Gell-Mann, P. Ramond and R. Slansky in Supergravity, ed. by P. van Nieuwenhuizen and D. Freedman (North Holland 1979); P.H. Frampton, Phys. Lett. 88B (1979) 219; ibid. 89B (1980) 352. P.H. Frampton and S. Nandi, Phys. Rev. Lett. 43 (1979) 1461; H. Georgi, Nucl. Phys. B156 (1979) 126.

39. M. Bowick and P. Ramond, Phys. Lett. 103B (1981) 338; R. Barbieri and D.V. Nanopoulos, Phys. Lett. 91B (1980)

369; <u>95B</u> (1980) 43; R. Barbieri, D.V. Nanopoulos and
D. Wyler, Phys. Lett. <u>103B</u> (1981) 433; S.M. Barr, Phys.
Rev. <u>D24</u> (1981) 1895; L. Ibonez, Nucl. Phys. <u>B193</u>
(1982) 317; G. Segre, Phys. Lett. <u>103B</u> (1981) 355.

40. E. Gildener, Phys. Rev. <u>D14</u> (1976) 1667; Phys. Lett. <u>92B</u>
(1980) 111; E. Gildener and S. Weinberg, Phys. Rev. <u>D15</u>
(1976) 3333; G.'t Hooft, Proceedings of the Advanced Study
Institute, Cargese 1979. Eds. G.'t Hooft et al. (Plenum
Press N.Y. 1980).

41. L. Susskind, Phys. Rev. <u>D20</u> (1979) 2169; S. Weinberg,
Phys. Rev. <u>D19</u> (1979) 1277.

42. Yu. Gol'fand and E.P. Likhtman, JETP Lett. (1971) 323;
P. Ramond, Phys. Rev. <u>D3</u> (1971) 2415; A. Neveu and J.
Schwartz, Nucl. Phys. <u>B31</u> (1971) 86; J. Wess and B. Zumino,
Nucl. Phys. <u>B70</u> (1974) 39.

43. S. Dimopoulos and S. Raby, ITP preprint (1981). S. Dimopoulos, S. Raby and F. Wilczek ITP preprint 81-31 (1981).

44. S. Dimopoulos, Nucl. Phys. <u>B168</u> (1980) 69; M.E. Peskin,
Nucl. Phys. <u>B175</u> (1950) 197; J. Preskill, Nucl. Phys. <u>B177</u>
(1981) 21; P. Binétruy, S. Chadha and P. Sikivie, CERN
preprints TH-3122/LAPP-TH-49 and TH-3163/LAPP-TH-47 (1981).

45. S. Dimopoulos and J. Ellis, Nucl. Phys. <u>B182</u> (1981) 505.

46. R. Holdom, Phys. Rev. <u>D24</u> (1981) 1441.

47. H. Georgi and S.L. Glashow, Phys. Rev. Lett. <u>47</u> (1981) 1511;
H. Georgi and I. Mac Arthur, Nucl. Phys. B, to be published.

48. S. Coleman and J. Mandula, Phys. Rev. <u>159</u> (1967) 1251;
R. Haag, J. Lopuszanski and M. Sohnius, Nucl. Phys. <u>B88</u>
(1975) 257.

49. N. Sakai, Z. f. Phys. <u>C11</u> (1981) 153; S. Dimopoulos and H.
Georgi, Nucl. Phys. <u>B193</u> (1981) 150.

50. S. Dimopoulos and S. Raby, Nucl. Phys. <u>B192</u> (1981) 353;
M. Dine, W. Fischler and M. Srednicki, Nucl. Phys. <u>B189</u>
(1981) 575.

51. L.E. Ibáñez and G.G. Ross, Phys. Lett. <u>105B</u> (1981) 439.

52. L. Alvarez-Gaumé, M. Claudson and M.B. Wise, Nucl. Phys.
<u>B207</u> (1982) 96; C.R. Nappi and B.A. Ovrut, Phys. Lett.

113B (1982) 175; M. Dine and W. Fischler, Phys. Lett. 110B (1982) 227.

53. J. Ellis, L.E. Ibáñez and G.G. Ross, Phys. Lett. 113B (1982) 283; CERN preprint TH 3382 (1982).

54. For a general review see P. Fayet and S. Ferrara, Phys. Reports 32C (1977) 250.

55. J. Ellis, C.H. Llewellyn Smith and G.G. Ross, Phys. Lett. 114B (1982) 227.

56. D.V. Nanopoulos and K. Tamvakis, Phys. Lett. 110B (1982) 449; M. Srednick, Nucl. Phys. B202 (1982) 327.

57. L. Giradello, M.T. Grisaru and P. Salomonson, Nucl. Phys. B178 (1981) 331.

58. E. Witten, Phys. Lett. 105B (1981) 267.

59. L. O'Raifeartaigh, Nucl. Phys. B96 (1975) 331; P. Fayet, Phys. Lett. 58B (1975) 67.

60. P. Fayet and J. Iliopoulos, Phys. Lett. 51B (1974) 461.

61. P. Fayet, Phys. Lett. 69B (1977) 489.

62. N. Dragon and M. Schmidt, Univ. of Heidelberg preprint HD-THEP-82-24 (1982).

63. U. Ellwanger, to be published (1983).

64. S. Deser and B. Zumino, Phys. Rev. Lett. 38 (1977) 1433.

65. S. Dimopoulos and S. Raby, Los Alamos preprint LA-UR-82-1282 (1982); T. Banks and V. Kaplunovsky, Tel. Aviv preprint TAUP 1028-82 (1982).

66. V. Polchinski and L. Susskind, SLAC-Pub-2924 (1982), J. Polchinski, SLAC-Pub-2931-T.

67. A. Masiero, D.V. Nanopoulos, K. Tamvakis and T. Yanagida, Phys. Lett. 115B (1982) 380; H. Georgi, Phys. Lett. 108B (1982) 283; B. Grinstein, Harvard preprint HUTP-82/A014 (1982).

68. E. Cremmer, B. Julia, J. Scherk, S. Ferrara, L. Giradello and P. van Nieuwenhuizen, Nucl. Phys. B147 (1979) 105; E. Witten and J. Bagger, Princeton University preprint (1982).

69. J. Ellis and D.V. Nanopoulos, Phys. Lett. 116B (1982) 33.

70. M.K. Gaillard, L.J. Hall, B. Zumino, F. del Aguila, J. Polchinski and G.G. Ross, Lawrence Berkeley Preprint LBL-15215 (1982).

71. H.P. Nilles, M. Srednicki and D. Wyler, CERN preprint TH 3432 (1982).

72. R. Barbieri, S. Ferrara and C.A. Savoy, Phys. Lett. 119B (1982) 343; J. Ellis, D.V. Nanopoulos and K. Tamvakis, CERN preprint TH 3418 (1982); R. Arnowitt, Pran Nath and A. Chamseddine, Phys. Rev. Lett. 49 (1982) 970.

73. J. Polonyi, Univ. of Budapest Report No. KFKI-1977-93(1977).

74. H.P. Nilles, M. Srednicki and D. Wyler, CERN preprint TH 3461 (1982); A.B. Lahanas, CERN preprint TH 3467 (1982).

75. L.C. Ibáñez, Phys. Lett. 118B (1982) 73; Univ. Autonoma de Madrid preprint FTUAM/82-8; J. Ellis, D.V. Nanopoulos and K. Tamvakis, Phys. Lett. 121B (1983) 123, L. Alvarez-Gaumé, J. Polchinski and M.B. Wise, Harvard preprint HUTP-82/AO63 (1982); L.E. Ibáñez and C. Lóbez, Univ. Autonoma de Madrid preprint FTUAM/83-2 (1983).

76. G. Ross, in preparation.

77. P. Fayet in "Unification of the Fundamental Particle Interactions", eds. S. Ferrara, J. Ellis and P. van Nieuwenhuizen (Plenum Press, N.Y. 1980) 587.

78. For a recent review see P. Fayet, Proceedings of the seventeenth Recontre de Moriond, Les Arcs (1982) 483.

79. L. Ibáñez and G.G. Ross. Phys. Lett. 105B (1981) 439; M.B. Einhorn and D.R.T. Jones, Nucl. Phys. B196 (1982) 475; W. Marciano and G. Senjanowich.

80. C.H. Llewellyn Smith and J. Wheater, Phys. Lett. 105B (1981) 486; A. Salam and W. Marciano, Nucl. Phys. B159 (1981) 442.

81. D.V. Nanopoulos and K. Tamvakis, Phys. Lett. 113B (1982) 151.

82. N. Sakai and T. Yanagida, Nucl. Phys. B197 (1982) 533; S. Weinberg, Phys. Rev. D26 (1982) 287.

83. J. Ellis, D.V. Nanopoulos and S. Rudaz, Nucl. Phys. B202 (1982) 43; S. Dimopoulos, S. Raby and F. Wilczek UMHE 81-64.

84. J. Ellis and G.G. Ross, Phys. Lett. 117B (1982) 397.

85. C. Oakley and G.G. Ross, Phys. Lett. to be published (1983).

86. L. Hall and I. Hinchcliffe, LBL preprint LBL-14806.

87. S. Dimopoulos and H. Georgi, Phys. Lett. 117B (1982) 287.

Acta Physica Austriaca, Suppl. XXV, 249–250 (1983)
© by Springer-Verlag 1983

GUTS, PHASE TRANSITIONS, AND THE EARLY UNIVERSE[+]

by

Q. SHAFI

ICTP - Trieste

LECTURE I

Motivation for going beyond SU(3)×SU(2)×U(1)

Introduction to GUTs

Phase Transitions in Gauge Theories

The standard hot big bang model and cosmological puzzles

The Inflationary Scenario

LECTURE II

Dirac Magentic Monopole and Electric charge quantization

Monopoles in spontaneously broken gauge theories

Primordial GUT monopoles

Magnetic Monopoles and Nucleon Decay

LECTURE III

Motivation for going beyond SU(5)

[+] Lectures given at the XXII. Internationale Universitätswochen für Kernphysik, Schladming, Austria, February 23-March 5, 1983.

Strong CP problem and the "almost" invisible axion
Axions as dark matter in the universe and galaxy formation
The need for an intermediate mass scale in GUTs
Family Symmetry

LECTURE IV

Strings in the very early universe
Density perturbations from strings and galaxy formation
Cosmological Constraints on Supersymmetric GUTs
Conclusions

Acta Physica Austriaca, Suppl. XXV, 251–281 (1983)
© by Springer-Verlag 1983

STOCHASTIC QUANTIZATION[+]

by

John R. KLAUDER

Bell Laboratories
Murray Hill,USA

ABSTRACT

In this introductory survey to stochastic quantization
we outline this new approach for scalar fields, gauge
fields, fermion fields, and condensed matter problems such
as electrons in solids and the statistical mechanics of
quantum spins.

I. INTRODUCTION

The concepts of stochastic quantization have
attracted much attention since their introduction and application to Euclidean-space non-Abelian gauge field theories
[1]. It was originally believed that this formalism automatically selected the Landau gauge and offered a method
for computing scattering amplitudes without the need for
gauge fixing terms and the attendant Faddeev-Popov ghost
fields. It is now realized that this belief is not entirely
true (see Sec. 3). In addition, the formalism itself is of

[+]Lectures given at the XXII. Internationale Universitätswochen
für Kernphysik,Schladming,Austria,February 23 - March 5,1983.

considerable interest in a broad range of problems which
have little or nothing to do with the original motivating
gauge field problems. Indeed, we feel that the ideas behind
stochastic quantization have potential value for numerical
studies of various problems, such as those dealing with
fermions or quantum spins, and we orient our presentation to
some degree with such computations in mind.

Broadly speaking, the concepts underlying stochastic
quantization provide an alternative way to evaluate integrals
by analyzing associated Langevin equations (i.e., stochastic
differential equations). For the sake of illustration, and
with little of the necessary qualifications, let us consider the various integrals

$$I_p \equiv \int x^p f(x) \, dx \qquad (1.1a)$$

for $p = 1, 2, \ldots$, where

$$f(x) \equiv e^{-S(x)} / \int e^{-S(y)} \, dy, \qquad (1.1b)$$

assuming always that $\int \exp[-S(y)] \, dy$ neither vanishes nor
diverges. Here although we use the notation of a single
variable we can imagine as well that x is a D-dimensional
variable.

Now let $g(x, \tau)$ denote a function of an additional
variable τ (to be called "auxiliary time", or more simply
"time") with the property that

$$\lim_{\tau \to \infty} g(x, \tau) = f(x) \, . \qquad (1.2)$$

One among many ways to generate such a function is as a
solution to the partial differential equation

$$\frac{\partial g(x, \tau)}{\partial \tau} = \frac{b^2}{2} \frac{\partial}{\partial x} \left[\frac{\partial S(x)}{\partial x} + \frac{\partial}{\partial x} \right] g(x, \tau) \qquad (1.3)$$

for arbitrary $b > 0$ and initial condition $g(x, 0) = g_0(x)$ such that $\int g_0(x)dx = 1$. The formal analogy of this equation with a Fokker-Planck equation suggests the introduction of the analogous Langevin equation, here given by

$$\frac{dx(\tau)}{d\tau} \equiv \dot{x}(\tau) = -\frac{b^2}{2}\frac{\partial S(x(\tau))}{\partial x(\tau)} + b\xi(\tau) \quad . \tag{1.4}$$

Here ξ denotes a Gaussian white noise of unit power, a generalized stochastic variable for which

$$<\xi(\tau)> = 0 \quad ,$$
$$<\xi(\tau_1)\xi(\tau_2)> = \delta(\tau_1 - \tau_2) \quad , \tag{1.5}$$

with all higher-order cumulants vanishing, where $<\cdot>$ denotes average with respect to the noise ensemble. As a relation among distributions the Langevin equation is properly interpreted as a weak equation requiring smearing with suitable test functions. For most cases $b = \sqrt{2}$ or $b = 1$ are satisfactory choices for the scale factor.

By design the tendency for any normalized distribution to relax to $f(x)$ for large τ implies that

$$\lim_{\tau\to\infty} <x^p(\tau)> = \int x^p f(x)dx \quad . \tag{1.6}$$

Furthermore an ergodic-like property of the ensemble of solutions implies that

$$\lim_{T\to\infty} \frac{1}{T}\int_0^T x^p(\tau)d\tau = \int x^p f(x)dx \quad . \tag{1.7}$$

These two equations are the sought-for equivalents and alternatives to calculate the integrals I_p.

To see these properties most easily let us introduce

$$h(x,\tau) \equiv g(x,\tau)e^{S(x)/2} \tag{1.8}$$

which is readily seen to satisfy the Schrödinger-like
equation

$$\frac{\partial h(x,\tau)}{\partial \tau} = - Hh(x,\tau) \tag{1.9a}$$

where

$$H \equiv - \frac{b^2}{2} \frac{\partial^2}{\partial x^2} + V(x) \ , \tag{1.9b}$$

$$V(x) \equiv \frac{b^2}{4} \left[\frac{1}{2}\left(\frac{\partial S(x)}{\partial x}\right)^2 - \frac{\partial^2 S(x)}{\partial x^2}\right] \ . \tag{1.9c}$$

Since we may also express V as

$$V(x) \equiv \frac{b^2}{2} \frac{\partial^2 e^{-S(x)/2}/\partial x^2}{e^{-S(x)/2}} \tag{1.10}$$

it is clear that

$$h_0(x) = ce^{-S(x)/2} \tag{1.11}$$

for arbitrary nonzero c is an eigenfunction of H with
eigenvalue zero. A general solution h has the representation

$$h(x,\tau) = \sum_{n=0}^{\infty} c_n h_n(x) e^{-E_n\tau} \ , \tag{1.12}$$

assumed for simplicity to be a discrete sum, where $E_0 \equiv 0$.
We now require that h_0 be a nondegenerate eigenfunction and
that

$$\lim_{\tau \to \infty} e^{-E_n\tau} = 0 \tag{1.13}$$

for all $n \geq 1$. In this case

$$\lim_{\tau \to \infty} g(x,\tau) = f(x) \tag{1.14}$$

as desired provided only that $c_o \neq 0$.

The condition for nondegeneracy of the ground state suffices also to ensure the required ergodic property. Let

$$C_o(s) \equiv \langle e^{i\int_o^\infty s(\tau)x(\tau)d\tau} \rangle \tag{1.15}$$

denote the characteristic functional for the process for suitable test functions s. If we introduce

$$s_u(\tau) \equiv s(\tau-u) \tag{1.16}$$

then

$$C(s) \equiv \lim_{u\to\infty} C_o(s_u) \tag{1.17}$$

characterizes a stationary characteristic functional. Furthermore it follows from the uniqueness of the ground state of H that

$$\lim_{u\to\infty} C(s+r_u) = C(s)C(r) \tag{1.18}$$

for all test functions r and s. Consequently the variable

$$A_T \equiv \frac{1}{T} \int_o^T e^{i\int r_u(\tau)x(\tau)d\tau} du \tag{1.19}$$

satisfies the relation

$$\lim_{T\to\infty} \langle \{A_T - C(r)\} e^{i\int s(\tau)x(\tau)d\tau} \rangle = 0 \tag{1.20}$$

for all s. In this sense we may say that

$$\lim_{T\to\infty} \frac{1}{T} \int_o^T e^{i\int r(\tau-u)x(\tau)d\tau} du = \langle e^{i\int r(\tau)x(\tau)d\tau} \rangle , \tag{1.21}$$

which establishes the sought for ergodic property.

Complex Distributions

A fundamental and key point to emphasize here is that
although we have argued implicitly as if S were real and
therefore f>0, our discussion extends as well to cases
for which S is <u>complex</u> leading in turn to complex f, g and
h and a complex solution $x(\tau)$ of the Langevin equation. The
complex nature of S will arise when we discuss fermions and
quantum spins. The formalism sketched above will still
apply whenever h_o is a nondegenerate state with $E_o=0$ and
Re $E_n > 0$ for $n \geq 1$. The validity of the formalism for complex S applies to those cases where exp(-S) vanishes on a set
of dimension D-2 or less, where D is the dimension of x in
the many-dimensional case.

When exp(-S) vanishes the Langevin equation acquires
singularities in the drift coefficients. For most cases of
interest (and all our examples) the Langevin equation
with a singularity in the complex variable z at a finite
point, say at z = 0, locally reduces, say for $|z| \leq 10|z_0| <<1$,
to the generic form

$$\dot{z}(\tau) = \frac{(p-1)}{2z(\tau)} + \xi(\tau) \tag{1.22}$$

where p = 2,3,4,..., and ξ is standard white noise (real).
Here z_0 denotes a nonzero complex initial value at $\tau = \tau_o$
for this equation. In terms of polar coordinates

$$z(\tau) \equiv r(\tau)e^{i\theta(\tau)} \tag{1.23}$$

the radius r satisfies the Langevin equation (Itô form)

$$\dot{r}(\tau) = \frac{(p-1)}{2} \frac{\cos^2\theta(\tau)}{r(\tau)} + \cos\theta(\tau)\xi(\tau) \quad . \tag{1.24}$$

The solution to this equation may be realized with help of the stochastic time

$$\tau^* = \int_{\tau_o}^{\tau} \cos^2\theta(\sigma)\,d\sigma + \tau_o \qquad (1.25)$$

as

$$r(\tau) = [\sum_{m=1}^{p} W_m^2 (\tau^*)]^{1/2} \quad , \qquad (1.26)$$

where $W_1, \ldots, W_p$ denote p independent standard Wiener processes conditioned at $\tau = \tau_o = \tau^*$ so that

$$r(\tau_o) \equiv r_o = |z_o| > 0 \quad . \qquad (1.27)$$

It follows from this realization that $r(\tau) > 0$ for $\tau \geq \tau_o$ holds with probability one [2]. This relation extends the usual Bessel process where $|\cos\theta| = 1$ and $\tau^* = \tau$. Further information about the solution of the local Langevin equation is given by decomposing z into Cartesian coordinates, $z = x + iy$, and noting that

$$\dot{y}(\tau) = - \frac{(p-1)}{2} \frac{y(\tau)}{r^2(\tau)} \quad , \qquad (1.28)$$

a purely deterministic equation. Hence

$$y(\tau) = \exp\left[- \left(\frac{p-1}{2}\right) \int_{\tau_o}^{\tau} \frac{d\sigma}{r^2(\sigma)}\right] y_o \quad . \qquad (1.29)$$

To complete the solution we note that

$$\cos^2\theta(\tau) = 1 - y^2(\tau)/r^2(\tau) \quad . \qquad (1.30)$$

These equations generate well-behaved self-consistent solutions in the local vicinity of a singularity of the type indicated. In the examples that arise later we refer to such singularities as <u>standard</u> ones.

2. SCALAR FIELDS

In a familiar way the correlation functions of a Euclidean-space, covariant scalar field theory may be described by formal functional integrals of the form

$$<<\phi(x_1) \ldots \phi(x_m)>> = N \int \phi(x_1) \ldots \phi(x_m) e^{-W(\phi)} D\phi , \qquad (2.1)$$

where $x \epsilon R^n$, $D\phi \equiv \pi_x d\phi(x)$, N is a formal normalization constant, and W denotes the Euclidean action, for example, given by

$$W(\phi) = \int \{1/2[(\nabla\phi)^2 + m_o^2 \phi^2] + \lambda_o \phi^4\} d^n x \qquad (2.2)$$

for a quartic self-interacting field in an n-dimensional space time.

Following conventional practise we choose the scale factor $b = \sqrt{2}$ for this example. We interpret $N \exp(-W)$ as the equilibrium distribution of a nonequilibrium statistical problem characterized by the Fokker-Planck equation

$$\frac{\partial G(\phi,\tau)}{\partial \tau} = \{\int \frac{\delta}{\delta\phi(x)} [\frac{\delta W(\phi)}{\delta\phi(x)} + \frac{\delta}{\delta\phi(x)}] d^n x\} G(\phi,\tau) . \qquad (2.3)$$

In turn this ensemble is characterized by the Langevin equation [1]

$$\frac{\partial \phi(x,\tau)}{\partial \tau} = - \frac{\delta W(\phi)}{\delta\phi(x,\tau)} + \eta(x,\tau) \qquad (2.4)$$

where η is a generalized Gaussian stochastic process characterized by the ensemble averages

$$\langle \eta(x,\tau) \rangle = 0 \quad ,$$

$$\langle \eta(x,\tau) \eta(y,\sigma) \rangle = 2\delta(x-y)\delta(\tau-\sigma) \quad . \tag{2.5}$$

Note that the factor $\sqrt{2}$ has been absorbed into this generalization of white noise. The solution of the Langevin equation subject to the distribution of initial values at $\tau = 0$ being given by $G(\phi,0) \equiv G_0(\phi)$ exhibits the property that

$$\lim_{\tau \to \infty} \langle \phi(x_1,\tau) \ldots \phi(x_m,\tau) \rangle$$

$$= \langle\langle \phi(x_1) \ldots \phi(x_m) \rangle\rangle \quad . \tag{2.6}$$

Alternatively, for almost all solutions it follows that

$$\lim_{T \to \infty} \frac{1}{T} \int_0^T \phi(x_1,\tau) \ldots \phi(x_m,\tau) d\tau$$

$$= \langle\langle \phi(x_1) \ldots \phi(x_m) \rangle\rangle \quad . \tag{2.7}$$

For the $(\phi^4)_n$ model the Langevin equation reads

$$\frac{\partial \phi(x,\tau)}{\partial \tau} = (\Delta - m_0^2)\phi(x,\tau) - 4\lambda_0 \phi^3(x,\tau) + \eta(x,\tau) \quad . \tag{2.8}$$

Strictly speaking Langevin equations of this form are weak equations and their solutions yield multi-time correlation functions

$$\langle \phi(x_1,\tau_1) \ldots \phi(x_m,\tau_m) \rangle \quad . \tag{2.9}$$

By a suitable choice of $G_0(\phi)$ or by looking sufficiently far in the future we can restrict our attention to stationary multi-time correlation functions. As an example consider the free field ($\lambda_0 = 0$) problem with unit mass ($m_0 = 1$) which satisfies the Langevin equation

$$\frac{\partial \phi_o(x,\tau)}{\partial \tau} = (\Delta - 1)\phi_o(x,\tau) + \eta(x,\tau) \quad . \tag{2.10}$$

The desired stationary solution is a Gaussian process with zero mean and distributional covariance given by

$$<\phi_o(x,\tau)\phi_o(y,\sigma)> = \frac{2}{(2\pi)^{n+1}} \int \frac{e^{ik\cdot(x-y)-i\omega(\tau-\sigma)}}{\omega^2 + (k^2+1)^2} \, d^n k d\omega . \tag{2.11}$$

The equal-time limit reduces to

$$<\phi_o(x,\tau)\phi_o(y,\tau)> = <<\phi_o(x)\phi_o(y)>>$$

$$= \frac{1}{(2\pi)^n} \int \frac{e^{ik\cdot(x-y)}}{k^2 + 1} \, d^n k \quad , \tag{2.12}$$

which is the usual expression. The short-distance behavior of the equal-time, two-point function is $[(x-y)^2]^{-(n-2)/2}$ as usual. For the multi-time function the short-distance, short-time behavior is effectively represented by

$$[A|\tau-\sigma| + (x-y)^2]^{-(n-2)/2} \tag{2.13}$$

for some A, $0<A<\infty$. This time dependence leads to a certain regularizing behavior which we now illustrate.

Let us attempt to solve the nonlinear Langevin equation by a perturbation expansion in λ_o about the free solution ϕ_o. The first appearance of the nonlinear term is as ϕ_o^3, which we propose to define as a normal-ordered product since ϕ_o is a Gaussian variable. The stochastic variable

$$Y \equiv \int h(x,\tau) : \phi_o^3(x,\tau) : d^n x d\tau \tag{2.14}$$

is well defined provided that

$$<Y^2> = 6 \int h(x,\tau) h(y,\sigma) < \phi_o(x,\tau) \phi_o(y,\sigma) >^3 d^n x d^n y d\tau d\sigma \qquad (2.15)$$

is finite. In turn this requires the local integrability of

$$\int (A|\tau| + x^2)^{-3(n-2)/2} d^n x d\tau$$

$$\leq \text{const.} \int (x^2)^{-3(n-2)/2+1} d^n x \quad , \qquad (2.16)$$

which converges (at $x = 0$) if $n<4$. In the absence of the τ integration the convergence criterion instead becomes $n < 3$. Recall that $n < 3$ is the criterion for finite mass renormalization while $n < 4$ is the criterion for super renormalizability for this model. Thus while the conventional Heisenberg operator field equation for this model has finite parameters only when $n < 3$, it is remarkable to note that the Langevin equation contains finite parameters and is well defined whenever $n < 4$! [3]. This example illustrates the regularizing power of the auxiliary time τ.

Further discussion of scalar field models may be found in Refs. [3] and [4].

3. GAUGE FIELDS

There are by now a number of articles that address the question of stochastic quantization for gauge fields [5], and as a consequence we shall present only the barest outline of this topic. Our discussion closely follows that of Namiki, et al. [6].

For an Abelian Euclidean-space gauge field the action is given by

$$S = 1/4 \int F_{\mu\nu}^2(x) d^n x \qquad (3.1a)$$

where

$$F_{\mu\nu}(x) = \partial_\mu A_\nu(x) - \partial_\nu A_\mu(x) \quad . \tag{3.1b}$$

Gauge invariance of S implies that

$$\int e^{-S} DA = \infty \tag{3.2}$$

and thus that expressions for the mean

$$<<I(A)>> = \int I(A) e^{-S} DA / \int e^{-S} DA \tag{3.3}$$

are ill-defined as the quotient of divergent factors. The remedy for this problem is gauge fixing with the attendant Faddeev-Popov ghost fields especially for non-Abelian gauge theories [7]. Nevertheless one could imagine an integrable distribution $G(A,\tau)$, the solution of a Fokker-Planck equation, which has the tendency that

$$<I(A,\tau)> = \int I(A) G(A,\tau) DA$$

$$\xrightarrow[\tau \to \infty]{} <<I(A)>> \quad , \tag{3.4}$$

thereby yielding the desired results by another route. The ensemble described by this Fokker-Planck equation may be characterized instead by the Langevin equation which for the Abelian case reads (with the scale factor $b = \sqrt{2}$)

$$\frac{\partial A_\mu(x,\tau)}{\partial \tau} = (\Delta \delta_{\mu\nu} - \partial_\mu \partial_\nu) A_\nu(x,\tau) + \eta_\mu(x,\tau) \quad , \tag{3.5}$$

where η_μ is a Gaussian noise source with

$$<\eta_\mu(x,\tau)> = 0 \quad ,$$

$$<\eta_\mu(x,\tau)\eta_\nu(y,\sigma)> = 2\delta_{\mu\nu}\delta(x-y)\delta(\tau-\sigma) \quad . \tag{3.6}$$

In Fourier space the Langevin equation becomes

$$\frac{\partial A_\mu(k,\tau)}{\partial \tau} = -k^2(\delta_{\mu\nu} - k_\mu k_\nu/k^2)A_\nu(k,\tau) + \eta_\mu(k,\tau) \quad, \qquad (3.7a)$$

which expressed in transverse and longitudinal components,

$$A_\mu^T(k,\tau) \equiv (\delta_{\mu\nu} - k_\mu k_\nu/k^2)A_\nu(k,\tau) \quad, \qquad (3.7b)$$

$$A_\mu^L(k,\tau) \equiv (k_\mu k_\nu/k^2)A_\nu(k,\tau) \quad, \qquad (3.7c)$$

becomes simply

$$\frac{\partial A_\mu^T(k,\tau)}{\partial \tau} = -k^2 A_\mu^T(k,\tau) + \eta_\mu^T(k,\tau) \quad, \qquad (3.8a)$$

$$\frac{\partial A_\mu^L(k,\tau)}{\partial \tau} = \eta_\mu^L(k,\tau) \quad. \qquad (3.8b)$$

The vanishing of the drift for the longitudinal component is a direct consequence of the gauge invariance of S, namely that $\delta S/\delta A_\mu^L$ vanishes. While the equation for A_μ^T possesses a damping term, and thus for large τ no longer depends on the initial distribution, this is not the case for A_μ^L. In particular

$$<A_\mu^L(k,\tau)> = \overline{A_\mu^L(k,0)} \equiv (k_\mu/k^2)\overline{\Phi(k)} \qquad (3.9)$$

holds for all $\tau > 0$ where the overbar denotes an average in the initial distribution, here an average over the scalar function Φ. The natural choice is to take $\Phi \equiv 0$ which leads to the Landau gauge propagator since for large τ

$$<A_\mu(k,\tau)A_\nu(k',\tau)> = \delta(k+k') \left[\frac{1}{k^2}(\delta_{\mu\nu} - \frac{k_\mu k_\nu}{k^2}) + 2\tau \frac{k_\mu k_\nu}{k^2}\right] \quad. \qquad (3.10)$$

The term proportional to τ-which superficially could lead to nonrenormalizability - may be ignored due to gauge invariance. Thus it may seem that stochastic quantization automatically selects the Landau gauge without the need for a specific gauge-fixing term. However if $\Phi \neq 0$ then the propagator above is modified by the additional term

$$\frac{k_\mu}{k^2} \frac{k'_\nu}{k'^2} \Phi(k,k') \qquad\qquad (3.11)$$

where $\Phi(k,k') \equiv \overline{\Phi(k)\Phi(k')}$. It is not difficult to arrange a distribution of the Φ field, in the spirit of white noise, so that $\Phi(k,k') = -\alpha\delta(k+k'), \alpha>0$. In this case, for large τ, the propagator becomes

$$<A_\mu(k,\tau)A_\nu(k',\tau)> = \delta(k+k')\{\frac{1}{k^2}[\delta_{\mu\nu} -(1-\alpha)\frac{k_\mu k_\nu}{k^2}] + 2\tau \frac{k_\mu k_\nu}{k^2}\}$$

$$(3.12)$$

reflecting the nature of α as a gauge parameter. Thus we see that the choice of initial state distribution just corresponds to the choice of gauge. Nevertheless it is true that no gauge fixing term needs to be added to the action. Moreover we emphasize that gauge invariant correlation functions converge to the desired form as τ diverges, as for example,

$$\lim_{\tau\to\infty} <F_{\mu\nu}(x,\tau)F_{\alpha\beta}(y,\tau)> = <<F_{\mu\nu}(x)F_{\alpha\beta}(y)>> \quad . \qquad (3.13)$$

Non-Abelian Fields

For the non-Abelian case

$$S = 1/4\int F_{\mu\nu}{}^a(x)^2 d^n x \qquad\qquad (3.14)$$

and

$$F_{\mu\nu}^a(x) = \partial_\mu A_\nu^a(x) - \partial_\nu A_\mu^a(x) - gf^{abc}A_\mu^b(x)A_\nu^c(x) \quad . \tag{3.15}$$

As before gauge invariance of S renders $\int \exp(-S)DA$ divergent so that the expression

$$<<I(A)>> = \int I(A)e^{-S}DA / \int e^{-S}DA \tag{3.16}$$

is ill defined. The usual remedy involves gauge fixing with the attendant Faddeev-Popov ghost fields.

From the point of view of stochastic quantization we can imagine developing the result in a perturbation expansion in g about the Abelian case (g=0) and starting with $A_\mu(k,0)=0$. Since the Abelian case requires no gauge fixing term we may anticipate that none is necessary for the non-Abelian case. This belief is in fact true.

Let us write the Langevin equation for the non-Abelian field in momentum space:

$$\frac{\partial A_\mu^a(k,\tau)}{\partial\tau} = -(k^2\delta_{\mu\nu} - k_\mu k_\nu)A_\nu^a(k,\tau) + Y_\mu^a(k,\tau) \tag{3.17}$$

where

$$Y_\mu^a(k,\tau) \equiv \eta_\mu^a(k,\tau)$$

$$+[g/(2\pi)^{n/2}]\int d^n k_1 d^n k_2 \delta(k-k_1-k_2)V_{\mu\alpha\beta}^{abc}(k,-k_1,-k_2)A_\alpha^b(k_1,\tau)A_\beta^c(k_2,\tau)$$

$$+[g^2/(2\pi)^n]\int d^n k_1 d^n k_2 d^n k_3 \delta(k-k_1-k_2-k_3)W_{\mu\nu\alpha\beta}^{abcd}A_\nu^b(k_1,\tau)A_\alpha^c(k_2,\tau)A_\beta^d(k_3,\tau) \, .$$

$$\tag{3.18}$$

Here

$$V_{\mu\alpha\beta}^{abc}(k,k_1,k_2) = -\frac{i}{2}f^{abc}[(k-k_1)_\beta\delta_{\mu\alpha}+(k_1-k_2)_\mu\delta_{\alpha\beta}+(k_2-k)_\alpha\delta_{\mu\beta}],$$
$$\tag{3.19a}$$
$$W_{\mu\nu\alpha\beta}^{abcd} = -\frac{1}{6}[f^{abe}f^{cde}(\delta_{\mu\alpha}\delta_{\nu\beta}-\delta_{\mu\beta}\delta_{\nu\alpha})$$

$$+ f^{ace}f^{bde}(\delta_{\mu\nu}\delta_{\alpha\beta} - \delta_{\mu\beta}\delta_{\nu\alpha})$$

$$+ f^{ade}f^{cbe}(\delta_{\mu\alpha}\delta_{\nu\beta} - \delta_{\mu\nu}\delta_{\alpha\beta})] . \tag{3.19b}$$

An iterative solution to this nonlinear equation leads to modification of the free field correlation functions. One expects that terms in the correlation functions that grow with τ, for large τ, are connected with gauge variant expressions and drop out of gauge invariant averages. However for internal lines the growing terms in the longitudinal propagator will make important contributions which are just those corresponding to the Faddeev-Popov ghost effects in the conventional theory appropriate to the Landau gauge. A detailed calculation [6], carried out to second order in g confirms that the propagator, for large τ,

$$<F^a_{\mu\nu}(x,\tau)F^b_{\alpha\beta}(y,\tau)> \tag{3.20}$$

coincides with the expression obtained from the conventional theory plus gauge fixing and ghost contributions for the Landau gauge. Consequently it appears entirely reasonable to assert that the perturbation theory of the stochastic quantization of non-Abelian gauge fields without gauge fixing or ghost fields leads to results in agreement with the conventional theory as modified by gauge fixing and ghost field contributions.

4. FERMION FIELDS

In this and the next section we take up ways to deal with fermions. Here, following a suggestion of K.G. Wilson[8], we initially eliminate the fermions and then use the Langevin equation for the generally complex effective action that emerges.

To first present the idea in an extremely elementary form consider the second quantized Hamiltonian operator

$$H = N_1 + 2N_2 + 2N_1 N_2 - 2N_1 N_3 \tag{4.1}$$

where $N_j = \psi_j^\dagger \psi_j$ and $\{\psi_i, \psi_j^\dagger\} = \delta_{ij}$. In this simple model "particles" 1 and 2 repel while 1 and 3 attract; the interaction between 2 and 3 is ignored. Since $N_j^2 = N_j$ for fermions H can be rewritten in the form

$$H = N_1 + N_2 + N_3 + (N_1 + N_2)^2 - (N_1 + N_3)^3 \ . \tag{4.2}$$

For this example the partition function

$$Z = \text{Tr}(e^{-\beta H})$$

$$= \frac{1}{4\pi\beta} \int \text{Tr} \exp[-\beta(N_1 + N_2 + N_3) - i\phi(N_1 + N_2) - \chi(N_1 + N_3)]$$

$$\times \ e^{-(\phi^2 + \chi^2)/4\beta} d\phi d\chi$$

$$= \frac{1}{4\pi\beta} \int (1 + e^{-\beta - i\phi - \chi})(1 + e^{-\beta - i\phi})(1 + e^{-\beta - \chi})$$

$$\times \ e^{-(\phi^2 + \chi^2)/4\beta} d\phi d\chi. \tag{4.3}$$

Here we arrive at a <u>generally complex</u> integrand due to the need to introduce an imaginary effective coupling constant for the $\phi(N_1 + N_2)$ interaction. The resultant complex distribution $\exp(-S)$ vanishes whenever $\chi = -\beta$ and $\phi = \pm\pi, \pm3\pi, \ldots$. As a set of equally spaced points in the two-dimensional ϕ, χ space the general complex Langevin equation formalism presented in Sec. 1 is applicable.

For this example we choose the scale factor b=1, and we let ξ and η denote two independent, standard white noise sources. The Langevin equations

$$\dot{\phi}(\tau) = -\frac{1}{2}\frac{\partial S}{\partial\phi(\tau)} + \xi(\tau) \quad , \tag{4.4a}$$

$$\dot{\chi}(\tau) = -\frac{1}{2}\frac{\partial S}{\partial\chi(\tau)} + \eta(\tau) \quad , \tag{4.4b}$$

become, for the case at hand,

$$\dot{\phi} = -\frac{i/2}{1+e^{1/2+i\phi+\chi}} - \frac{i/2}{1+e^{1/2+i\phi}} + \xi \quad , \tag{4.5a}$$

$$\dot{\chi} = -\frac{1/2}{1+e^{1/2+\chi}} - \frac{1/2}{1+e^{1/2+i\phi+\chi}} + \eta \quad . \tag{4.5b}$$

The drift terms in these equations exhibit singularities, and it is important to assess their character. For example, for a complex ϕ given by $\phi=\pi+\zeta+i/2$, $|\zeta|<<1$, it follows that the ϕ equation is adequately approximated by

$$\dot{\zeta} = -\frac{i/2}{1-e^{i\zeta}} + \xi = \frac{1}{2\zeta} + \xi \quad , \tag{4.6}$$

which is just one of the standard singularities discussed in Sec. 1 (with p=2). The other singularities in these equations are of a similar nature.

Finally we observe that ϕ and χ correlation functions are related to fermion statistical averages. For example, it follows that

$$\lim_{T\to\infty}\frac{1}{T}\int_0^T\phi(\tau)d\tau = -2i\beta\,\frac{Tr[(N_1+N_2)e^{-\beta H}]}{Tr[e^{-\beta H}]} \quad . \tag{4.7}$$

Although this is a very simple model it is worth emphasizing that a typical Monte-Carlo approach (e.g., Metropolis algorithm) to evaluate the partition function is impossible because the distribution is complex. On the other hand, the Langevin equation and its numerical simulation works even when the distribution is complex. This is an illustration of an advantage which we feel has wide application.

General Case

Even when one deals with more complicated fermion problems the situation is conceptually similar. Consider the Hamiltonian

$$H = \sum \psi_\alpha^\dagger \hat{K}_{\alpha\beta} \psi_\beta + \sum \psi_\alpha^\dagger \psi_\gamma^\dagger W_{\alpha\gamma,\delta\beta} \psi_\delta \psi_\beta \tag{4.8}$$

where α,β denote spatial or lattice sites, internal indices, ect.,

$$\{\psi_\alpha, \psi_\beta^\dagger\} = \delta_{\alpha\beta} \ ,$$

with all other anticommutators vanishing, and $\hat{K}_{\alpha\beta}^* = \hat{K}_{\beta\alpha}$, $W_{\alpha\gamma,\delta\beta}^* = W_{\beta\delta,\gamma\alpha}$ ($=W_{\delta\beta,\alpha\gamma}$). We may decompose $W_{\alpha\gamma,\delta\beta}$ according to

$$W_{\alpha\gamma,\delta\beta} = \sum_m \lambda_m U_{\alpha\beta}^m U_{\gamma\delta}^m - \sum_n g_n V_{\alpha\beta}^n V_{\gamma\delta}^n \tag{4.9}$$

where $\lambda_m > 0$, $g_n > 0$, $U_{\alpha\beta}^{m*} = U_{\beta\alpha}^m$ and $V_{\alpha\beta}^{n*} = V_{\beta\alpha}^n$. In turn we may write H as

$$H = \sum \psi_\alpha^\dagger K_{\alpha\beta} \psi_\beta + \sum_m \lambda_m (\psi_\alpha^\dagger U_{\alpha\beta}^m \psi_\beta)^2 - \sum_n g_n (\psi_\alpha^\dagger V_{\alpha\beta}^n \psi_\beta)^2 \tag{4.10}$$

where

$$K_{\alpha\beta} \equiv \hat{K}_{\alpha\beta} - \sum_m \lambda_m U_{\alpha\delta}^m U_{\delta\beta}^m + \sum_n g_n V_{\alpha\delta}^n V_{\delta\beta}^n \quad . \tag{4.11}$$

The λ-interactions are repulsive while the g-interactions are attractive. We consider the general case in which neither the λ nor g sum is empty.

It follows in this case that the partition function can be written as

$$Z = Tr(e^{-\beta H})$$

$$= N \int Tr(T \exp\{-\sum \int \psi_\alpha^\dagger [K_{\alpha\beta} + i\phi_m(t) U_{\alpha\beta}^m + \chi_n(t) V_{\alpha\beta}^n] \psi_\beta dt\})$$

$$\times \exp\{-\frac{1}{4}\sum \int [\phi_m^2(t)/\lambda_m + \chi_n^2(t)/g_n] dt\} \Pi D\phi_m D\chi_n \quad , \qquad (4.12)$$

where we let one summation sign do the work of many, the thermal time t runs between 0 and β, and the operators are time ordered by the operator T. The trace involving the resultant quadratic fermion Hamiltonian may be represented as a determinant according to standard results.

Recall, for fermion field functional integrals [9], that

$$Tr(T \exp\{-1/2\sum \int [\psi_\alpha^\dagger, \Lambda_{\alpha\beta}(t) \psi_\beta] dt\})$$

$$= N \int \exp\{-\sum \int [\bar{\psi}_\alpha(t) \partial_t \psi_\alpha(t) + \bar{\psi}_\alpha(t) \Lambda_{\alpha\beta}(t) \psi_\beta(t)] dt\} \Pi D\bar{\psi}_\alpha D\psi_\alpha , \qquad (4.13)$$

where on the right-hand side one deals with Grassmann variables $\psi_\alpha, \bar{\psi}_\beta$ which satisfy antiperiodic boundary conditions, e.g., $\psi_\alpha(\beta) = -\psi_\alpha(0)$, etc. This integral is formally evaluated as

$$M det[\partial_t + \Lambda_{\alpha\beta}(t)] , \qquad (4.14)$$

where the indicies of the determinant are α, β and the thermal time. The factor M is a formal normalization clearly given by

$$M = \text{Tr}(1) / \det [\partial_t] \quad . \tag{4.15}$$

To see this machinery in action consider the elementary case

$$\text{Tr}\{\exp(-1/2\beta[\psi^\dagger,\psi]\Lambda)\} \tag{4.16}$$

with one index and a time independent Λ, which is then represented by

$$N\int\exp\{-\int[\bar{\psi}\partial_t\psi + \bar{\psi}\Lambda\psi]dt\}D\bar{\psi}D\psi \quad . \tag{4.17}$$

Introduce the Fourier decomposition

$$\psi(t) = \sum\psi_q e^{-i\pi(2q+1)(t/\beta)} \tag{4.18}$$

which respects the antiperiodic condition. In these variables the functional integral is given by

$$N\int\exp\{-\sum\beta[\Lambda-i\pi(2q+1)/\beta]\bar{\psi}_q\psi_q\}\Pi d\bar{\psi}_q d\psi_q$$

$$= 2 \frac{\Pi[\beta\Lambda-i\pi(2q+1)]}{\Pi[-i\pi(2q+1)]}$$

$$= 2 \prod_{m=1}^{\infty}\{ 1+\beta^2\Lambda^2/[\pi^2(2m+1)^2]\}$$

$$= 2 \cosh(\beta\Lambda/2) \quad , \tag{4.19}$$

which is readily seen to yield the correct answer in this case.

For the case at hand it follows that the partition function Z may be represented by

$$Z = \text{Tr}(e^{-\beta H})$$

$$= N \int \det[\partial_t + K_{\alpha\beta} + i \sum \phi_m(t) U^m_{\alpha\beta} + \sum \chi_n(t) V^n_{\alpha\beta}] e^{-F}$$
$$\times \exp\{-1/4 \sum \int [\phi^2_m(t)/\lambda_m + \chi^2_n(t)/q_n] dt\} \Pi D\phi_m D\chi_n \quad . \tag{4.20}$$

where

$$F \equiv 1/2 \sum \int [K_{\alpha\alpha} + i \sum \phi_m(t) U^m_{\alpha\alpha} + \sum \chi_n(t) V^n_{\alpha\alpha}] dt \quad . \tag{4.21}$$

This final representation for Z is the one of interest since it is purely functional. It is this complex effective action that one wishes to study by means of associated complex Langevin equations. It is straightforward to augment the functional integral to include additional interactions with boson fields. For any particular case of interest the appropriate Langevin equations may be deduced by following the general principles outlined previously.

5. ELECTRONS AND QUANTUM SPINS

In this section we first recall the construction of path integrals in terms of coherent states, and then develop the associated Langevin equation for them. We shall consider coherent states for canonical and for spin variables.

The canonical coherent states [10] are defined for all real p and q by

$$|p,q\rangle \equiv e^{i(pQ-qP)} |0\rangle , \tag{5.1}$$

where $[Q,P]=i$, and $|0\rangle$ is the normalized harmonic-oscillator ground state characterized by $(Q+iP)|0\rangle=0$. The overlap of two such states is given by

$$\langle p_2, q_2 | p_1, q_1 \rangle = \exp\{\tfrac{i}{2}(q_2 p_1 - p_2 q_1) - \tfrac{1}{4}[(p_2-p_1)^2 + (q_2-q_1)^2]\} \tag{5.2}$$

and it never vanishes. These states are complete, in fact overcomplete, and admit a resolution of unity in the form

$$1 = \int |p,q\rangle\langle p,q| \, (dpdq/2\pi) \qquad (5.3)$$

when integrated over the entire phase space R^2. Indeed these states are so overcomplete that an operator H is fully characterized by its diagonal matrix elements [10], i.e., by the function

$$H(p,q) \equiv \langle p,q|H|p,q\rangle \quad . \qquad (5.4)$$

Operators other than unity also admit a diagonal expansion such as

$$H = \int h(p,q) |p,q\rangle\langle p,q| \, (dpdq/2\pi) \quad , \qquad (5.5)$$

and it follows by taking diagonal matrix elements that we may express h by

$$h(p,q) = e^{-1/2(\partial^2/\partial p^2 + \partial^2/\partial q^2)} H(p,q) \quad . \qquad (5.6)$$

We confine ourselves to operators H for which $h(p,q)$ is a reasonable function.

If ε is a small parameter then it follows that

$$e^{-\varepsilon H} = \int e^{-\varepsilon h(p,q)} |p,q\rangle\langle p,q| \, (dpdq/2\pi) \qquad (5.7)$$

with an error that is of order ε^2. Choose N to be a large integer such that $\beta \equiv N\varepsilon$, and multiply the preceding formula times itself N times. A trace of the resultant expression, and a limit as $\varepsilon \to 0$ leads to the important formula [11]

$$Z = \text{Tr}(e^{-\beta H})$$

$$= \lim_{\varepsilon \to 0} \int \dots \int_{\ell} \prod_{1}^{N} \langle p_{\ell+1}, q_{\ell+1} | p_\ell, q_\ell \rangle e^{-\varepsilon h(p_\ell, q_\ell)} \, (dp_\ell dq_\ell/2\pi) \qquad (5.8)$$

where

$$p_{N+1}, q_{N+1} \equiv p_1, q_1.$$

For N large but finite this expression provides an approximate evaluation of the partition function. It is straightforward to write down the Langevin equations appropriate to this complex distribution (we choose the scale factor b=1). These equations are

$$\dot{q}_\ell = \frac{i}{4}(p_{\ell-1} - p_{\ell+1}) - \frac{1}{4}(2q_\ell - q_{\ell+1} - q_{\ell-1}) - \frac{\varepsilon}{2} \frac{\partial h(p_\ell, q_\ell)}{\partial q_\ell} + \xi_\ell, \quad (5.9a)$$

$$\dot{p}_\ell = \frac{i}{4}(q_{\ell+1} - q_{\ell-1}) - \frac{1}{4}(2p_\ell - p_{\ell+1} - p_{\ell-1}) - \frac{\varepsilon}{2} \frac{\partial h(p_\ell, q_\ell)}{\partial p_\ell} + \eta_\ell, \quad (5.9b)$$

where ξ_ℓ, η_ℓ, $1 \leq \ell \leq N$, are all independent standard white noise terms.

For M distinguishable particles the preceding Langevin equations are simply extended by the addition of a particle label to read

$$\dot{q}_\ell^m = \frac{i}{4}(p_{\ell-1}^m - p_{\ell+1}^m) - \frac{1}{4}(2q_\ell^m - q_{\ell+1}^m - q_{\ell-1}^m) - \frac{\varepsilon}{2} \frac{\partial h_\ell}{\partial q_\ell^m} + \xi_\ell^m, \quad (5.10a)$$

$$\dot{p}_\ell^m = \frac{i}{4}(q_{\ell+1}^m - q_{\ell-1}^m) - \frac{1}{4}(2p_\ell^m - p_{\ell+1}^m - p_{\ell-1}^m) - \frac{\varepsilon}{2} \frac{\partial h_\ell}{\partial p_\ell^m} + \eta_\ell^m, \quad (5.10b)$$

where $1 \leq \ell \leq N$ and $1 \leq m \leq M$, and all noise terms are independent, standard white noise sources. Here $h_\ell \equiv h(p_\ell, q_\ell)$ is associated to

$$H(p_\ell, q_\ell) \equiv \langle p_\ell, q_\ell | H | p_\ell, q_\ell \rangle \qquad (5.11)$$

where

$$|p_\ell, q_\ell\rangle \equiv \bigotimes_{m=1}^{M} |p_\ell^m, q_\ell^m\rangle. \qquad (5.12)$$

Electrons

When we deal with nonrelativistic electrons we must augment the above M-particle formulation to include indistinguishability and the exclusion principle. This may be accomplished simply by replacing the distinguishable-particle coherent-state overlap

$$\prod_{m=1}^{M} <p_{\ell+1}^{m}, q_{\ell+1}^{m} | p_{\ell}^{m}, q_{\ell}^{m}> \tag{5.13}$$

each time it appears by the Slater determinant

$$(M!)^{-1} \det <p_{\ell+1}^{m}, q_{\ell+1}^{m} | p_{\ell}^{n}, q_{\ell}^{n}> \tag{5.14}$$

where the determinant is over the indicies m and n. The modification of the coherent state path integral is evident, and the new form of the M-particle Langevin equations incorporating the exclusion principle is given by

$$\dot{q}_{\ell}^{m} = \frac{1}{2} \frac{\partial}{\partial q_{\ell}^{m}} [\ln \det <p_{\ell+1}^{n}, q_{\ell+1}^{n} | p_{\ell}^{s}, q_{\ell}^{s}>$$

$$+ \ln \det <p_{\ell}^{n}, q_{\ell}^{n} | p_{\ell-1}^{s}, q_{\ell-1}^{s}>] - \frac{\varepsilon}{2} \frac{\partial h_{\ell}}{\partial q_{\ell}^{m}} + \xi_{\ell}^{m} \quad , \tag{5.15a}$$

$$\dot{p}_{\ell}^{m} = \frac{1}{2} \frac{\partial}{\partial p_{\ell}^{m}} [\ln \det <p_{\ell+1}^{n}, q_{\ell+1}^{n} | p_{\ell}^{s}, q_{\ell}^{s}>$$

$$+ \ln \det <p_{\ell}^{n}, q_{\ell}^{n} | p_{\ell-1}^{s}, q_{\ell-1}^{s}>] - \frac{\varepsilon}{2} \frac{\partial h_{\ell}}{\partial p_{\ell}^{m}} + n_{\ell}^{m} \quad . \tag{5.15b}$$

These equations exhibit singularities in the drift terms due to a vanishing of the determinant when two or more particles are at the same phase-space point. In all cases such singularities are of the standard type discussed in Sec. 1.

The system of Langevin equations developed above for a system of M fermions appears particularly well suited to analyze the two-dimensional (surface problem) electrons in the quantized Hall effect [12]. This application will be discussed in a separate publication.

Quantum Spins

To deal with quantum spins we forego the canonical coherent states in favor of the spin coherent states defined for all points θ, ϕ on the unit sphere by [13]

$$|\theta,\phi\rangle \equiv e^{-i\phi S_3} e^{-i\theta S_2}|0\rangle \qquad (5.16)$$

for an irreducible spin representation where $S_1^2 + S_2^2 + S_3^2 = s(s+1)$, $s \geq 1/2$, and $|0\rangle$ is chosen so that $S_3|0\rangle = s|0\rangle$. The overlap of two such states is given by

$$\langle\theta_2,\phi_2|\theta_1,\phi_1\rangle = [\cos\frac{\theta_2-\theta_1}{2}\cos\frac{\phi_2-\phi_1}{2} + i\cos\frac{\theta_2+\theta_1}{2}\sin\frac{\phi_2-\phi_1}{2}]^{2s}$$

$$(5.17)$$

and this expression vanishes only when θ_2,ϕ_2 and θ_1,ϕ_1 refer to diametrically opposite points. These states admit a resolution of unity in the form

$$1 = N_s \int |\theta,\phi\rangle\langle\theta,\phi| d\Omega \ , \qquad (5.18)$$

where $N_s \equiv (2s+1)/4\pi$ and $d\Omega \equiv \sin\theta d\theta d\phi$, when integrated over the unit sphere. Indeed the overcompleteness of the spin coherent states suffices to determine an operator H uniquely by its diagonal matrix elements, i.e., by the function

$$H(\theta,\phi) = \langle\theta,\phi|H|\theta,\phi\rangle \ . \qquad (5.19)$$

An operator H also possesses a diagonal representation,

$$H = N_s \int h(\theta,\phi)|\theta,\phi\rangle\langle\theta,\phi| d\Omega \ . \qquad (5.20)$$

To find h we note that

$$H(\theta',\phi') = N_s \int h(\theta,\phi) |<\theta',\phi'|\theta,\phi>|^2 d\Omega \quad , \tag{5.21}$$

where

$$(2s+1)|<\theta',\phi'|\theta,\phi>|^2 = (2s+1)2^{-2s}[1+\cos\beta]^{2s}$$

$$\equiv \sum_{\ell=0}^{2s} (2\ell+1) r_\ell^{-1} P_\ell(\cos\beta)$$

$$= 4\pi \sum r_\ell^{-1} Y_{\ell m}(\theta',\phi') Y_{\ell m}^*(\theta,\phi) \quad . \tag{5.22}$$

Here

$$\cos\beta \equiv \cos\theta'\cos\theta + \sin\theta'\sin\theta\cos(\phi'-\phi) \quad , \tag{5.23}$$

P_ℓ denotes the Legendre polynomials, and $Y_{\ell m}$ the spherical harmonics. It follows from these formulas that

$$r_0 = 1 \quad ,$$
$$r_{\ell+1} = \frac{(2s+\ell+2)}{(2s-\ell)} r_\ell \tag{5.24}$$

which determines r_ℓ recursively for all $\ell \leq 2s$ as needed. Consequently H and h are related by their spherical harmonic expansions for $\ell \leq 2s$ according to

$$H(\theta,\phi) = \sum H_{\ell m} Y_{\ell m}(\theta,\phi) \quad ,$$

$$h(\theta,\phi) = \sum H_{\ell m} r_\ell Y_{\ell m}(\theta,\phi) \quad . \tag{5.25}$$

For example, for s=1/2 all H can be written in the form $H = a + \vec{b} \cdot \vec{S}$. In this case $h(\theta,\phi) = a + 3\vec{b} \cdot <\theta,\phi|\vec{S}|\theta,\phi>$; the factor 3 is just r_1 for s = 1/2.

With the nature of the diagonal representation clarified we proceed in analogy with the canonical coherent state case[14].

For small ε we can write

$$e^{-\varepsilon H} = N_s \int e^{-\varepsilon h(\theta,\phi)} |\theta,\phi><\theta,\phi| d\Omega \qquad (5.26)$$

which is correct up to an order of ε^2. Hence for $\beta \equiv N\varepsilon$ it follows that

$$Z = \text{Tr}(e^{-\beta H})$$

$$= \lim_{\varepsilon \to 0} \int .. \int \prod_{\ell=1}^{N} <\theta_{\ell+1},\phi_{\ell+1}|\theta_\ell,\phi_\ell> e^{-\varepsilon h(\theta_\ell,\phi_\ell)} (N_s d\Omega_\ell) \qquad (5.27)$$

where

$$\theta_{N+1},\phi_{N+1} \equiv \theta_1,\phi_1 \quad .$$

For N large but finite this formula provides an approximate evaluation of Z. Here the complex distribution $\exp(-S)$ is determined by

$$S \equiv -\sum \ln<\theta_{\ell+1},\phi_{\ell+1}|\theta_\ell,\phi_\ell> + \varepsilon \sum h(\theta_\ell,\phi_\ell) \quad . \qquad (5.28)$$

Associated with this example is the complex analogue of the Fokker-Planck equation which reads (b=1)

$$\frac{\partial G}{\partial \tau} = \frac{1}{2}\sum \{ \frac{1}{\sin\theta_\ell} \frac{\partial}{\partial\theta_\ell}[\sin\theta_\ell (\frac{\partial S}{\partial\theta_\ell} + \frac{\partial}{\partial\theta_\ell})]$$

$$+ \frac{1}{\sin^2\theta_\ell} \frac{\partial}{\partial\phi_\ell}(\frac{\partial S}{\partial\phi_\ell} + \frac{\partial}{\partial\phi_\ell}) \} G \qquad (5.29)$$

reflecting the geometry of the sphere appropriate to the phase space for this example. The associated Langevin equation is given by [14]

$$\dot{\theta}_\ell = \frac{1}{2} \cot\theta_\ell - \frac{1}{2} \frac{\partial S}{\partial\theta_\ell} + \xi_\ell \, , \qquad (5.30a)$$

$$\dot{\phi}_\ell = -\frac{1}{2} \frac{1}{\sin^2\theta_\ell} \frac{\partial S}{\partial\phi_\ell} + \frac{1}{\sin\theta_\ell} \eta_\ell \, , \qquad (5.30b)$$

where ξ_ℓ, η_ℓ denote independent, standard white noise terms. These equations have two sources of singularities. One type comes from the zero of the overlap of two spin coherent states; this is just a standard singularity of the type discussed in Sec. 1. The other type is induced by a coordinate singularity at $\theta=0$ and $\theta=\pi$. This type of singularity is entirely unphysical and can be removed by an appropriate change of coordinates.

To obtain the Langevin equation for a system with M spin degrees of freedom it suffices to append an index m to the preceding equations to give

$$\dot{\theta}_\ell^m = \frac{1}{2} \cot \theta_\ell^m - \frac{1}{2} \frac{\partial S}{\partial \theta_\ell^m} + \xi_\ell^m \quad , \tag{5.31a}$$

$$\dot{\phi}_\ell^m = - \frac{1}{2} \frac{1}{\sin^2\theta_\ell^m} \frac{\partial S}{\partial \phi_\ell^m} + \frac{1}{\sin\theta_\ell^m} \eta_\ell^m \tag{5.31b}$$

where now

$$S \equiv - \sum \ln\langle \theta_{\ell+1}^m, \phi_{\ell+1}^m | \theta_\ell^m, \phi_\ell^m \rangle + \epsilon \sum h_\ell \tag{5.32}$$

in which $h_\ell \equiv h(\theta_\ell, \phi_\ell)$, a function associated with $H(\theta_\ell, \phi_\ell) \equiv \langle \theta_\ell, \phi_\ell | H | \theta_\ell \ \phi_\ell \rangle$ and where

$$|\theta_\ell, \phi_\ell\rangle \equiv \bigotimes_{m=1}^{M} |\theta_\ell^m, \phi_\ell^m\rangle \quad . \tag{5.33}$$

The numerical solution to these M-spin Langevin equations should permit the calculation of correlation functions for various quantum spin problems.

Lastly we observe that fermi degrees of freedom may be formulated as quantum spin problems when it is recalled that the Jordan-Wigner representation for fermi degrees of freedom[15] is formulated essentially as a quantum spin problem for spin s=1/2. However this approach may not be practical for computer simulation since large numbers (made from many factors of 3) may arise.

ACKNOWLEDGEMENTS

It is a pleasure to thank W.F. Brinkman, H. Ezawa, A.T. Ogielski and K.G. Wilson for discussions pertaining to the subject matter of these lecture notes.

REFERENCES

1. G. Parisi and Wu Yong-Shi, Sci. Sinica $\underline{24}$ (1981) 483.
2. See. e.g., H. McKean, Stochastic Integrals (Academic,1969).
3. J.R. Klauder and H. Ezawa, Prog. Theor. Phys. $\underline{69}$ (1983) 664.
4. See, e.g., Y. Nakano, "One-Time Characteristic Functional in the Stochastic Quantization" and "Hidden Super-symmetry in the Stochastic Quantization", University of Alberta preprints; W. Grimus and H. Hüffel, "Perturbation Theory from Stochastic Quantization of Scalar Fields", CERN preprint.
5. See, e.g., G. Parisi and Wu Yong-Shi, Sci. Sinica $\underline{24}$ (1981) 483; D. Zwanziger, Nucl. Phys. $\underline{B192}$ (1981) 259; L. Baulieu and D. Zwanziger, $\underline{B193}$ (1981) 163; J. Alfaro and B. Sakita, "Stochastic Quantization and Large N Limit of U(N) Gauge Theory", City College preprint.
6. M. Namiki, I. Ohba, K. Okano and Y. Yamanaka, "Stochastic Quantization of Non-Abelian Gauge Field - Unitary Problem and Faddeev-Popov Ghost Effects", Waseda University preprint.
7. L. Faddeev and V. Popov, Phys. Lett. $\underline{25B}$ (1967) 29.
8. K.G. Wilson (private communication).
9. See, e.g., F.A. Berezin, The Method of Second Quantization (Academic, 1966).
10. See, e.g., J.R. Klauder and E.C.G. Sudarshan, Fundamentals of Quantum Optics, (W.A. Benjamin, 1968), Chapter 7.
11. M. Ciafaloni and E. Onofri, Nucl. Phys. $\underline{B151}$ (1979) 118;

    E. Onofri, in Functional Integration, ed. J.-P. Antoine and E. Tirapegui (Plenum, 1980).

12. W.F. Brinkman (private communication).

13. See, e.g., J.R. Klauder, J. Math. Phys. $\underline{4}$ (1963) 1058; $\underline{23}$ (1982) 1797.

14. J.R. Klauder, "A Langevin Approach to Fermion and Quantum Spin Correlation Functions", Bell Labs preprint.

15. P. Jordan and E. Wigner, Z. Physik $\underline{47}$ (1928) 631.

Acta Physica Austriaca, Suppl. XXV, 283–356 (1983)
© by Springer-Verlag 1983

LATTICE QCD[+]

by

P. HASENFRATZ

CERN

INTRODUCTION

The surprising message from the deep inelastic e-p
scattering experiment at SLAC more than ten years ago was
that under the photon microscope the energetic proton seems
to be built up from pointlike elements, from quark partons.
The parton struck by the photon behaves like a free particle.
Strong interactions become weak at large energies and this
feature singled out QCD almost uniquely as the candidate
theory of strong interactions.

QCD is asymptotically free - at high enough energies it
becomes perturbative. This opened the way for perturbation
theory, and almost all the feasible calculations have been
completed in a remarkable short time. A typical prediction
for a deep inelastic process can be characterized in a crude
way as

$$\sim 1 + \sim\frac{1}{\ln Q^2/\Lambda^2} + \sim\frac{1}{Q^2/m^2} \quad ,$$

where ~ 1 is the lowest order result predicting scaling, the
second term is the calculable, perturbative correction leading to weak scaling violations, while the last term is a

---

[+] Lectures given at the XXII. Internationale Universitätswochen
für Kernphysik,Schladming,Austria,February 23-March 5,1983.

non-perturbative correction, m being some kind of mass (quark mass, proton mass, $<k_\perp^2>$,...). Q is the characteristic momentum transfer in the process. For large Q^2 the perturbatively non-calculable last term is small. It was a painful observation, however, that these non-perturbative corrections are not really under control at present day energies.

Non-perturbative methods are needed. Even more so since we want to understand and calculate the hadron masses and widths and other spectroscopical data as well - all beyond the reach of perturbation theory. The idea of defining QCD - more generally gauge theories on a lattice has been put forward by Wilson in 1974 [1]. Not only the formulation was given that time, but new notions and different approximations were raised also. In a few years time the lattice established itself as a sensible regularization of QCD, and the new techniques proved to be powerful enough to attack exciting non-perturbative questions (confinement, string tension, mass gap, ...) in the theory [2].

What is the greatest thing about the lattice approach? It opened the way towards new, non-perturbative techniques ... yes, this is true. It opened the way towards the rigorous results derived in statistical physics ... yes, that is also true. But, what one might enumerate first is: the lattice approach created a new standard in strong interactions. There is a way to check (and hopefully control) the systematical and statistical errors and the approximations.

1. GENERAL INTRODUCTION TO LATTICE GAUGE THEORIES

1.1 Going to Euclidean Space. The Normal Relation Between Quantum Field Theory and Classical Statistical Physics

The relation between the problems of quantum field theory and those in classical statistical physics played a very important role in all the ideas and methods going beyond perturbation theory in QCD.

In the path integral formulation of quantum mechanics[3] the amplitude of propagation between the points of x_a, t_a and x_b, t_b is given by

$$K(x_b,t_b;x_a,t_a) = \sum_{\text{paths}} e^{i\int_{t_a}^{t_b} L(x,\dot{x})dt} \quad , \quad (1.1)$$

where the summation is over all paths connecting the initial and final space-time points. The weight of a path is given by e^{iS}, where S is the classical action of the path.

The path integral formulation can be generalized to systems with many or infinite degrees of freedom. In quantum field theory the vacuum functional is defined analogously

$$Z = \sum_{\substack{\text{field} \\ \text{configurations}}} e^{i\int dt \int d^3x L(\phi,\partial_\mu \phi)} \quad . \quad (1.2)$$

Here, again, we might consider the amplitude that the field is equal to $f_a(x)$ and $f_b(x)$ at time t_a and t_b respectively (by summing over the connecting configurations). More often, however, $t_a \to -\infty$, $t_b \to +\infty$ is taken, some boundary condition is prescribed (the nature of which is expected to be irrelevant, like that of the spacial boundary conditions for large systems), and the system is probed by the introduction of external sources, taking $L \to L + F\phi$.

Both Eqs.(1.1) and (1.2) are formal until some sensible definition is not given to the summations. This definition can be given easily if the field theory is treated perturbatively (when functional integrals of Gaussian nature are considered only). The usual diagrammatic expansion is obtained in this case.

The vacuum expectation value of the time ordered product of the fields (Green's function) is obtained in the path integral formulation as the expectation value of the product of the fields, using the measure

$$\sim \frac{1}{Z} e^{i \int d^4 x L(\phi, \partial_\mu \phi)} \quad . \tag{1.3}$$

This measure is complex. In order to establish a relation with classical statistical physics an analytic continuation is done in the time coordinate

$$x_o \to -ix_4 \quad , \tag{1.4}$$

which defines the model in four-dimensional Euclidean space. A real, normalized measure is obtained

$$\sim \frac{1}{Z} e^{-\int d^4 x L_E(\phi, \partial_\mu \phi)} \quad , \tag{1.5}$$

where L_E is the Euclidean Lagrangian, where all the indices enter symmetrically with positive metric diag $(1,1,1,1)$. Specifically, for QCD the Euclidean vacuum functional is given as (the zeroth component of a vector field becomes imaginary also):

$$Z = \sum_{\substack{\text{all Euclidean} \\ \text{gauge field configurations}}} e^{-\frac{1}{g^2} \int d^4 x \sum_{a=1}^{8} \sum_{\mu,\nu=1}^{4} \frac{1}{4} F^a_{\mu\nu} F^a_{\mu\nu}} \quad , \tag{1.6}$$

which can be interpreted as the partition function of a four-dimensional classical Yang-Mills system, g^2 plays the role of temperature in this analogy.

The possibility of the analytic continuation to imaginary times can be established easily in perturbation theory[4] (Appendix A) and there are rigorous theorems beyond that[5].

In practice, the results obtained in Euclidean space can be used directly to obtain physical predictions, and an explicit analytic continuation back to Minkowski space is unnecessary. Take for instance the free propagator in Euclidean space

$$G(x) = \frac{1}{(2\pi)^4} \int d^4p \, e^{ipx} \frac{1}{p^2+m^2} \quad . \tag{1.7}$$

For large $|x| = \sqrt{x_m x_\mu}$ $G(x)$ behaves as

$$G(x) \sim e^{-m|x|} \quad . \tag{1.8}$$

An analytic continuation back to Minkowski space would lead to the usual oscillatory behaviour for time-like separations. However, in order to recover the mass of the particle from the asymptotic behaviour of the propagator - a problem often encountered in non-perturbative studies - we can use (1.8) directly as well.

1.2 Gauge Theory on the Lattice

By going to Euclidean space the quantum field theoretical problem is transformed into the statistical physics of classical fields. This is not an easy problem either. In particular, it is plagued by the same kind of divergences as the original system. Regularization is needed. We want to discuss non-perturbative phenomena, therefore this regularization should be more general than a special prescription for Feynman diagrams.

The lattice regularization satisfies not only this criterion, but opens the way towards the non-perturbative techniques of statistical physics.

The continuous Euclidean space is replaced by discrete lattice points. Consider a regular hyper-cubic lattice [6]. The lattice unit is denoted by "a", while a lattice point by $n_\mu = (n_1, n_2, n_3, n_4)$ (integers). The Fourier transform of a function $f(n)$ defined on the lattice points is given as

$$\tilde{f}(p) = a^4 \sum_n e^{i(p \cdot n)a} f(n) \quad . \tag{1.9}$$

$\tilde{f}(p)$ is periodic over $p_\mu = 2\pi/a$, therefore the momentum values can be constrained within the first Brillouin zone

$$- \frac{\pi}{a} \leq p_\mu \leq \frac{\pi}{a} \quad . \tag{1.10}$$

The inverse transform is given by

$$f(n) = \frac{1}{(2\pi)^4} \int\!\!\int\!\!\int\!\!\int_{-\pi/a}^{\pi/a} d^4p \; \tilde{f}(p) \; e^{-i(p.n)a} \quad . \tag{1.11}$$

As we see from Eqs. (1.10) and (1.11), the lattice provides for a cut-off in momentum space:

$$\text{cut-off momentum} = \pi/a \quad . \tag{1.12}$$

Given the continuum Lagrangian, there is a natural way to define a field theory on the lattice: scalar fields are defined on the points, vector fields (characterized by a position and a direction) on the links of the lattice[7]. Derivatives are replaced by discrete differences

$$\partial_\mu f(x) \rightarrow \Delta_\mu f(x) = \frac{1}{a} (f(n+\hat{\mu}) - f(n)), \tag{1.13}$$

where $\hat{\mu}$ is the unit vector along the μ direction.

In the case of gauge fields, however, the requirement of gauge invariance complicates the matter.

Consider the case of electrodynamics. In Maxwell's theory Gauss' law says: the electric flux coming through a closed surface is equal to the charge inside. There is no charge without its accompanying Coulomb field. It is the same in QED.

Let us denote the vector potential and the electric field by $\hat{A}_r(\underline{x})$ and $\hat{E}_r(\underline{x})$ respectively, $r = 1,2,3$. The electric field is the momentum conjugate to the vector potential. They satisfy the usual commutation rules, the Hilbert space is restricted to physical states (satisfying Gauss' law). Consider the operator

$$e^{i\int d^3x \, \hat{A}_r(\underline{x}) \, C_r(\underline{x})} \quad , \tag{1.14}$$

where $C_r(\underline{x})$ are three c-number functions. By acting on a state with this operator, the value of the electric field is increased by $\underline{C}(\underline{x})$. Really

$$\hat{E}_i(\underline{y}) \, e^{i\int d^3x \hat{A}_r(\underline{x}) \, C_r(\underline{x})} = e^{i\int d^3x \hat{A}_r(\underline{x}) \, C_r(\underline{x})} (E_i(\underline{y}) + C_i(\underline{y})) \tag{1.15}$$

Consider now a source and a point of strength g at the points 1 and 2, respectively. An electric flux of g should be led from the point 1 to 2. It is the requirement of Gauss law. It is a kinematical requirement. Gauss law does not tell us, what distribution the electric flux takes. It only requires the conservation of the electric flux. A special choice could be to let the electric flux propagate along an infinitesimal flux tube between 1 and 2.

In this case c_r is a transversal δ function along the path and the situation is described by

$$(\text{source at 1}) \; e^{ig\int_1^2 \hat{\underline{A}} d\underline{s}} \; (\text{sink at 2}) \quad . \tag{1.16}$$

Let 1 and 2 be two neighbouring points separated by a small distance "a" along the μ direction. We get:

$$\phi^+(\underline{x}) \, e^{iga\hat{A}_\mu(x)} \phi(\underline{x} + a\hat{\mu}) \quad . \tag{1.17}$$

From small "a" we can expand (1.17) in a Taylor series, and the terms linear in "a" give the usual covariant derivative. For finite "a", however, we must keep the complete expression to preserve fluxes (or equivalently, to preserve gauge invariance).

These considerations suggest that the basic variables of the lattice formulation are not the vector potentials themselves, but their exponentiated forms

$$U_{n\mu} = e^{iga\,A_{n\mu}} : \underset{n \qquad\qquad n+\hat{\mu}}{\rule{2cm}{0.4pt}\bullet} \qquad\qquad (1.18)$$

associated to the directed link with endpoints n and $n + \hat{\mu}$. The oppositely oriented link is associated with $U^{+}_{n\mu}$. $U_{n\mu}$ is called - for obvious reasons - a string bit. $U_{n\mu}$ is an element of the local symmetry group, U(1) here, SU(3) for QCD:

$$(U_{n\mu})_{st} = (e^{iga\,A^{b}_{n\mu}T^{b}})_{st}\,, \qquad \begin{array}{l} b = 1,2,\ldots,8 \\[4pt] s,t = 1,2,3 \ ; \end{array} \qquad (1.19)$$

where T^{b} are the SU(3) generators.

If there are no sources and sinks, that is we are dealing with a pure gauge theory, the string bits must form closed loops, like smoke rings.

In the simplest case the smoke rings run along the smallest loops of the lattice: around the plaquettes. Keeping the symmetry between the four directions of the Euclidean lattice one arrives to the action[1,8]

$$S_{W} = \text{const.} \sum_{\text{plaquettes}} (\text{Tr } U_{p} + \text{Tr } U^{+}_{p}) \ , \qquad (1.20)$$

where

$$U_{p} = U_{n\mu}\,U_{n+\hat{\mu},\nu}\,U^{+}_{n+\hat{\nu},\mu}\,U^{+}_{n\nu} \ , \qquad P : \ \begin{array}{c} n+\hat{\nu} \qquad\ n+\hat{\mu}+\hat{\nu} \\ \boxed{} \\ n \qquad\qquad n+\hat{\mu} \end{array} \qquad (1.21)$$

U_{p} and U^{+}_{p} describe oppositely oriented loops. Trace (Tr) is taken to get numbers from matrices. Gauge invariance can be expressed as an invariance under local symmetry transformations - like in the continuum. S_W is invariant under the independent rotations of colour reference frames at every lattice n by an SU(3) matrix V_n:

$$U_{n\mu} \rightarrow V_n \ U_{n\mu} V^+_{n+\hat{\mu}} \quad . \tag{1.22}$$

This action respects maximally the symmetries of the lattice: it is invariant under 90° rotations and lattice translations. It is invariant also under parity transformations and charge conjugation.

The constant in eq.(1.20) should be chosen to reproduce the continuum action in the classical $a \rightarrow 0$ continuum limit. By using repeatedly the Baker-Hausdorff formula

$$e^A \ e^B \ = e^{A+B+ \frac{1}{2} [A,B]+ \ ...} \tag{1.23}$$

one obtains [9]

$$U_p \equiv e^{iga^2 \ F_{n,\mu\nu}} = 1 + iga^2 \ F_{n,\mu\nu} + \frac{1}{2!} \ (iga^2)^2 \ F^2_{n,\mu\nu}+ \ ... \ , \tag{1.24}$$

where

$$F_{n,\mu\nu} = \sum_{a=1}^{8} F^a_{n,\mu\nu} \ T^a \ ,$$

and

$$F^a_{n,\mu\nu} = \Delta_\mu A^a_{n\nu} - \Delta_\nu A^a_{n\mu} - gf^{abc}(A^b_{n\mu}A^c_{n\nu}+ a \ \Delta_\nu A^b_{n\mu}A^c_{n\nu} +$$

$$+ aA^b_{n\mu}\Delta_\mu A^c_{n\nu}- \frac{1}{2} a \ A^b_{n\mu}\Delta_\nu A^c_{n\mu} - \frac{1}{2}a \ \Delta_\mu A^b_{n\nu}A^c_{n\nu} + \frac{1}{2}a^2 \ \Delta_\nu A^b_{n\mu}\Delta_\mu A^c_{n\nu})+$$

$$+ 0(g^2,a^2) \quad . \tag{1.25}$$

$F^a_{n,\mu\nu}$ is gauge covariant on the lattice. In the $a \to 0$
classical continuum limit it reduces to the continuum field
strength tensor (the first three terms in eq.(1.25). By using
eq.(1.24) and Tr $T^a T^b = 1/2(\delta_{ab})$, the continuum action is
obtained from eq.(1.20), if const. $= -1/g^2$ is chosen.

Let us write down Wilson's action again, now with the
correct normalization

$$S_W = - \frac{1}{g^2} \sum_{\text{plaquettes}} (\text{Tr } U_p + \text{Tr } U_p^+) \quad . \tag{1.26}$$

In the partition function we sum over the field configurations, that is we integrate over the SU(3) group on each
link of the lattice:

$$Z = \prod_{\text{links}} \int dU_\ell \, e^{-S_W} \quad , \tag{1.27}$$

where dU is the invariant Haar measure having the properties

$$\int dU = 1 \quad ,$$

$$\int dU \, f(U) = \int dU \, f(U_o U) \quad , \tag{1.28}$$

where U_o is an arbitrary element of the group.

1.3 The Continuum Limit of Lattice Field Theories

The lattice is only a regularization. It should be re-
movied at the end of the calculation by taking $a \to 0$ limit.

This limit is subtle. The action in (1.26) violates
Euclidean rotation invariance (corresponding to Lorentz invariance in Minkowski space) - it is invariant only under 90°
rotations. In the continuum limit the full symmetry should be
recovered.

To achieve this goal it is not enough to consider
correlations over distances r, which are much larger than a.

For a generic g^2 in (1.26) the asymptotic decay of correlation functions will show a direction dependence, defining masses [via relations analogous to (1.8)] which are direction dependent also. It is like in a solid crystal which behaves differently along different directions.

The only dimensionful parameter in pure gauge QCD is the lattice unit a. The lattice predictions are always dimensionless numbers, the correct dimensions are restored with the help of a. For instance, a mass prediction will have the form

$$m = \frac{1}{a} f(g) \ . \tag{1.29}$$

In a naive a → 0 limit, all the masses would go to infinity. Saying differently, for a generic g, f(g) is a number of O(1). Therefore the mass is ∿1/a ∿cut-off.

The order to keep the mass finite as a → 0, g, the bare coupling constant, must also be tuned:

$$\left. \begin{array}{l} a \to 0 \\ g \to g^* \end{array} \right\} \ , \qquad \text{where} \qquad f(g^*) = 0 \ . \tag{1.30}$$

In this limit (am) → 0, 1/am (≡ correlation length in lattice units) →∞ . Therefore:

In the continuum limit, the lattice problem approaches a continuous phase transition point.

Close to this critical point, there will be strong fluctuations over many lattice units in the system washing away the original lattice structure, and leading to the possibility of restoring Euclidean rotation invariance.

Let us illustrate these points on an exactly solvable simple system, on the two-dimensional Ising model: classical spins coupled ferromagnetically on a quadratic lattice. The partition function is defined as:

$$Z = \sum_{\{S_i=\pm1\}} e^{K \sum_{<ij>} S_i S_j} \quad , \tag{1.31}$$

where $\sum_{<ij>}$ denotes a summation over nearest neighbour couplings, and K is proportional to the inverse temperature $1/T$.

Consider the high temperature phase. The correlation function $<S_0 S_n>$ decays exponentially as $a|n| \equiv r \to \infty$ in an angle dependent way [$\sin\alpha = n_1/|n|$ here]:

$$<S_o S_n> \underset{r\to\infty}{\sim} e^{-m(\alpha)r} \quad . \tag{1.32}$$

for instance,

$$\frac{m(\alpha=0^{\circ})}{m(\alpha=45^{\circ})} = \frac{1}{\sqrt{2}} \frac{\ln(V\frac{1+V}{1-V})}{\ln(\frac{2V}{1-V^2})} \quad , \tag{1.33}$$

where $V =$ th K[10]. For small K (high temperature) this ratio is $1/\sqrt{2}$, rotation invariance is strongly violated. By decreasing the temperature the situation is improving, and rotation symmetry is restored exactly by approaching the critical point (Curie point) $V_c = \sqrt{2}-1$ (Tab.1).

V	$m(\alpha=0^{\circ})/m(\alpha=45^{\circ})$
0.	$1/\sqrt{2} \approx 0.707$
0.05	0.890
0.1	0.929
0.2	0.972
0.3	0.992
$\sqrt{2}-1$ 0.414	1

Table 1: Restoration of rotation symmetry in d=2 Ising model as the critical point is approached.

In the relation analogous to (1.29)

$$m(\alpha) = \frac{1}{a} f(T,\alpha) \tag{1.34}$$

$f \to \, \sim (T-T_c)^\nu$, $\nu = 1$, independently of α, and a finite mass is obtained if a is going to be zero as $(T-T_c)^\nu$ in this limit.

What is the critical coupling g^* for QCD? g^* is the bare coupling constant of the continuum theory. The bare coupling constant describes the interactions on momentum scales $\sim$cut-off. In an asymptotically free theory this coupling becomes vanishingly small as the cut-off is increased: $g^* = 0$.

In order to find our asymptotically free theory in the continuum limit of lattice QCD, this continuum limit should occur when the coupling is turned towards zero.

1.4 Strong Coupling Limit, Confinement

Lattice gauge theories can be solved exactly in the other extreme limit, when $g \to \infty$. (This is like the high temperature limit in statistical physics.)

The model exhibits confinement at strong coupling: the potential energy between two heavy quark sources increases linearly for large separations. This potential energy can be obtained by studying special gauge invariant loop expectation values <W>[1,11](Fig.1). This loop ("Wilson loop") describes the propagation of a q-q̄ pair at rest over a "time-period" T. Really, the spacelike parts of the loop describe the creation and annihilation of a gauge invariant q-q̄ source as it is discussed in Section 1.2. The presence of an external current leads to an extra term in the continuum action:

$$\int d^4 x \, ig \, V_\mu A_\mu \tag{1.35}$$

where V_μ is the current of the external source. For a point-

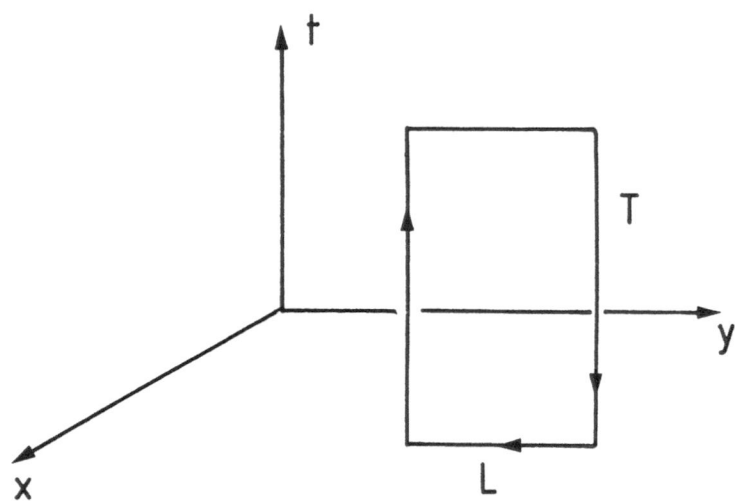

Fig.1

like source it gives a factor in the path integral:

$$e^{ig\int_C dx_\mu A_\mu} \rightarrow \prod_C U \quad ,$$ (1.36)

which gives the "timelike" part of the Wilson loop.

Let us denote by $E(L)$ the energy of a source-sink system separated at a distance L. We expect

$$<W> \underset{T\to\infty}{\sim} e^{-T\,E(L)} \quad ,$$ (1.37)

which leads to an area decay if the sources are confined by a linear potential $E(L) = \sigma L$ (σ is called the string tension). If, for large distances, the binding potential becomes vanishingly small, the energy is independent of L, the decay of the Wilson loop expectation value is governed by the perimeter of the loop.

Consider therefore

$$<W> = \frac{\int DU\ W\ e^{1/g^2 \sum_p (\mathrm{Tr}\ U_p + \mathrm{Tr}\ U_p^+)}}{\int DU\ W\ e^{1/g^2 \sum_p (\mathrm{Tr}\ U_p + \mathrm{Tr}\ U_p^+)}} \quad ,$$ (1.38)

and expand the Boltzmann factor in powers of $1/g^2$. Using the group integrals

$$\int dU \; U = \int dU \; U^+ = \int dU \; UU = \int dU \; U^+U^+ = 0,$$

$$\int dU \; U_{ij} \; U^+_{k\ell} = \frac{1}{3} \, \delta_{i\ell}\delta_{jk} \; , \tag{1.39}$$

[which can be deduced from (1.28) and from $UU^+ = 1$], we see that in lowest order a U_ℓ variable of the Wilson loop should meet a U^+_ℓ from the Boltzmann factor, which leads to the

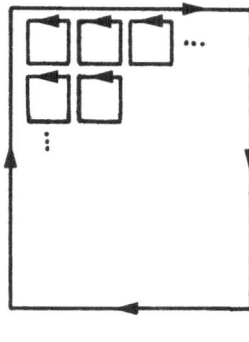

Fig.2

arrangement of Fig.2. Therefore in leading order

$$<W> \underset{g \to \infty}{\sim} (\frac{1}{3g^2})^{LT} = e^{-LT\ln(3g^2)} \; . \tag{1.40}$$

W decays exponentially with the area of the loop, signalling confinement. For $g \to \infty$ the model confines. This conclusion is not spoiled by higher order corrections since it can be shown rigorously that the strong coupling expansion has a finite radius of convergence[12].

After restoring the dimensions, the string tension is given by

$$\sigma \underset{g \to \infty}{=} \frac{1}{a^2} \ln(3g^2) \tag{1.41}$$

Unfortunately, in this limit the model has not much to do with continuum QCD. The masses are much larger than the cut-off, rotation invariance is badly violated (the string tensio would increase by rotating the sources away from the axes).

Thus g should be changed towards g = 0, where the continuum limit is to be found. It is a long way to go, and we might meet surprises. If we want an asymptotically free, confining theory at the end, a deconfining phase transition must not be among them.

1.5 Renormalizability, Dimensional Transmutation

In calculating the string ension in the strong coupling limit the leading contribution is obtained by constructing a minimal surface over the Wilson loop (Fig. 2). It describes the propagation of an infinitely narrow, straight flux tube spanned between the sources.

In higher orders this surface begins to fluctuate, the flux tube acquires a finite transversal width. The dimensionless function f(g) in

$$\sigma^{1/2} = \frac{1}{a} f(g) \quad , \qquad (f(g) \underset{g \to \infty}{\to} (\ln 3g^2)^{1/2}) \quad , \qquad (1.42)$$

becomes smaller as g is decreased – the string begins to thaw. The relation between the cut-off and $\sigma^{1/2}$, or between the width of the tube and the lattice distance becomes more and more acceptable physically (Fig.3).

In the continuum limit the lattice distance a is small compared to the characteristic physical distances, (like the radius of the flux tube), the lattice becomes fine grained. In this limit the lattice is expected to play no important role anymore. The size of the cut-off becomes irrelevant. We expect that it is possible to change the lattice distance a and the coupling g together keeping the physical properties of the system unchanged. This is the usual requirement of

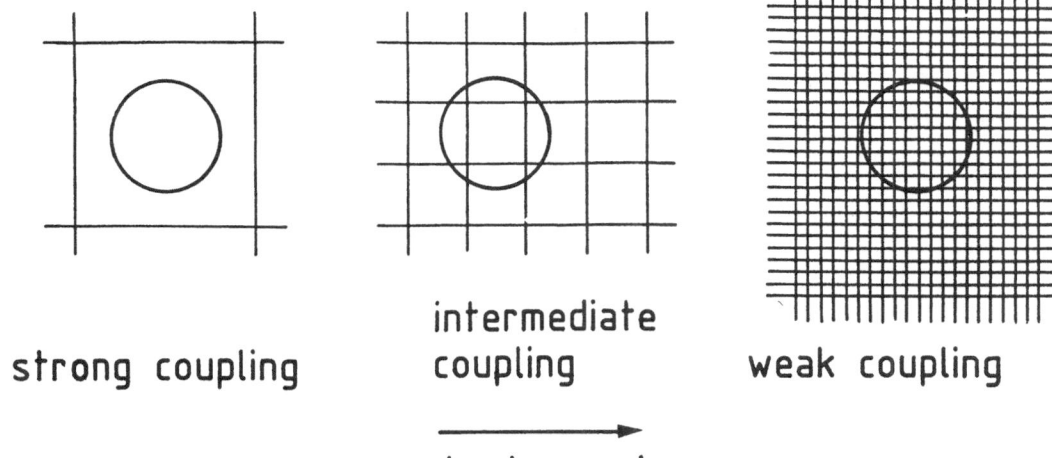

strong coupling   intermediate coupling   weak coupling

g is decreasing

Fig.3

renormalizability.

This requirement uniquely determines the coupling constant dependence of any masses (and similar dimensionful quantities) in this limit. If m is a physical mass, cut-off independence implies

$$a \frac{d}{da} m \underset{\substack{(a \to 0 \\ g \to 0)}}{=} 0 \quad , \tag{1.43}$$

which is a first order differential equation for the function
f of eq.(1.29):

$$-f(g) + \frac{d}{dg} f(g) a \frac{d}{da} g(a) = 0 . \tag{1.44}$$

Of course, not only masses, but other physical quantities
should also become independent of the cut-off, for instance,
results related to high energy deep inelastic processes.
The requirement of cut-off independence determines the function
$-\beta(g) \equiv (1/a)(d/da)g(a)$ there, in perturbation theory. $\beta(g)$
is the usual β function

$$\beta(g) = -\beta_0 g^3 - \beta_1 g^5 - \beta_2 g^7 + \ldots \quad . \tag{1.45}$$

The first two coefficients, β_0 and β_1 are universal[13]: they are the same in the continuum (in any renormalization scheme) and in the lattice perturbation theory.

Apart from an integration constant, the function f (and m) is completely determined by eqs. (1.44) and (1.45) when $g \to 0$:

$$m = c \frac{1}{a} e^{-\frac{1}{2\beta_0 g^2}} (\beta_0 g^2)^{-\frac{\beta_1}{2\beta_0^2}} [1 + O(g^2)]$$

$$\equiv c \Lambda^{latt} . \tag{1.46}$$

Λ^{latt} is a cut-off independent mass parameter. This mass parameter sets the scale for QCD. It is an external parameter, like the hyperfine coupling $\alpha = 1/137$ in QED.

The dimension of physical masses and lengths is carried by Λ^{latt}[14], and the non-perturbative content of the theory lies in the value of the connecting constants C_i:

$$\sqrt{\sigma} = c_\sigma \cdot \Lambda^{latt}$$

$$m_{gap} = c_g \cdot \Lambda^{latt} \tag{1.47}$$

...

On the lattice dimensionless quantities are obtained originally, like (ma). According to eq.(1.46) this quantity should behave in a definite way as $g \to 0$. In a semi-logarithmic plot (ma) falls almost linearly as the function of $1/g^2$, with a definite slope $(-1/2\beta_0)$. Observing this behaviour ("scaling") is a sign of entering a coupling constant regime, where cut-off independence holds.

1.6 Weak Coupling Perturbation Theory. Freedom in Choosing the Lattice Action

In the limit $g \to 0$, the factor $1/g^2$ in front of the action [eq.(1.26)] becomes very large, which tends to suppress the fluctuations in the plaquette variable. Apart from the gauge freedom (which can be handled by fixing the gauge) the U matrices will be close to 1:

$$U_{n\mu} = e^{iga\,A_{n\mu}} \simeq 1 + iga\,A_{n\mu} + \ldots \tag{1.48}$$

Using this expansion, a systematic perturbation theory (in g) can be constructed.

Lattice perturbation theory is rather involved, because ever more new types of interaction terms enter as the order of the calculation is increased. Expanding the Wilson action we obtain[15]:

$$
\begin{aligned}
S_W = a^4 \sum_n \sum_{\mu,\nu} \{ &-\tfrac{1}{4}(\Delta_\mu A_{n\nu} - \Delta_\nu A_{n\mu})^2 + g f^{abc}[A^a_{n\mu} A^b_{n\nu} \Delta_\mu A^c_{n\nu} \\
&+ \tfrac{1}{2} a\, A^a_{n\mu} \Delta_\nu A^b_{n\mu} \Delta_\mu A^c_{n\nu}] - \tfrac{1}{4} f^{abt} f^{cdt} A^a_{n\mu} A^b_{n\nu} A^c_{n\mu} A^d_{n\nu} \\
&+ \tfrac{1}{2} f^{abt} f^{cdt} a\, \Delta_\mu A^a_{n\nu} A^b_{n\mu} A^c_{n\mu} A^d_{n\nu} + \ldots \} \quad .
\end{aligned}
\tag{1.49}
$$

Although the new type of vertices are all proportional to a^k (k positive), they cannot be neglected, since they contain extra derivatives and fields leading to divergencies cancelling the a^k factor in front.

Nevertheless, it is true that these extra contributions can be absorbed into the definition of the coupling and fields and the lattice and continuum formulations of QCD lead to identical physical predictions in perturbation theory[16]. In this case the lattice is just one of the possible regularizations (a rather inconvenient one). It is a special scheme, like the dimensional regularization with minimal subtraction, the MOM scheme, and so on. In different renormalization or regularization schemes the Λ parameter is defined as

$$\Lambda = (\text{some mass } \mu) \cdot e^{-\frac{1}{2\beta_0 g^2(\mu)}} (\beta_0 g^2(\mu))^{-\frac{\beta_1}{2\beta_0^2}} [1 + O(g^2(\mu)],$$

$$(1.50)$$

where $g^2(\mu)$ is the coupling corresponding to this mass. For instance

μ	mass, introduced in n-dim. reg. with min. subtr.	Λ^{DR}
m_{PV}	PV regulator mass	Λ^{PV}
M	where $\Gamma^{(2)}$ and $\Gamma^{(3)}$ are fixed (MOM scheme)	Λ^{MOM}
$1/a$	on the lattice	Λ^{latt} .

Let us choose these masses to be very large (then the corresponding couplings are very small), and equal to each other. Then the ratio between the Λ parameters of schemes A and B is given by

$$\Lambda_A/\Lambda_B = e^{-\frac{1}{2\beta_0}(\frac{1}{g_A^2} - \frac{1}{g_B^2})} .$$

$$(1.51)$$

The difference in the exponent is a constant for weak couplings, and can be determined in perturbation theory at the one-loop level

$$\frac{1}{g_A^2} - \frac{1}{g_B^2} = C_{AB} + O(g^2) .$$

$$(1.52)$$

After calculating C_{AB}, the ratio between the scales is known and the predictions can be converted from one scheme to the other. For example, the relation between the scale parameter of the Wilson action (1.26) Λ^{latt}_{W} and Λ^{MOM} is given by [15,17]

$$\Lambda^{MOM}/\Lambda^{latt}_{W} = \begin{array}{ll} 57.5 & (SU(2)) \\ 83.5 & (SU(3)) \end{array} . \tag{1.53}$$

In particular, the different lattice formulations of QCD can be converted the same way. There is a large freedom in choosing the lattice action. Even after satisfying the requirement of gauge invariance and of other symmetries, there are many possibilities to write down a lattice action which reproduces the continuum action in the a → 0 classical continuum limit. Different functions of Tr U_p, can be chosen and/or other type of loops (rectangualrs, or more complicated) can be included. In perturbation theory these formulations are identical. Additionally, a one-loop calculation predicts the relation between their scales, connecting trivially their non-perturbative predictions.

Does the full non-perturbative theory respect these considerations? We can not answer this question in relation with the lattice and continuum formulations, since there are no quantitative non-perturbative methods in the continuum. However, different lattice formulations can be compared and the basic assumption of universality can be investigated (Section 5).

2. NON-PERTURBATIVE METHODS IN THE GAUGE SECTOR

By going to Euclidean space and using lattice regularization, the non-perturbative methods developed and tested in statistical physics become available for the study of quantum field theoretical problems.

Although the non-perturbative techniques we shall discuss are all familiar in statistical physics, there are special new features due to the fact that a d = 4 locally invariant gauge system is a rather unusual statistical problem.

The essential elements of strong coupling expansion, Monte Carlo simulation, mean field techniques and variational Monte Carlo studies will be discussed briefly. Some of these methods have not become quantitative yet in QCD. However, they influence our intuition and they are part of the everyday vocabulary.

2.1 Strong Coupling Expansion

As we discussed earlier, lattice QCD can be solved exactly in the strong coupling, $g \to \infty$ limit. Gauge theories exhibit confinement in this limit. Unfortunately, continuum QCD is to be recovered in the other extreme region, when $g \to 0$ (which is "weak coupling", but not perturbative!).

The strong coupling expansion is a systematic expansion in $1/g^2$, starting from the exact, confining solution at $g=\infty$. The idea is to derive a reasonable long series for the physical quantity in question, and then to extrapolate this power series towards the continuum point at $g = 0$[18]. The technique is in complete analogy with the high temperature expansion in statistical physics.

Consider, for definiteness, the calculation of the string tension, which is related to the expectation value of large Wilson loops $<W(C)>$, eq.(1.38). The Boltzmann factor is expanded in powers of $1/g^2$:

$$e^{1/g^2 \sum_p (\text{Tr } U_p + cc)} = \Pi_p e^{1/g^2 (\text{Tr } U_p + cc)}$$

$$= \Pi_p \{1 + 1/g^2 (\text{Tr } U_p + cc) + \frac{1}{2!} (1/g^2)^2 (\text{Tr } U_p + cc)^2 + \ldots\} \quad .$$

$$(2.1)$$

After resolving the brackets, the result is a sum of terms, each term being the product of N_p factors (N_p is the number of plaquettes of the lattice) of the form:

$$1/\ell_p! \ (1/g^2)^{\ell_p} \ (\text{Tr } U_p + cc)^{\ell_p} \ , \qquad \ell_p = 0,1,\dots \qquad . \qquad (2.2)$$

Up to any given order in $1/g^2$, the number of terms is finite (on a finite lattice).

This expansion is inserted into (1.38) and then the group integrals are performed. This last step is simplified significantly by replacing the expansion in (2.1) by the character expansion. The Boltzmann factor

$$e^{1/g^2(\text{Tr } U_p + cc)}$$

is defined over equivalence classes:

$$e^{1/g^2(\text{Tr}(VU_pV^{-1})+cc)} = e^{1/g^2(\text{Tr } U_p + cc)} \ , \qquad (2.3)$$

where V is an arbitrary group element, therefore it can be expanded in terms of the characters χ_r of the group

$$e^{1/g^2(\text{Tr } U_p + cc)} = \sum_r \nu_r \beta_r(1/g^2) \chi_r(U_p) \quad . \qquad (2.4)$$

Here the summation is over the irreducible representations r, ν_r is a conveniently chosen number (=dimension of the representation r). The basic orthogonality relation is given by

$$\int dU \ \chi_r(SU) \bar{\chi}_s(TU) = \frac{\delta_{rs}}{\nu_r} \chi_r(ST^{-1}) \ , \qquad S,T \epsilon G \ . \qquad (2.5)$$

As an explicit example, consider the strong coupling expansion of the string tension in a d=3, U(1) gauge theory. The expectation value of an N×N (N→∞) Wilson loop is calculated (Fig.4). The character expansion (2.4) has the simple form:

$$e^{1/g^2(U_p+cc)} = \sum_{r=-\infty}^{\infty} \beta_r(1/g^2)(U_p)^r \ , \qquad (2.6)$$

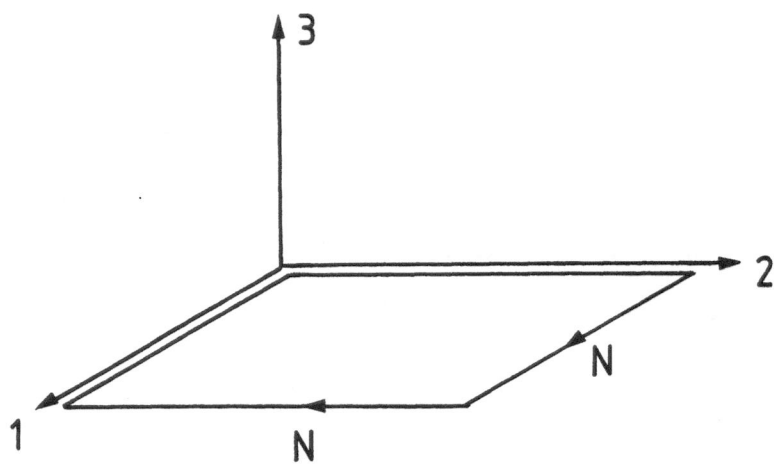

Fig.4

where $U_p = e^{i\alpha_p}$, $\alpha_p \in (-\pi, \pi)$, and

$$\beta_{-r}(1/g^2) = \beta_r(1/g^2) = I_r(2/g^2)$$

$I_r(x)$ being the rth modified Bessel function. Equation (2.6) can be written as

$$I_0(2/g^2)\{1 + \frac{I_r(2/g^2)}{I_0(2/g^2)} \; (U_p + U_p^{-1}) + \frac{I_2(2/g^2)}{I_0(2/g^2)} \; (U_p^2 + U_p^{-2}) + \ldots\}$$

$$(2.7)$$

for small $1/g^2$,

$$\frac{I_r(2/g^2)}{I_0(2/g^2)} \sim (1/g^2)^r \quad .$$

2.1.1 Lowest order

As we discussed previously, the lowest order contribution is obtained by taking the "1" of (2.7) on all the plaquettes, except on those covering the Wilson loop, where

$$\frac{I_1(2/g^2)}{I_0(2/g^2)} \, U_p$$

is taken (Fig.2). The resulting group integrals are trivial and we obtain

$$<W> \; = \; (\frac{I_1}{I_0})^{N^2} \quad \to \quad \sigma = -\ln(\frac{I_1}{I_0}) \quad , \tag{2.8}$$

where the (dimensionless) tension is denoted by σ.

2.1.2 Next order

The minimal covering surface is distorted by pushing it out of the plane, as given in Fig.5.

The contribution is $2N^2(I_1/I_0)^{N^2+4}$, and we get

$$<W> \; = \; (\frac{I_1}{I_0})^{N^2}(1+ 2N^2(\frac{I_1}{I_0})^4) \quad \to \quad \sigma = -\ln(\frac{I_1}{I_0}) - 2(\frac{I_1}{I_0})^4 \quad . \tag{2.9}$$

The string begins to fluctuate, the tension becomes smaller.

2.1.3 Next-to-next order

There are three types of contributions giving:

$$(N_p - 2N^2) \; 2 \; (\frac{I_1}{I_0})^{N^2+6} \qquad \text{(Fig.6a)} \tag{2.10}$$

$$4N^2 (\frac{I_1}{I_0})^{N^2+6} \qquad \text{(Fig.6b)} \tag{2.11}$$

and

$$2N^2 (\frac{I_1}{I_0})^{N^2+4} (\frac{I_2}{I_0}) \qquad \text{(Fig.6c)} \tag{2.12}$$

Here N_p in (2.10) is the total number of plaquettes on the

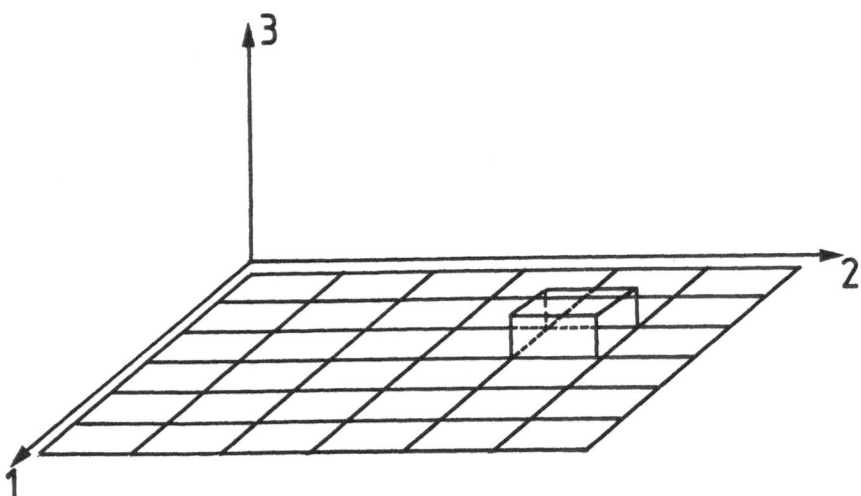

Fig.5

lattice, and this extensive contribution is cancelled by the
normalizing partition function in the denominator of (1.38)

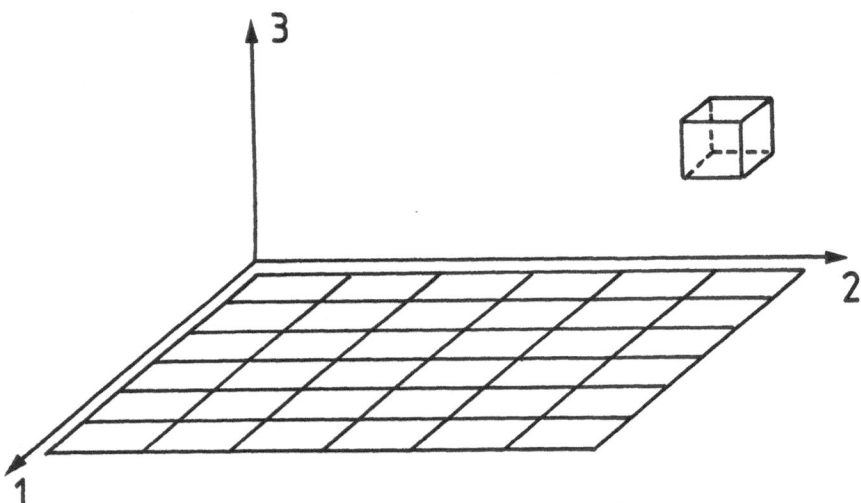

Fig.6a

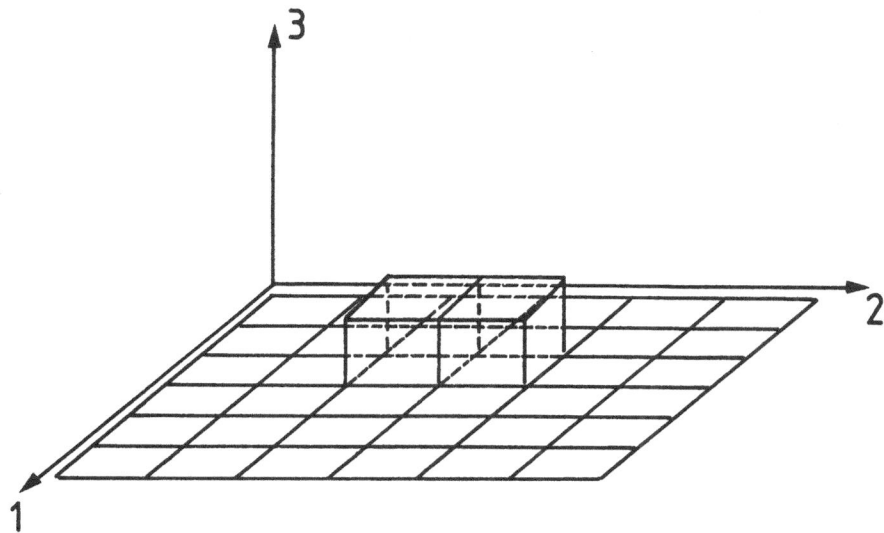

Fig.6b

$$Z = 1 + 2\ N_p (\frac{I_1}{I_o})^6 \tag{2.13}$$

from these type of graphs. Terms, proportional to the perimeter of the loop are neglected in (2.11). Collecting everything we get for the string tension

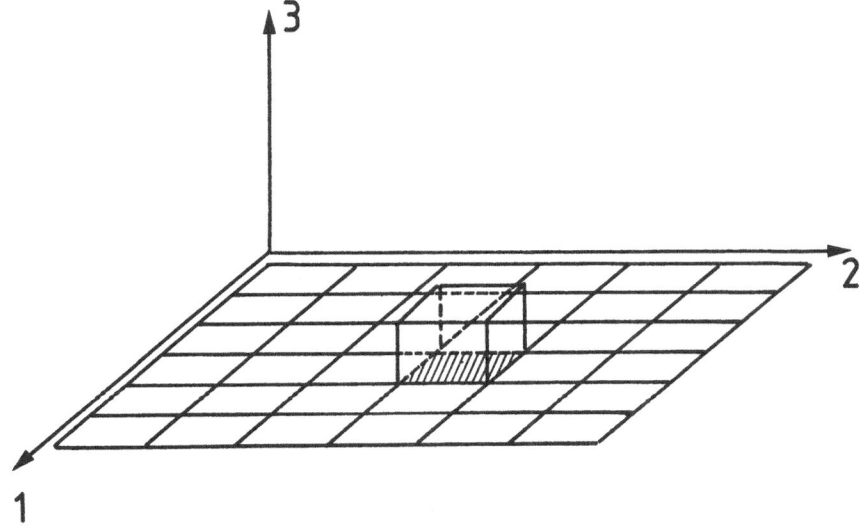

Fig.6c

$$\sigma = -\ln(\frac{I_1}{I_o}) - 2(\frac{I_1}{I_o})^4 - 2(\frac{I_1}{I_o})^4 (\frac{I_2}{I_o}) \quad . \tag{2.14}$$

By this order we met all the main features of the expansion: the correct, order by order exponentialization

$$W \sim e^{-(\text{area})\sigma} \quad , \tag{2.15}$$

the cancellation of the extensive contributions and the way the higher characters enter.

Although the procedure is straightforward, the calculation becomes more and more laborous in higher orders. In QCD applications, the method waits for new ideas. The series derived for the string tension and for different glueball masses are "hand-made" and rather short. The analytic continuation towards the continuum limit is ambiguous[19].In the case of the string tension the probable presence of the roughening transition[20](where the flux tube begins to fluctuate like a string in the continuum theory) constitutes an extra problem for this continuation.

2.2 Monte Carlo Simulation

At present, the Monte Carlo simulation[21] is most powerful technique for the quantitative study of lattice QCD[22].

Our problem is to calculate expectation values like

$$<A> = \frac{\prod_\ell \int dU_\ell A(U) e^{-S(U)}}{\prod_\ell \int dU_\ell e^{-S(U)}} \quad . \tag{2.16}$$

On a finite lattice, this is a well defined, multidimensional integral. For instance, on a 10^4 lattice there are 4.10^4 links, on every link there is a $U_{n\mu}$ matrix defined by three integration variables in the case of SU(2). That defines a 120.000 dimensional integral. Monte Carlo simulation is a method to evaluate this huge integral numerically.

In calculating expectation values like that in (2.16),

one has to sum over (in the case of a continuous group: one has to integrate over) all the configurations. Random sampling, that is generating configurations randomly, does not work. Due to the Boltzmann factor e^{-S}, the integrand changes very rapidly, and with random sampling most of the time irrelevant configurations would be generated. The relevant configurations should be generated with higher probability: importance sampling.

Let us choose a configuration $\{U\}$ with a probability $P\{U\}$. After generating M configurations we have

$$\langle A \rangle \simeq \bar{A} = \frac{\sum\limits_{\nu=1}^{M} A(\{U\}_\nu) P^{-1}(\{U\}_\nu) e^{-S(\{U\}_\nu)}}{\sum\limits_{\nu=1}^{M} P^{-1}(\{U\}_\nu) e^{-S(\{U\}_\nu)}} \quad . \tag{2.17}$$

A natural possibility is to generate a configuration according to the corresponding Boltzmann factor:

$$P_{eq}(\{U\}) \sim e^{-S(\{U\})} \quad . \tag{2.18}$$

In this case (2.17) gives:

$$\bar{A} = \frac{1}{M} \sum_{\nu=1}^{M} A(\{U\}_\nu) \quad . \tag{2.19}$$

Of course, the correctly normalized P_{eq} is not known a priori. A Markov process is constructed such that after many iterations the probability of creating a configuration is just the equilibrium probability P_{eq}.

This Markov process is defined by giving the transition probability $W(\{U\}_\nu \to \{U\}_{\nu'})$ from one configuration to another. W is required to satisfy

a)

$$\sum_{\nu'} W(\{U\}_\nu \to \{U\}_{\nu'}) = 1$$

b) any finite-action configuration should be reachable (at least after a number of steps).

c) detailed balance condition

$$P_{eq}(\{U\}_\nu) W(\{U\}_\nu \rightarrow \{U\}_{\nu'}) = P_{eq}(\{U\}_{\nu'}) W(\{U\}_{\nu'} \rightarrow \{U\}_\nu) \ . \quad (2.20)$$

It is easy to show (at least on a non-rigorous level) that this transition probability drives the system towards equilibrium.

Even within the requirements above, there is a large freedom in choosing the transition probabilities. In most of the lattice gauge theory calculations, the Metropolis and the Heat Bath methods were used.

2.2.1 Metropolis method

First, an initial configuration is chosen (for instance, a completely ordered or disordered state). Then local changes are performed: a link is chosen (randomly or regularly), and the gauge matrix U_ℓ of this link is replaced by U'_ℓ by some random process. After this replacement, the change of the action δS is calculated. If δS is negative, the change is accepted, if δS is positive, the $-\delta S$ change is accepted with the conditional probability $e^{-\delta S}$. This occasional acceptance simulates the thermal fluctuations. Without it, the system would be driven towards a dead, lowest-action state.

2.2.2 Heat bath method

Successively a local heat bath is touched to the links which are updated. That is, a new U'_ℓ is chosen (all the other U's are fixed) with a probability which is proportional to the Boltzmann factor. Of all the possible Monte Carlo algorithms which vary only a single link at a time, this method leads to equilibrium in the least number of iterations.

In both methods, the subsequent configurations are

different along a single link only, therefore these configurations are highly correlated. Usually, A is calculated on configurations which are separated by many updates. Additionally,
at the beginning, the system is not in equilibrium, therefore
a certain number of configurations should not be used in
calculating averages.

If the configurations over which A is calculated are
independent, the statistical error is given by

$$\delta A = \{\frac{1}{M(M-1)} \sum_{\nu=1}^{M} (W(\{U\}_\nu) - \bar{W})^2\}^{1/2} \quad .$$ (2.21)

The statistical error decreases like $\sim M^{-1/2}$ for large M, which
is very slow - a notorious point of Monte Carlo studies.

The configurations are generated according to the Boltzmann factor e^{-S}. This probability distribution is independent
of the operator A considered. For instance, these configurations
do not "know" whether and where quark sources will be inserted
later. How will a bag or flux tube be formed then? By random
fluctuations, there will be configurations in the generated
sequence which are relevant in the wave-functional of a q-$\bar{q}$
source. Those are enhanced, the other ones are suppressed
in the averaging process. It is clear, however, that we are
looking for a needle in a haystack if the Wilson loop is large.
It would be nice to advise a generating procedure where the
probability distribution would know about the presence of
sources in an essential way[23].

2.3 Mean Field Techniques

The mean field method [1,24-26] is an old, well-known
procedure in statistical physics. Its limitations are also
well known. For spin models the long-distance properties (for
instance, critical indices) are predicted incorrectly by mean
field theory in low dimensions. It is a quick, easy method to
obtain some information on the system and in many cases it

works surprisingly well in predicting the existence and the place of phase transitions.

In lattice gauge theories there is a recent, renewed interest in mean field techniques[25,26]. The original "one-shot" method has been extended to a systematic expansion, and the problem related to gauge invariance has been understood [25].

Consider first the naive procedure which later became the first approximation of a systematic expansion. In the mean field analysis, one concentrates on a single link in the partition function $\overset{\bullet}{n}\rule[0.5ex]{1.5em}{0.4pt}\overset{\bullet}{n+\hat{\mu}}\ U_{n\mu}$, and the effect of all the other links is replaced by a mean field value M, which is taken to be proportional to the unit matrix. In this

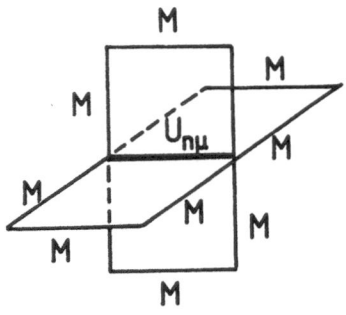

Fig. 7

approximation the action is reduced to (Fig.7)

$$-S(U) = \frac{\beta}{N} \sum_p 1/2(\text{Tr } U_p + cc) + 2(d-1)\beta M^3 \frac{1}{N} \frac{1}{2}(\text{Tr } U_p + cc)$$

$$+ \text{(terms, independent of } U_{n\mu}) ,$$

$$(2.22)$$

where $\beta = 2N/g^2$ for SU(N), d is the space-time dimension. Consistency requires

$$\frac{1}{N} \langle \frac{1}{2}(\text{Tr } U_p + cc) \rangle = M , \qquad\qquad (2.23)$$

leading to the equation

$$M = \underbrace{\frac{\partial}{\partial \alpha} \ln \left(\int dU \ e^{\alpha \frac{TrU+cc}{2N}} \right)}_{F(\alpha)} \bigg|_{\alpha = 2(d-1)\beta M^3} \qquad (2.24)$$

The function F is 0 and 1 for $\alpha = 0$ and ∞ respectively, and it is changing smoothly between these values. For small β, (2.24) has only the trivial $M = 0$ solution (Fig. 8a). There is a certain $\beta = \beta_0$, however, above which a non-trivial, $M \neq 0$ solution exists also (Fig. 8b).

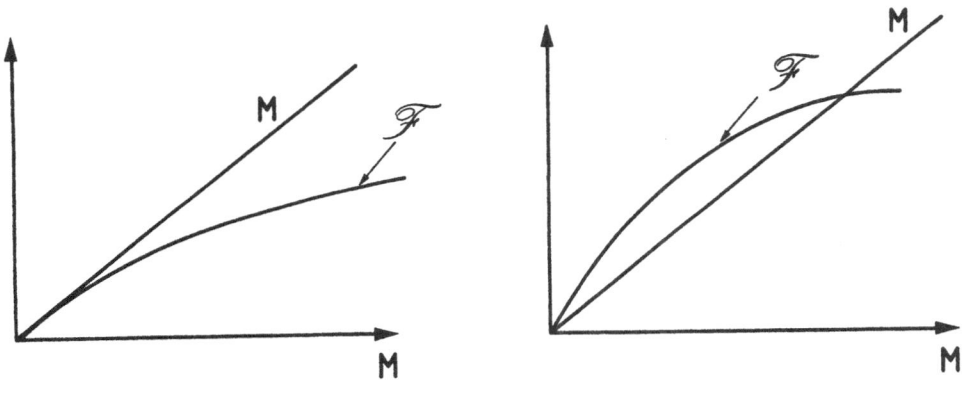

Fig. 8a   Fig. 8b

The transition point should be determined by comparing the free energies. At large β, the free energy with $M \neq 0$ is larger. The crossing with the $M = 0$ free energy happens at $\beta_c > \beta_0$ in general.

For simple gauge models (like $Z_2, Z_3, \ldots$) which have a deconfining phase transition as β is increased, β_c agrees quite well with the known phase transition points. For SU(2) and SU(3), where no deconfining phase transition is expected to occur, it is hoped (and there are indications for that) that the phase transition, predicted by the lowest approximation, disappears as corrections are included systematically. One might note that for SU(2) β_c is close to the point where an abrupt change towards the continuum behaviour is observed in Monte Carlo studies.

However, there is an apparently basic theoretical objection against the mean field method. According to a rigorous theorem, local gauge symmetry cannot be broken spontaneously, $\langle U_{n\mu} \rangle = 0$ is true for any coupling[27]. In a system with global symmetry there is a possibility for degenerated ground states separated by an infinite potential barrier. In the case of a local gauge symmetry there are degenerate states close to each other in phase space. By applying a gauge transformation which is different from unity only at a single point of the lattice, a new configuration is created. This configuration has the same action as the original one, they are indentical everywhere except on 2d links, and they are separated by a finite potential barrier. The vacuum functional is a superposition of these states. On the links effected, the gauge matrix can be changed arbitrarily by the gauge transformation and $\langle U_{n\mu} \rangle = 0$ occurs unavoidably.

The starting point of the mean field procedure is to assume $\langle U_{n\mu} \rangle = M$ which contradicts Elitzur's theorem for $M \neq 0$. A nice, recent result is the observation that the mean field method can be reconciled with Elitzur's theorem the same way as the soliton quantization steps restore translation invariance[25].

In quantizing a field theory around a soliton solution the tree graph approximation is given by the classical, localized solution itself, violating translational invariance. In the correct quantization procedure (the usual way is to introduce collective co-ordinates related to the centre-of-mass motion of the soliton) translation invariance is restored already at the one-loop level.

The mean field method can be extended into a systematic

saddle point approximation, the old mean field result providing the zeroth order result. The mean field solution $U_{n\mu} = M$
and all its gauge transformed forms are equivalent saddle
points, which should be integrated over, restoring gauge
invariance.

The possibility of a novel systematic expansion method
for lattice QCD is an exciting prospect. Unfortunately it
turned out that in the strong coupling region this expansion
essentially coincides with the usual strong coupling expansion,
while in the weak coupling regime it is a rearranged weak
coupling expansion.

Although the theoretical status of the mean field method
improved significantly, it has not become yet a new quantitative method in lattice QCD.

2.4 Variational Methods Combined with Monte Carlo Simulations

The variational method is very effective in quantum
mechanics in determining low-lying energy levels. In field
theory, even staying within the simplest type of ansätze, the
calculation of the necessary matrix elements is very difficult.
It has been suggested by Wilson[28] to use Monte Carlo simulation for this problem. This method has been extensively
applied in glueball mass calculations[29].

The glueballs are hypothetical particles formed solely from
gluons, without valence quarks. They are colourless and fla-
vourless objects. Their existence is predicted (almost) unavoidably by QCD. The glueball with lowest mass is called
the mass gap of the theory.

In searching for the lowest mass with given quantum
numbers the usual procedure is to create an excitation with
the given quantum numbers and to follow its propagation over
large distances. Let be θ a local operator having the required quantum numbers. Then

$$\langle\theta(\underline{0},\ t)\theta(\underline{0},\ 0)\rangle_c \underset{t\to\infty}{\sim} e^{-mt}\ , \tag{2.25}$$

where m is the lowest mass in the channel. θ is arbitrary to a large extent. The only a priori requirement is that $\theta|0\rangle$ should have a non-zero component in the exact wave function of the lowest-lying state. Of course, by hitting the vaccum with θ, not only the lowest-lying state, but excited single particle states, possible two, three,... particle states are also created. However, they have a faster exponential decay and at large enough t the propagation of the relevant state is observed.

Assume the lowest glueball is 0^{++}. Then a natural possibility is to take

$$\theta(t) = \sum_{\substack{\text{orientations} \\ \text{and } \underline{x}}} (\square(\underline{x},t) + \square(\underline{x},t))\ , \tag{2.26}$$

where $\square(\underline{x},t)$ is the product of U matrices around the plaquette at $(\underline{x},t)$. Summing over $\underline{x}$ assures that the excitations created by $\theta(t)$ are in the rest frame $(\underline{p} = 0)$.

In early glueball calculations the correlation

$$\langle\theta(t)\theta(0)\rangle - \langle\theta(t)\rangle\langle\theta(0)\rangle \tag{2.27}$$

was measured by Monte Carlo simulation. The main problem is, however, that the signal decreases rapidly with increasing t, and for $t > 3$ completely disappears into the noise (within acceptable computer time limits).

Combining this method with variational considerations significantly improved the situation. The operator θ is taken to be a combination of different loop products with unknown coefficients.

$$\theta = c_1\sum(\square+\square) + c_2\sum(\square+\square) + c_3\sum(\square+\square) + \ldots\ . \tag{2.28}$$

The unknown variational parameters are determined by the usual minimalization procedure:

$$\min[- \frac{1}{t} \ln \frac{<0|\theta(0)e^{-Ht}\theta(0)|0>}{<0|\theta(0)\theta(0)|0>}] =$$

$$= \min[- \frac{1}{t} \ln \frac{<0|\theta(0)\theta(t)|0>}{<0|\theta(0)\theta(0)|0>}] \quad . \tag{2.29}$$

If the basis is large enough in (2.28), then even t = 1 will result in a good value for the mass gap. In practice, a few operators are taken and t = 0,1,2 is measured by Monte Carlo simulation.

3. HOW THE LATTICE WORKS: STRING TENSION IN SU(2)

The way the string tension has been measured in the SU(2) lattice gauge theory including the effort to control the systematic errors, to check the restoration of rotation symmetry, universality and so on, shows nicely the lattice at work.

The string tension in an SU(2) lattice gauge theory has been measured first by Creutz[22,30]using Monte Carlo simulation. Assuming that the expectation value of large Wilson loops can be parametrized as

$$<W> = e^{-[\text{tension} \cdot \text{area} + b \cdot \text{perimeter} + c]} , \tag{3.1}$$

the tension can be obtained by taking appropriate ratios of Wilson loop expectation values. A possibility is to take

$$\chi(I) = -\ln \frac{<W(I,I)><W(I-1,I-1)>}{<W(I,I-1)><W(I-1,I)>} , \tag{3.2}$$

or[31]

320

$$K(I) = -\ln \frac{\langle W(I,I)\rangle}{\langle W(I+1,I-1)\rangle} \qquad , \qquad (3.3)$$

and so on. Here $W(I,I)$ is a Wilson loop of size $I \times I$. It is expected that

$$\chi(I), \; K(I) \xrightarrow[I \to \infty]{} \text{(dimensionless) string tension.} \qquad (3.4)$$

By measuring $\chi(I)$ $(K(I))$ for different values of I at a given $\beta \equiv 4/g^2$,

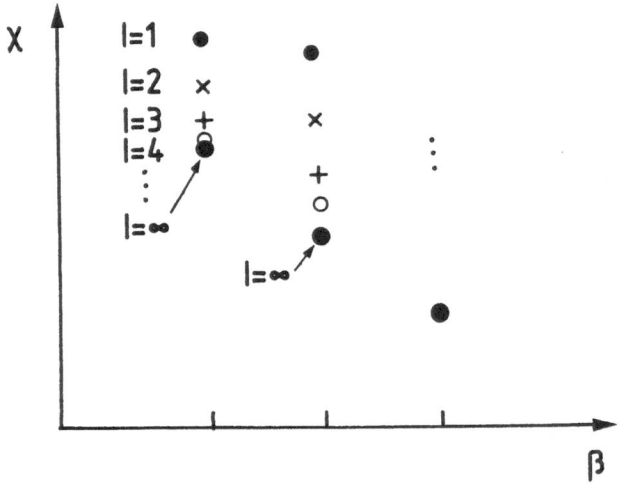

Fig. 9

we expect a convergence towards a number as I is increased (Fig.9). By repeating the calculation for different values of β, the limiting points represent the (dimensionless) tension as the function of the coupling. In the continuum limit, renormalization group arguments dictate a very definite behaviour (scaling) [Section 1.5, Eq. (1.46)], and if this behaviour is confirmed, the relation between the tension and the QCD scale parameter can be determined.

The data points on a 10^4 lattice by Creutz are given in Fig.10. The statistical errors are suppressed in this Figure. Observe that $\chi(3)$ measured at $\beta = 2.3$ and 2.5 is consistent with the renormalization group behaviour (dotted line).

On the other hand, there is not much sign of a convergence towards a limiting point as I is increased - clearly there is a need to measure χ for larger separations of the quark sources (-larger Wilson loops). If $\chi(3)$ can be taken as representing well the true limiting points at $\beta = 2.3$ and 2.5, the relation between the tension and the scale parameter is predicted to be [Eq.(1.47)]

$$\Lambda^{\text{latt}} = (0.013 \pm 0.002) \, (\text{tension})^{1/2} \, . \tag{3.5}$$

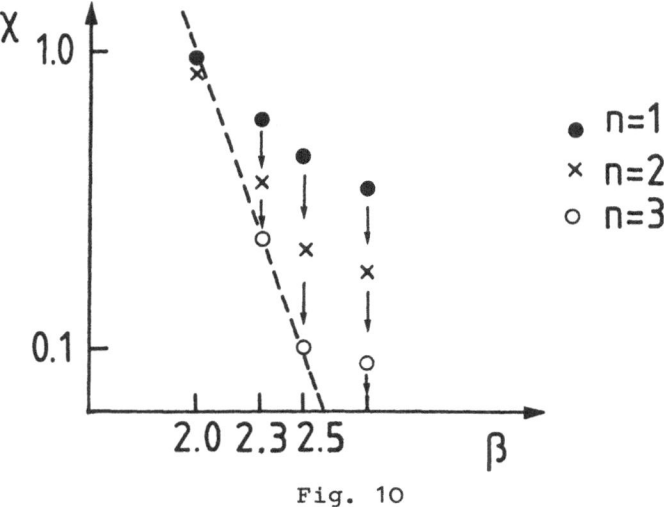

Fig. 10

Bhanot and Rebbi[31] increased the lattice size to 16^4 and studied the potential up to five lattice distances. Their result is given in Fig. 11. There is a significant

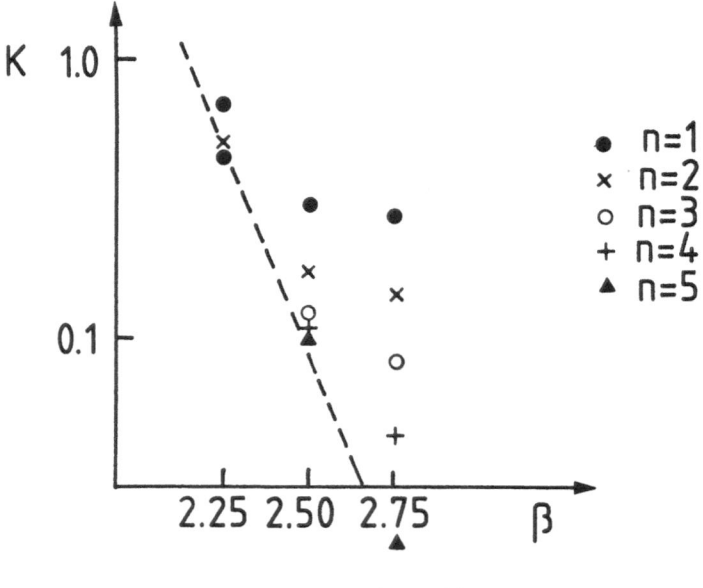

Fig. 11

improvement[+]. There is a sign of convergence at β = 2.25
and 2.5.

The limiting points are consistent with the renormalization group behaviour and the numerical result agrees with
(3.5) within the statistical errors.

These results tell us that beyond β≈2.25 the continuum
behaviour sets in and that the potential is essentially
linear after a few lattice distances in this coupling constant
region.

However, we cannot think in terms of the value of the
bare charge and in terms of lattice units. We can think in
terms of Fermi's and MeV's. There are certain questions which
we should always ask for and answer in interpreting a lattice
result:

[+] The case is less convincing if the statistical errors are
also included. The errors are quite large for I=4 and 5.

A) Resolution? Take the point $\beta = 2.5$. At this coupling $a^2\sigma \simeq 0.1$. Using the experimental value of the tension $\sqrt{\sigma} \simeq 400$ MeV, one obtains

$$a\big|_{\beta=2.5} \stackrel{\sim}{\sim} 0.16 \text{ fermi} \quad . \tag{3.6}$$

The characteristic distances in hadron physics are of the order of ~ 1 fm. A resolution of 0.16 fm is not extremely good, but acceptable. It is not against our intuition that the continuum behaviour already sets in at this resolution.

B) Separation? A separation of 4a, 5a corresponds to ~ 0.6–0.8 fm. This is a rather small distance, but either by taking the usual Coulomb + linear form of the potential, or by taking the potential derived in the bag model for heavy quarks, we obtain a precocious linear behaviour. Therefore, the lattice prediction saying that at 0.6–0.8 fm the potential is essentially linear, is not against our expectation.

C) $\Lambda^{latt} = (5.2 \pm 1.0)$MeV? Equation (3.5) predicts this small value for Λ^{latt}. We do not know very well Λ from deep inelastic experiments, but there is an order of magnitude problem here. However, the lattice is a strange regularization and the value of Λ is scheme-dependent (Section 1.6). Actually, $\Lambda^{latt} = 5.2$ MeV corresponds to $\Lambda^{MOM} = 280$ MeV[15], which is quite reasonable. Of course, we do not know Λ^{MOM} precisely, especially not in an SU(2) world without quarks. But this is an acceptable number.

D) Rotation invariance (Lorentz invariance)? Rotation invariance is expected to be restored only in the continuum limit. The string tension was measured along the co-ordinate axes. We would like to see that the results to not depend on that.

In Fig. 12a, the equipotential lines of two heavy

quarks at β = 2.0 are given, as it was measured by Lang
and Rebbi[32]. According to the tension calculations,
β = 2.0 is not in the continuum limit yet, and really,
the equipotential lines clearly remember the lattice
structure.

The coupling β = 2.25 is expected to be in the
continuum regime. Figure 12b gives the equipotential lines
at this coupling.

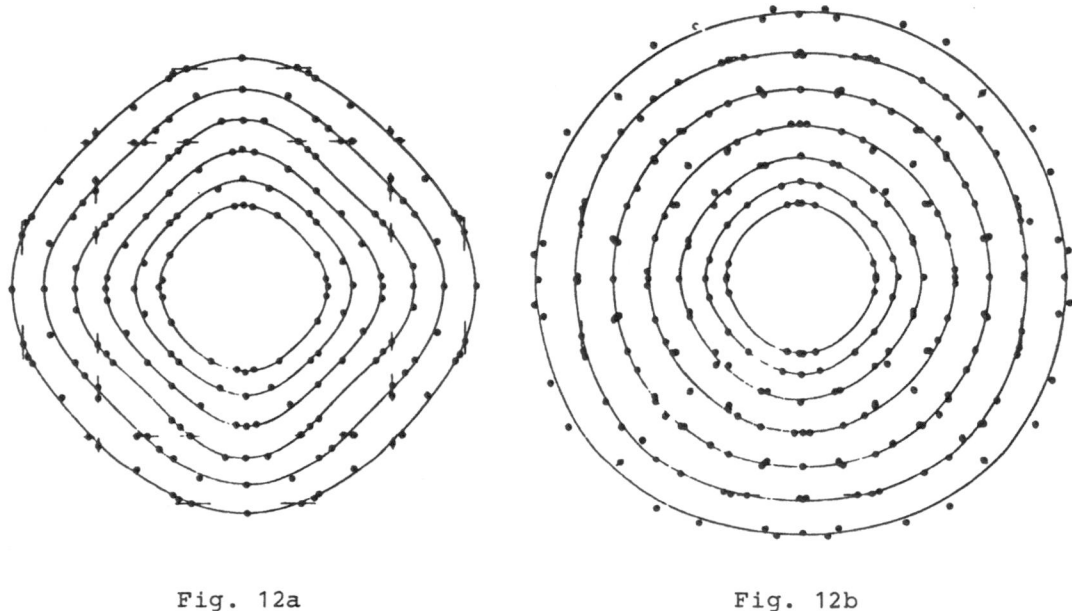

Fig. 12a   Fig. 12b

The restoration of rotation invariance is quite reassuring.

E) Universality? Let us postpone the discussion on this
point to Section 5.

The story of the string tension in the SU(2) gauge
theory illustrates nicely the strength of the lattice method
and its present limitations. Although there is a lot to
be done there, I think it is a piece of work we can be
proud of.

4. LATTICE QCD AT FINITE PHYSICAL TEMPERATURE

The exciting problems of cosmology and heavy ion collisions[33] call for the investigation of QCD under extreme conditions: when the system is at finite temperature, or when a non-zero matter density is present. The question of introducing chemical potential on the lattice will be discussed at the end of these lectures. Let us consider only finite temperature now.

There has been a long-standing conjecture that a phase transition must take place between the high temperature and low temperature phases of a non-Abelian gauge field theory[34]. In a simplified picture, at low temperatures the hadrons form isolated islands in the highly non-perturbative confining vacuum "soup". Inside the hadrons the quarks and gluons can propagate; it is close to a Fock vacuum. Owing to the fluctuations, the number of these islands is increasing with the temperature, and above a certain temperature the islands form regions which are connected over macroscopic distances.

How to introduce finite temperature in a quantum field theory?

Consider first a quantum mechanical particle, described by the Hamiltonian H. Consider the behaviour of this simple system under the influence of a heath bath of temperature $T \equiv 1/\beta$. The partition function is given by

$$Z = \text{Tr} (e^{-\beta H}) \quad . \tag{4.1}$$

The trace can be calculated in any basis, take a co-ordinate basis, for instance:

$$Z = \int dx \ \langle x|e^{-\beta H}|x\rangle \quad . \tag{4.2}$$

This is just the propagation kernel [Eq.(1.1)] continued to the imaginary time $-i\beta$. Therefore, in a path integral representation, the partition function is given by

$$Z = \sum_{\substack{\text{periodic} \\ \text{path}}} e^{-\int_0^\beta dx_4 L_E(x)} \quad , \tag{4.3}$$

where the summation is over all paths satisfying $x(\beta) = x(0)$.

By generalization, it is easy to modify the Euclidean prescription of zero temperature Yang-Mills theory to include the effect of finite temperatures. The partition function is given as

$$Z = \prod_\ell \int dU_\ell \; e^{-S_E} \quad ,$$

$$-S_E = 1/g^2 \sum_p (\text{Tr } U_p + cc) \quad , \tag{4.4}$$

with the constraint:

    a) the lattice is finite in the fourth direction
       $N_\beta \times N_s \times N_s \times N_s$, where $(N_s a) \to \infty$, $(N_\beta a)$ is fixed,
       $N_\beta a = 1/T$.

    b) the gauge field is periodic in the fourth direction.

Having the partition function we can take derivatives with respect to β to obtain the internal energy and other thermodynamical functions. These functions are expected to behave singularly if a phase transition occurs as the temperature is changed.

Another possibility is to insert an isolated quark into the system. In the confining phase its (free) energy is expected to be infinite, while in the deconfining phase it is expected to be finite. An isolated quark source is represented by a loop running along the temperature direction and closing upon itself (Fig. 13). It is expected that:

$$\langle L \rangle = e^{-F_q \beta} \qquad \begin{cases} = 0 & \text{in the confining phase} \\ \neq 0 & \text{in the deconfined phase} \end{cases},$$

therefore <L> is an order parameter.

It has been shown quite a long time ago that lattice
QCD undergoes a thermal quark liberation in the strong coupling limit[35]. The question was (like in the case of confinement itself at zero temperature) what happens with

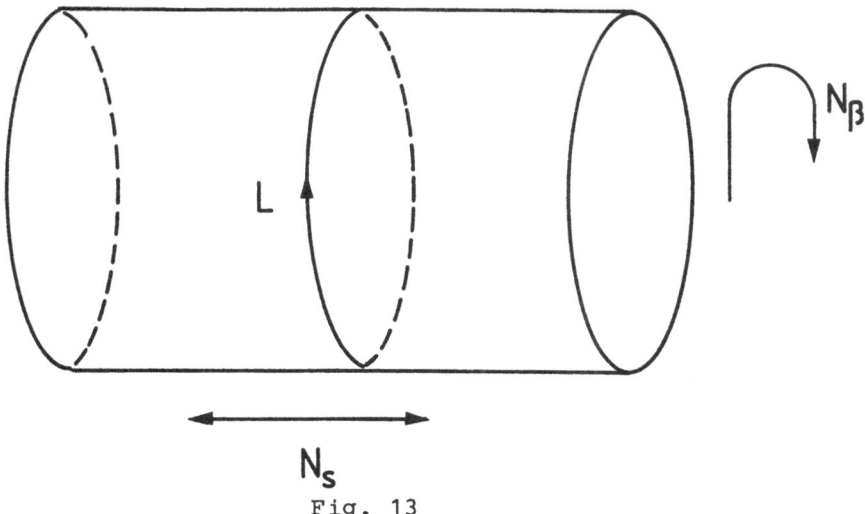

Fig. 13

this phase transition in the continuum limit, and if it
exists, what is the critical temperature.

The first Monte Carlo studies have been done for
SU(2) by Kuti, Polônyi and Szlachányi and by McLerran and
Svetitski[36]. The results confirmed the presence of a
deconfining phase transition in the continuum limit and predicted the critical temperature.

There are Monte Carlo measurements for SU(3) also
[37,38], and presumably they are the most reliable results
available at present for the SU(3) gauge theory. Observing
a phase transition is somewhat **easier** than measuring the
spectrum and additionally, in SU(3) the deconfining phase
transition is of first order (as the function of the temperature in the continuum limit)[38], therefore the finite size
effects are less severe.

5. UNIVERSALITY TESTS IN SU(2) LATTICE QCD

As we discussed in Section 1.6, there is a considerable freedom in defining a lattice action. The requirement of obtaining the continuum action in the classical continuum limit, and the requirement of gauge invariance, translation and 90° rotation invariance, parity and charge conjugation invariance are not very restrictive. The Wilson action in Eq. (1.26) is built up from the simplest gauge invariant products of the $U_{n\mu}$ matrices: Tr U_p [Eq. (1.21)]. Consider SU(2) gauge theories and define

$$U_p = e^{i\theta_p^a \tau^a} = \cos\theta_p + i\tau^a (\frac{\theta_p^a}{\theta_p}) \sin\theta_p \quad , \tag{5.1}$$

where $\theta_p^2 = \theta_p^a \theta_p^a$. Tr U_p, and therefore θ_p is gauge invariant. [Actually θ_p^a is proportional to the lattice field strength tensor defined in Eqs. (1.24) and (1.25).] In terms of θ_p the Wilson action can be written as:

$$S_W = -4/g_W^2 \sum_p \cos\theta_p \quad . \tag{5.2}$$

There are many other functions of θ_p satisfying the basic requirements. The Manton action is defined as [39]

$$S_M = 2/g_M^2 \sum_p \bar{\theta}_p^2 \quad , \qquad \bar{\theta}_p = \theta_p \, (\text{mod } 2\pi) \quad , \tag{5.3}$$

while the Villain action has the form[40]:

$$e^{-S_V} = \prod_P \{ \sum_{\ell=0}^{\infty} (\ell+1) \frac{\sin(\ell+1)\theta_p}{\sin\theta_p} e^{-\ell(\ell+2)\frac{g_V^2}{8}} \} \quad . \tag{5.4}$$

(The Villain action corresponds to a simple, special choice of β_r in the character expansion $e^{f(\text{Tr}U_p)} = \Sigma_r \beta_r \chi_r$). Additionally, other loop products, higher representations, different lattice structures, and so on, can be considered.

The independence of the physical predictions on the

specific action chosen is a basic requirement for this approach to make sense. It is expected (and can be checked explicitly in perturbation theory) that the different formulations define identical renormalized theories after adjusting a single parameter (the coupling constant) appropriately[+]. This expectation is based on the accumulated knowledge and experience concerning universality in statistical physics.

Consider the actions defined in Eqs.(5.2)-(5.4). In renormalized perturbation theory the three models are identical in what concerns the physical predictions. The only difference, predicted by perturbation theory, is in their respective scale parameter. One-loop calculation gives the following result (exact in the $g \to 0$ continuum limit)

$$\Lambda_M^{latt} / \Lambda_W^{latt} = 3.07 \quad , \tag{5.5}$$

$$\Lambda_M^{latt} / \Lambda_V^{latt} = 2.45 \quad . \tag{5.6}$$

On the other hand, Monte Carlo studies give the following nonperturbative predictions[43]

$$(\text{string tension})^{1/2} = \begin{cases} (83 \pm 14)\Lambda_W^{latt} & \text{Wilson action} \\ (16.2 \pm 0.5)\Lambda_M^{latt} & \text{Manton action} \\ (48.5 \pm 2.6)\Lambda_V^{latt} & \text{Villain action} \end{cases} \tag{5.7}$$

[+] For Green's functions, which are not renormalization group invariant, wave function renormalization is needed also. It is assumed that the symmetries of the specific lattice formulation reduce the number of dimension-four, gauge invariant operator combinations to 1. If this condition is not satisfied (like in the Hamiltonian formualtion[41], where the absence of the symmetry of 90° rotations between space and time allows two combinations: $F^a_{oi}F^a_{oi}$ and $F^a_{ik}F^a_{ik}$) then a careful tuning process is required in a multidimensional parameter space in order to get a correct Lorentz and gauge invariant continuum theory[42].

Compare the theoretical prediction with these results. The tension should be independent of the action, giving

	TH	"EXP"
$\Lambda_M^{latt}/\Lambda_W^{latt}$	3.07	5.14 ± 0.87
$\Lambda_M^{latt}/\Lambda_V^{latt}$	2.45	2.99 ± 0.19

$$(5.8)$$

There is some discrepancy beyond the statistical errors. A possibility is that the theoretical calculation should include the next loop correction[+]

$$\Lambda^{latt} = \frac{1}{a} e^{-\frac{1}{2\beta_0 g^2}} (\beta_0 g^2)^{-\frac{\beta_1}{2\beta_0^2}} \{1 - \frac{\beta_0 \beta_2 - \beta_1^2}{2\beta_0^3} g^2 + O(g^4)\} \, ,$$

$$(5.9)$$

where β_2, the third coefficient of the β function, is not universal.

The continuum limit is defined as $g \to 0$, this is the limit when scaling and universality is expected to hold. Is it consistent to keep g^2 terms in the bracket of (5.9)?

Yes, it is. Cut-off independence requires

$$a \frac{d}{da} \Lambda^{latt} = 0 + O(a) \qquad (5.10)$$

Continuum behaviour starts when the $O(a)$ correction is negligible in (5.10). This is an exponentially small correction in g^2:

$$a \sim e^{-\frac{1}{2\beta_0 g^2}} \qquad .$$

[+] There exists the possibility also that the statistical and systematic errors are underestimated in the Monte Carlo calculations.

There exists a coupling constant region, where $e^{-(1/2\beta_0 g^2)}$ is small, but the $O(g^2)$ correction cannot be neglected with respect to 1 in (5.9).

Unfortunately, no complete calculation exists for β_2. A few graphs - believed to be important- were calculated, and they work in the right direction[44].

Until we wait for a full calculation, there is another possibility: measure a different physical quantity, T_c for instance, and form the ratio:

$$\frac{(\text{tension})^{1/2}}{T_c} \quad - \quad \text{universal} \quad . \tag{5.11}$$

From this ratio the unknown corrections cancel. The Monte Carlo results for T_c are[+][45]

$$T_c = \begin{array}{ll} 42.8 \; \Lambda_W^{latt} & \text{Wilson action} \\[2mm] 10.5 \; \Lambda_M^{latt} & \text{Manton action} \\[2mm] 27.3 \; \Lambda_V^{latt} & \text{Villain action} \quad . \end{array} \tag{5.12}$$

Equations (5.7) and (5.12) give the following predictions for the ratio Eq.(5.11):

$$\frac{(\text{tension})^{1/2}}{T_c} = \begin{array}{ll} 1.94 \pm 0.33 & \text{Wilson action} \\[2mm] 1.54 \pm 0.05 & \text{Manton action} \\[2mm] 1.78 \pm 0.10 & \text{Villain action} \quad , \end{array} \tag{5.13}$$

which is consistent with universality within the statistical errors.

In the case of the mixed SO(3)-SU(2) action the situation is more problematic[46]. The question of universality is of basic importance and deserves further careful studies.

[+] Unfortunately, no errors are quoted by the authors.

6. NUMERICAL RESULTS IN SU(3) PURE GAUGE THEORY

Let us summarize briefly the numbers obtained in SU(3)
for the string tension, mass gap and critical temperature.
In all the cases the method was Monte Carlo simulation
(+variational method in the case of the mass gap).

As we discussed earlier, presumably the critical temperature measurement is the most reliable, although even
there the independence of the result on the number of lattice
units along the temperature direction would be highly required. The string tension data are less convincing than
they are for SU(2). The mass gap determinations were made
on very small lattices. In all the cases - even with the
available techniques and computers - there is a possibility for more convincing studies, and improved results are expected in the near future.

The present results are as follows[47]:

$$(\text{string tension})^{1/2} = (170 \pm 30) \; \Lambda^{\text{latt}} \tag{6.1}$$

$$m(0^{++}) = (280 \pm 50) \; \Lambda^{\text{latt}} \tag{6.2}$$

$$T_c = (84 \pm 2) \; \Lambda^{\text{latt}} \tag{6.3}$$

7. FERMIONS ON THE LATTICE

By introducing dynamical quarks, the number of questions
for which an answer might be attempted is significantly
increased. The basic question is whether QCD is the theory
of strong interactions. Is QCD capable of describing the
large wealth of spectroscopical data?

During the last year exciting new results were published
on hadron spectroscopy[48]. In a remarkable short time a
large number of predictions were obtained in lattice QCD including the effect of dynamical (valence) quarks.

The merit of these results is somewhat controversial [49]. Presumably much more effort is needed to clarify the problems and to perform spectroscopical calculations where the systematical and statistical errors are really under control. Concerning the existing results and possible problems, the reader is referred to the literature[48,49], as for the very interesting results concerning the nature of chiral symmetry breaking in QCD[50].

In the following we shall pick up a few introductory topics on the theoretical and technical problems of introducing fermions in lattice calculations.

At the end, a special problem will be discussed related to introducing finite particle density in lattice calculations, i.e., the problem of finite chemical potential on the lattice[51]. This is the only part of these lectures which contains new results.

7.1 Defining Fermions on the Lattice Naively, the Coupling Problem

Let us try to put fermions on the lattice following the usual recipe. In the continuum action of a free Dirac particle, the derivatives are replaced by differences giving the lattice action:

$$S = a^4 \sum_n \{ \sum_\mu \frac{1}{2a} [\bar{\psi}(n) \gamma_\mu \psi(n+\hat{\mu}) - \bar{\psi}(n+\hat{\mu}) \gamma_\mu \psi(n)] + m \, \bar{\psi}(n) \psi(n) \} \; . \quad (7.1)$$

By inserting the gauge matrices in a gauge invariant way and by adding the gauge field action, a candidate prescription of QED on the lattice is obtained

$$S = a^4 \sum_n \{ \sum_\mu \frac{1}{2a} [\bar{\psi}(n) \gamma_n U_{n\mu} \psi(n+\hat{\mu}) - \bar{\psi}(n+\hat{\mu}) \gamma_\mu U_{n\mu}^+ \psi(n)] +$$

$$+ m \, \bar{\psi}(n) \psi(n) \} + S_{gf} \quad , \quad\quad\quad\quad\quad (7.2)$$

where $U_{n\mu} = e^{iagA_{n\mu}} \varepsilon\ U(1)$. For m = 0 the model is invariant under the transformations

$$\psi_n \to e^{i\alpha}\psi_n\ , \qquad\qquad \bar\psi_n \to \bar\psi_n\ e^{-i\alpha}\ ,$$

and

$$\psi_n \to e^{i\alpha\gamma_5}\psi_n\ , \qquad\qquad \bar\psi_n \to \bar\psi_n\ e^{i\alpha\gamma_5}\ . \qquad\qquad (7.3)$$

The symmetry group is $U(1)_{vector} \otimes U(1)_{axial\ vector}$. This symmetry is exact for any value of the lattice constant, therefore it is there in the a → 0 continuum limit also.

However, there is a general theorem due to Adler[52] claiming that although the classical theory is U(1) × U(1) invariant, there is no regularization which could respect both of these symmetries. In the quantum theory the chiral symmetry is necessarily explicitly broken. The axial vector current is not conserved[53], its divergence receives a non-zero contribution from the triangle graph

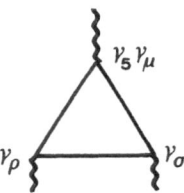

The above construction seemingly contradicts this theorem. The resolution of this paradox is the following. Though we wanted to describe the interaction of a single fermion with the electromagnetic field, actually our action describes 16 identical fermion species. Running around the triangle graph all of these fermions give a contribution to the axial anomaly, but - as it was shown by Karsten and Smit[54] - their contribution alternates in sign and adds up to zero.

It is easy to see that this action describes 16 fermion species. The free fermion part of the action

$\sim \bar{\psi}(n) \; \gamma_\mu \; \psi(n+\hat{\mu}) \; - \; \bar{\psi}(n+\hat{\mu}) \gamma_\mu \; \psi(n)$

gives the following propagator in momentum space:

$$\sim \frac{1}{\sum_\mu \gamma_\mu \sin p_\mu} \quad , \qquad \qquad p_\mu \; \epsilon \, (-\pi, \pi) \quad . \qquad \qquad (7.4)$$

This propagator has 16 poles at the points $p_\mu = (0,0,0,0)$, $(\pi,0,0,0), \ldots, (\pi,\pi,\pi,\pi)$.

Chiral symmetry implies species doubling on the lattice. If we want to describe a single fermion, the U(1) chiral symmetry must be explicitly broken. We should accept it, it cannot be otherwise due to Adler's theorem.

Similarly, for the general case with n_f massless fermions, we accept that the flavour singlet U(1) axial symmetry is explicitly broken, since it is a phenomenon which is general, independent of the lattice. However, we would like to keep the SU(n_f) axial symmetry. That is the point where the solutions suggested until now are not satisfactory.

7.2 Wilson's Method of Removing the Degeneracy

In order to avoid species doubling, Wilson suggested adding a new term to the action[55]. This term is chosen in such a way that:

- it gives large ($\sim 1/a$ cut-off) masses to the unwanted 15 fermion species. They disappear from the theory in the continuum limit;

- it goes to zero in the formal $a \to 0$ limit, therefore we expect that it will not affect the behaviour of the remaining single fermion at the end.

The action has the following form:

$$-S = \sum_n \{ -\bar{\psi}_\alpha^{a,i}(n) \, \psi_\alpha^{a,i}(n) + \sum_\mu K_i \; \bar{\psi}_\alpha^{a,i}(n) \, (1-\gamma_\mu)_{\alpha\beta} U_{n\mu}^{ab} \, \psi_\beta^{b,i}(n+\hat{\mu}) \; +$$

$$+ \sum_\mu K_i \; \bar{\psi}_\alpha^{a,i}(n+\hat{\mu})(1+\gamma_\mu)_{\alpha\beta} U_{n\mu}^{+ab} \psi_\beta^{b,i}(n) \} + 1/g^2 \sum_p (\mathrm{Tr} U_p + cc) ,$$

$$(7.5)$$

where $a,b;\alpha,\beta$ and i are colour, Dirac and flavour indices, respectively.

By finding the coefficient of the quadratic term $\bar{\psi}(n)\psi(n)$, K_i can be related to the bare (classical) quark mass m_i. In the limit $(m_i a) \to 0$ one obtains:

$$K_i = \frac{1}{8+2m_i a} , \qquad\qquad i = u,d,s,c,\ldots \quad .$$

$$(7.6)$$

There is no species doubling in this prescription. Unfortuantely, the new terms explicitly break chiral symmetry. Before adding these terms, the classical symmetry is (for massless quarks)

$$SU(n_f)_{vector} \times SU(n_f)_{axial} \times U(1)_{vector} \times U(1)_{axial} ,$$

which is reduced by the new terms to

$$SU(n_f)_{vector} \times SU(n_f)_{axial} \times U(1)_{vector} \times U(1)_{axial} .$$

This is the symmetry for any finite lattice distance a. We can only hope that the $SU(n_f)_{axial}$ symmetry will be recovered in the continuum limit (and it will be broken spontaneously).

The amplitude of moving a quark by one lattice unit is proportional to K_i. Hence the name hopping parameter. If $g = 0$, $K = 1/8$ defines a massless fermion. For $g \neq 0$, there are mass corrections (there is no chiral symmetry which would prevent the occurrence of mass counter terms) and the value of K giving a massless fermion is not 1/8, but receives perturbative and non-perturbative corrections. K_i should be renormalized. $K_i = K_i(g^2)$ is not known a priori.

7.3 Technical Problems and Methods

In a path integral formulation the fermion fields are
represented by anticommuting c numbers, by Grassmann variables. There is no effective way of representing them on a
computer. In every method the first step is to integrate over
the fermion fields.

The action is quadratic in the fermion fields. In a
concise notation it has the form:

$$S = \sum_{i,j} \bar{\psi}_i \, \Delta_{ij}(U) \psi_j + S_{gf} \quad , \tag{7.7}$$

where i,j represent all kind of indices. By integrating over
the fermion variables in the vacuum functional or in any
expectation value of fermion fields one obtains:

$$\int D\psi D\bar{\psi} \; e^{-S} = \det(\Delta(U)) e^{-S_{gf}} \quad ,$$

$$\int D\psi D\bar{\psi} \; \bar{\psi}_i \psi_j \; e^{-S} = \Delta_{ji}^{-1}(U) \det(\Delta(U)) e^{-S_{gf}} \quad , \tag{7.8}$$

...

The second example describes the propagation of a quark in
the background field U. This background field is generated
with a probability distribution governed by $S_{eff} = S_{gf} -$
$-\text{Tr}(\ell n\Delta(U))$. The second term in S_{eff} represents the effect
of virtual quark loops.

Until now, most of the results were obtained under the
approximation of neglecting the virtual quarks. The valence
quarks are treated dynamically. There are qualitative arguments (large N limit, Zweig rule,...) supporting this approximation. Including the virtual quark contribution would
require an updating process in Monte Carlo with S_{eff} - a
non-local action.

The hadron propagators are constructed from the quark
propagators. According to (7.8) this requires the inversion
of a very large matrix $\Delta_{ij}(U)$. There are different suggestions
for this problem.

A) Direct numerical methods

Δ_{ij} depends on the background gauge field and contains the hopping parameter K as an external parameter. One might try to invert this matrix numerically for a given value of K. In spectrum calculations the Gauss-Seidel iteration method was used but there are other methods suggested.

The matrix $\Delta(U)$ can be written as:

$$\Delta(U) = 1 - KB(U) \quad . \tag{7.9}$$

The equation

$$\chi = (1-KB(U)^{-1}\phi \tag{7.10}$$

can be rearranged as

$$\chi = KB(U)\chi + \phi \ , \tag{7.11}$$

and this form is ready for iteration.

B) Hopping parameter expansion

The hopping parameter expansion[55,56] is analogous to the high temperature expansion in statistical physics in many respects. The amplitude of moving a quark by one lattice unit is proportional to the hopping parameter K. An expansion in K is equivalent to an exapnsion in the length of quark paths in configuration space. In their propagation the quarks are constrained by the maximum order of the expansion; nevertheless they should still gather the essential information on the hadron's structure. This defines the conditions under which the expansion to a given order will be reliable. The size of the hadron in lattice units should not be too large, it has to be comparable to the regions covered by possible quark paths. Or, alternatively the lattice distance a cannot be too small.

The quark propagator can be written:

$$\Delta^{-1}_{ji}(U) = \sum_{\ell=\ell_{min}} C_\ell(U;j,i) K^\ell \quad . \tag{7.12}$$

In this method the coefficients $C_\ell(U;j,i)$ are calculated.

7.4 The Procedure of Extracting the Hadron Masses

The procedure is rather different for the methods A)
and B). Let us discuss them separately. In both methods
gauge field configurations are generated by MC using $S = S_{gf}$.
Next:

A) On a given background field $\{U_{n\mu}\}$ the quark propagator $\Delta^{-1}_{ij}(U)$ is calculated by some direct numerical method.
Then the meson and baryon propagators are built up from
the quark propagator. More precisely, the "time slice propagators" are constructed

$$D(n_t) = \sum_{\underline{n} \atop (n_t \text{ fixed})} D(\underline{n},n_t) \quad , \qquad n_t = 0,\pm1,\pm2,\ldots \quad , \tag{7.13}$$

corresponding to a hadron in the rest frame. The procedure
is repeated for different K values on the same background
field. After that, the next (independent) background field
is taken and so on.

For a given K (quark mass), the hadron propagator $D(n_t)$
is expected to decay exponentially. At shorter distances
excited states are also present. Their effect might be taken
into account by fitting the data points with a sum of ex-
ponentials.

B) On a given background field the (quark and) hadron
propagators are expanded in the hopping parameter K. By
taking the average over many background fields one obtains

$$D(n) = \sum_{\ell=\ell_{min}} C_\ell(n) K^\ell . \qquad (7.14)$$

The propagator is expected to have a pole in momentum space.
Close to the pole we have, in the continuum limit,

$$D(p_0 = iE, \quad f = 0; \quad K,g) \sim \frac{1}{E^2 - M^2(K,g)} . \qquad (7.15)$$

Consider now D as a function of K at a fixed E (and g). Let
us define K^* by the equation $E^2 = M^2(K^*,g)$. Then, by using
the expansion

$$M^2(K,g) = E^2 + \frac{\partial}{\partial K} M^2(K,g)|_{K=K^*} (K-K^*) + \ldots , \qquad (7.16)$$

one finds that a pole in momentum space implies a pole in K.

Of course, a finite expansion cannot produce a pole.
The situation is similar to the high temperature expansion,
where we have a power series (in 1/T), and we have some
knowledge about the singularity structure. In our case the
nearest singularity is expected to be a pole. Padé approxi-
mants might be used to identify this pole.

7.5 How the Parameters Are Fixed

The calculation is done at a given value of g. In order
to check the renormalization group behaviour the calculation
should be repeated for different g values.

Let us fix g and consider an idealized case. Assume that
the lowest (dimensionless) meson mass has been determined
as the function of K both in the pion and the rho channels
(Fig. 14). K_u^c is the hopping parameter value where the pion
is massless. It corresponds to massless quarks and to spontaneously broken chiral symmetry. Increase the quark mass
(decrease K) until the m_p/m_π ratio is predicted correctly.
This defines the physical value of K (and the quark mass).

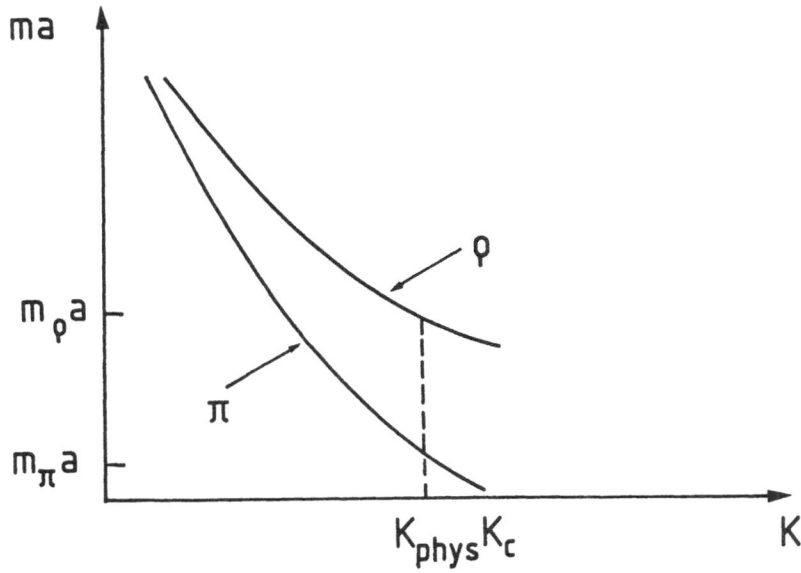

Fig. 14

Then the requirement m_π = 140 MeV gives the value of the lattice distance in physical units at this coupling.

The lattice unit "a" obtained this way should be consistent with the pure gauge theory results (we neglected the virtual quark loops!)

K_s^{phys} and K_c^{phys} are fixed similarly by using the experimental values of m_K and m_ψ, respectively. From now on everything is predicted!

8. CHEMICAL POTENTIAL ON THE LATTICE

Let me close these introductory lectures by a short seminar on the way the chemical potential is introduced on the lattice[51].

The exciting problems of cosmology and heavy ion collisions call for the introduction of finite temperature and particle density in (nonperturbative) QCD studies. The investigation of QCD under extreme conditions has a special flavour. Rather than attempting to explain and quantitatively derive old, well-known experimental data, we deal with a subject unexplored in laboratory experiments, and the results derived from QCD might influence future experiments.

There is a straightforward way of introducing finite temperature in lattice calculations. The continuum prescription for the partition function in terms of a functional integral can be taken over directly on the lattice. The partition function is defined as

$$Z = \int DU_\ell D\psi D\bar\psi \ e^{-S_E(U,\psi,\bar\psi)} \tag{8.1}$$

on a lattice, which is finite along the fourth (temperature) direction: $N_\beta a = 1/T$ is finite. The gauge and fermion fields satisfy periodic and antiperiodic boundary conditions respectively.

The lattice approach proved to be powerful in determining the critical temperature and other thermodynamical quantities of pure gauge QCD[36,37]. There are attempts at taking into account the effect of quarks[38].

On the other hand, the introduction of finite particle density is hindered by the problems of defining the chemical potential on the lattice in a satisfactory way. The naive generalization of the continuum prescription leads to quadratic divergences even for free fermions: in the continuum limit ($a \to 0$) the energy density ε is proportional to $(\mu/a)^2$ instead of the correct finite result $\varepsilon \sim \mu^4$ (for massless fermions), with μ being the chemical potential.

Let H be the Hamiltonian of the system and N some conserved quantity. In the following, we shall take N to be the fermion number, the generalization is straightforward. The partition function is given by

$$Z = e^{-\beta\Omega} = \text{Tr } e^{-\beta(H-\mu N)} \quad . \tag{8.2}$$

Going to functional integral representation, the presence
of a non-zero chemical potential implies a new term in the
Lagrangian: μ times the fourth component of the current
the space integral of which gives the conserved quantity N.

Let us take free fermions in the continuum. The action
is

$$\int_0^\beta \left(\frac{1}{2}\, \bar{\psi}(x)\gamma_\mu \overleftrightarrow{\partial}_\mu \psi(x) + m\bar{\psi}(x)\psi(x) + \mu\bar{\psi}(x)\gamma_4\psi(x)\right)d^4x \tag{8.3}$$

with the usual antiperiodic boundary conditions along the
fourth direction. The energy density is defined as

$$\varepsilon = -\frac{1}{V}\frac{\partial}{\partial\beta}\ln Z\Big|_{\beta\mu \text{ fixed}} \quad , \tag{8.4}$$

and is given by the following integral when $T \to 0$:

$$\varepsilon = \frac{4}{(2\pi)^4}\int d^4p\, \frac{(p_4+i\mu)^2}{(p_4+i\mu)^2+\underline{p}^2+m^2} \quad . \tag{8.5}$$

ε is a $(\text{mass})^4$ quantity. By power counting, the integral is
quartically divergent. There is a trivial part, the vacuum
energy at $\mu = 0$. By subtracting this term we get:

$$\varepsilon = -\frac{4}{(2\pi)^4}\int d^4p\, \left[\frac{\underline{p}^2+m^2}{(p_4+i\mu)^2+\underline{p}^2+m^2} - (\mu = 0)\right] \quad . \tag{8.6}$$

This integral is finite (not quadratically divergent) as it
should. By performing the p_4 integral first, two poles are
encountered in the p_4 plane both for $\mu \neq 0$ and $\mu = 0$ (Fig. 15).
The difference is zero, except when both poles lie on the
lower half plane for $\mu \neq 0$. This gives a θ function:

344

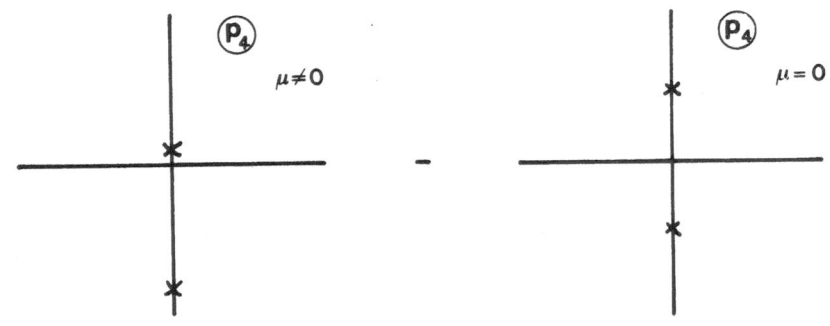

Fig. 15

$$\varepsilon = \frac{4}{(2\pi)^4} \pi \int d^3p \ \theta(\mu - \sqrt{\underline{p}^2 + m^2}) \sqrt{\underline{p}^2 + m^2} \ , \tag{8.7}$$

or

$$\varepsilon \big|_{m=0} = \frac{1}{4\pi^2} \mu^4 \ ,$$

which is the correct result.

The naive generalization of this prescription would lead to the lattice action

$$S = \sum_{x} \{\frac{1}{2} \sum_{\mu} (\bar{\psi}_x \gamma_\mu \psi_{x+\hat{\mu}} - \bar{\psi}_{x+\hat{\mu}} \gamma_\mu \psi_x) + ma\bar{\psi}_x\psi_x + \mu a \ \bar{\psi}_x \gamma_4 \psi_x\} \ , \tag{8.8}$$

and yields the following result for the energy density $(T \to 0)$

$$\varepsilon = a^{-4}\{ - \frac{1}{4\pi^4} \int_{-\pi}^{\pi} d^4q \ \frac{\sum\limits_{j=1}^{3} \sin^2 q_j + (ma)^2}{(\sin q_4 - i\mu a)^2 + \sum\limits_{j=1}^{3} \sin^2 q_j + (ma)^2}\} -$$

$$- a^{-4}\{\mu = 0\} \ . \tag{8.9}$$

What was a potential problem before, is present here explicitly: ε is quadratically divergent

$$\varepsilon \sim (1/a)^2 \, \mu^2 \, . \tag{8.10}$$

The problem is not related to the species doubling implied
by Eq.(8.3). The 16-fold degeneracy can be removed by Wilson's
prescription, but the problem discussed above, remains.
Replacing the current $\bar{\psi}_x \gamma_4 \psi_x$ by a point-split form does not
help either. In order to obtain a finite result, non-co-
variant counterterms should be introduced in Eq.(8.3).
Although their presence is understandable (there is no Euclidean symmetry for $\mu \neq 0$), it would be an awkward way to
proceed.

What is the reason that no similar problems occurred
in the continuum formulation? The key to understanding is
the observation that in the Euclidean formulation of thermodynamics the chemical potential acts like the fourth component of an imaginary, constant vector potential. In continuum QED, for instance, the chemical potential is introduced
exactly as a photon field:

$$e \, \bar{\psi}_x \, \gamma_\mu \, A_\mu \, \psi_x - i_\mu \, \bar{\psi}_x \, \gamma_4 \, \psi_x \tag{8.11}$$

For this reason an expansion in powers of μ is equivalent to inserting external, zero momentum photon lines
to the amplitude. For instance, the contribution of order
μ^2 to the thermodynamic potential $\Omega = -(\beta V)^{-1} \ln Z$ is proportional to

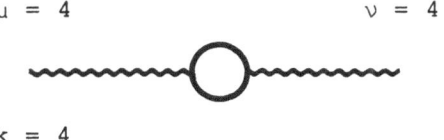

This graph is proportional to the (potentially quadratically
divergent) photon mass renormalization which, however, is
zero in a gauge invariant formulation[57] at zero temperature
(or, due to the plasmon effect, finite at finite temperature).

Similarly, the finiteness of contributions of the order μ^ℓ to the amplitude $\Gamma^{(n,m)}$ (n and m being the number of external photon and electron lines, respectively) follows from the renormalizability of the amplitude $\Gamma^{(n+\ell,m)}$. Although ℓ factors of the electromagnetic coupling constant e are replaced by the chemical potential μ and ℓ factors of wave function renormalization are also absent, the missing factors for charge and wave function renormalizations just cancel. This is again the consequence of gauge invariance[+].

In Eq.(8.3) this Abelian gauge invariance is violated: μ does not enter like the fourth component of a gauge field on the lattice. The correct solution is

$$S = a^3 \sum_x \{ ma\, \bar{\psi}_x \psi_x + \frac{1}{2} \sum_{j=1}^{3} (\bar{\psi}_x \gamma_j \psi_{x+\hat{j}} - \bar{\psi}_{x+\hat{j}} \gamma_j \psi_x)$$

$$+ 1/2(e^{\mu a}\bar{\psi}_x \gamma_4 \psi_{x+\hat{4}} - e^{-\mu a}\psi_{x+\hat{4}} \gamma_4 \psi_x) \} \quad . \tag{8.12}$$

Expand this form in terms of aμ. The first term is the kinetic term. The second term is the point-splitted current. However, there are higher order extra contributions - just the appropriate non-covariant counterterms we discussed before. Of course, we shall not expand in μa - like we do not expand $U_{n\mu}$ in terms of $A_{n\mu}$.

By this prescription, in Eq.(8.9) in the integrand $(\sin q_4 - i\mu)^2$ is replaced by $\sin^2(q_4 - i\mu)$ like in the continuum theory, and performing the q_4 integration we get

$$\varepsilon a^4 = \frac{1}{2\pi^3} \int_{-\pi}^{\pi} d^3 q\; \theta(e^{\mu a} - b - \sqrt{b^2+1}) \frac{b}{\sqrt{b^2+1}} \quad , \tag{8.13}$$

with

$$b^2 = \sum_{j=1}^{3} \sin^2 q_j \quad . \tag{8.14}$$

[+] The importance of the generalized gauge invariance in the renormalization of theories with finite chemical potentials is emphasized by Baluni[58].

Therefore, we see from Eq. (8.13) that in every corner of the Brillouin zone the q_4 integration leads in the $a \to 0$ limit to the expected, correct result for the momentum cut-off $\sim\theta(\mu-\sqrt{q^2+m^2})$, and the resulting energy density is 16 times the usual finite energy density of free fermions at zero temperature.

Using Wilson fermions, the degeneracy is removed and the factor 16 disappears for any $r \neq 0$ as it should ($0 < r \leq 1$ is the usual arbitrary parameter in the Wilson action).

Equation (8.12) can be immediately generalized to the case of QCD. At finite temperature and chemical potential the Wilson action with one flavour has the form

$$S = a^3 \sum_x \{\bar{\psi}_x \psi_x - K \sum_{j=1}^{3} [\bar{\psi}_x (r-\gamma_j) U_{x,j} \psi_{x+\hat{j}} + \bar{\psi}_{x+\hat{j}} (r+\gamma_j) U^+_{x,j} \psi_x] -$$

$$- K[e^{+\mu a}\bar{\psi}_x (r-\gamma_4) U_{x4} \psi_{x+\hat{4}} + e^{-\mu a}\bar{\psi}_{x+\hat{4}} (r+\gamma_4) U^+_{x4} \psi_x]\} +$$

$$+ \frac{2N}{g^2} \sum_P (1- \frac{1}{N} \text{Re Tr } U_p) \quad , \tag{8.15}$$

where the gauge fields U are periodic, while the fermion fields ψ, $\bar{\psi}$ are antiperiodic along the "temperature" direction. The prescription is very simple: the hopping parameter K, related to the quark propagation by one lattice unit along the positive (negative) imaginary time axis is replaced by $e^{\mu a}K(e^{-\mu a}K)$.

Considering the thermodynamic potential at finite temperature, there are quark paths wrapping around the lattice in the imaginary time direction. Only these paths can lead to chemical potential dependence - from ordinary closed the μ dependence cancels. This is understandable: ordinary loops describe virtual pair creation and annihilation, and the chemical potential of quarks and antiquarks is of opposite sign. It follows that it is not advisable to study this system at zero temperature exactly, even if we want to discuss the effect

of the finite particle density alone. The hopping parameter
expansion (and related iterative methods) breaks down in this
case. This fact can also be seen from the explicit result
for the energy density of free quarks. At zero temperature ε
is proportional to $\theta(\mu-m(K))$, and this distribution cannot
be expanded in terms of the hopping parameter. No similar
problem occurs at finite temperature.

ACKNOWLEDGEMENT

The author is grateful for the invitation to partici-
pate in this school and for the kind hospitality extended
to him during the time he stayed there.

APPENDIX A
ANALYTIC CONTINUATION TO IMAGINARY TIME IN PERTURBATION THEORY

Consider a Green's function in a scalar field theory
in Minkowski space

$$G(x_1,x_2\ldots,x_n) = \int \frac{d^4p_1}{(2\pi)^4} \ldots \int \frac{d^4p_n}{(2\pi)^4} e^{ip_1x_1} e^{ip_nx_n} \tilde{G}(p_1,\ldots,p_n).$$

$$(A.1)$$

$\tilde{G}$ is given as a sum of Feynman diagrams, built up from vertices and propagators. The propagators remember the time
ordered product of configuration space by their special $i\varepsilon$
prescription:

$$\frac{1}{q^{o2}-\underline{q}^2-m^2+i\varepsilon} \quad , \tag{A.2}$$

describing a pole slightly above and below the negative and
positive part of the real q^o axes respectively on the complex
q^o plane

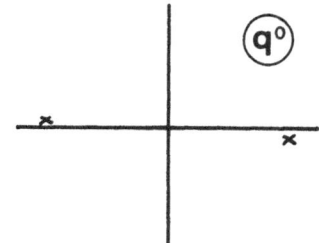

It follows that if all the zero component integration contours are rotated identically anticlockwise, these singularities are avoided. This leads to the possibility of defining the function

$$G(\phi x_1^o, \underline{x}_1; \ldots; \phi x_n^o, \underline{x}_n) \quad , \qquad \phi = \rho e^{i\theta} \qquad (A.3)$$

which is given by the right-hand side of (A.1), where all the zero component integration contours are rotated by $-\theta(-\pi < \theta < 0)$ and $x_1^o, \ldots, x_n^o$ are replaced by $x_1^o\phi, \ldots, x_n^o\phi$. This function is analytic in the lower ϕ plane and the original Green's function can be rediscovered as $\phi \to +1$.

After introducing the new integration variables $p_1^o \to p_1^o/\phi, \ldots$ (everywhere including the internal loop integrations), in the particular case of $\phi = -i$ we obtain

$$G(-ix_1^o, \underline{x}_1, \ldots, -ix_n^o, \underline{x}_n) \sim \int_{-\infty}^{\infty} \frac{d^4 p_n}{(2\pi)^4} \ldots \int_{\infty}^{\infty} \frac{d^4 p_n}{(2\pi)^4} e^{-ip_1 x_1} \ldots$$

$$\ldots e^{ip_n x_n} \tilde{G}(p_1^o/-i, \mathbf{p}_1, \ldots, p_n^o/-i, p_n) \qquad (A.4)$$

Therefore in the perturbative expansion the result of the Euclidean field theory right-hand side of Eq. (A.4) is analytically connected to the Minkowski result. They are uniquely determined by each other.

REFERENCES

1. K.G. Wilson, Phys. Rev. $\underline{D10}$ (1974) 2445.

2. A partial list of summary papers on the subject:
   J.-M. Drouffe and C. Itzykson, Phys. Reports $\underline{38C}$ (1978) 133.
   L.P. Kadanoff, Rev. Mod. Phys. $\underline{49}$ (1977) 267.
   J.P. Kogut, Rev. Mod. Phys. $\underline{51}$ (1979) 659.
   G. Parisi, XX Int. Conf. on High Energy Physics, Madison,
   Wisconsin, 1980 (Amer. Inst. Phys., New York, 1981),
   p. 1531.
   P. Hasenfratz, International Conference on High Energy
   Physics, Lisbon, 1981, Proceedings p. 620.
   C. Rebbi, XXI Int. Conf. on High Energy Physics, Paris,
   1982.
   P. Hasenfratz, in the Proceedings of the Johns Hopkins
   Workshop, Florence, 1982.
   J.B. Kogut, Illinois preprint, ILL-(TH)-82-46 (1982).

3. R.P. Feynman, Rev. Mod. Phys. $\underline{20}$ (1948) 367.
   R.P. Feynman and A.R. Hibbs, Quantum Mechanics and Path
   Integrals, New York, (Mc Graw-Hill Book Comp.) (1965).

4. J. Schwinger, Phys. Rev. $\underline{115}$ (1959) 721
   G.C. Wick, Phys. Rev. $\underline{96}$ (1954) 1124.

5. K. Osterwalder and R. Schrader, Commun. Math. Phys. $\underline{42}$
   (1975) 211.
   B. Simon, The $p(\phi)_2$ Euclidean (Quantum) Field Theory
   (Princeton University, Princeton, N.J.) (1974).
   E. Nelson, in Constructive Quantum Field theory, (Springer
   Verlag, Berlin) (1973).

6. The independence of the continuum limit on the lattice
   structure is an important question. Unfortunately not much
   has been done until now in this field. An example is discussed by
   W. Celmaster, Northeastern Univ. preprint NUB Nr. 2561 (1982).
   The exciting attempt of defining field theories on a
   random lattice will not be discussed here. We refer the
   interested reader to the literature
   N.H. Christ, R. Friedberg and T.D. Lee, Nucl. Phys. $\underline{B202}$

(1982) 89, B210 [FS6] (1982) 310,337.

7. Half-integer spin fermion fields will be defined on lattice
    points also (Section 7). This is not very natural in
    the above sense. The consequences of a different proce-
    dure are discussed in
    P. Becker and H. Joos, DESY preprint, DESY-82-031 (1982).
    T. Banks, Y. Dothan and D. Horn, Phys. Lett. 117B (1982)
    413.

8. For a detailed discussion on the non-Abelian case see
    R. Balian, J.-M. Drouffe and C. Itzykson, Phys. Rev. D10
    (1974) 3376.

9. B.E. Baaquie, Phys. Rev. D16 (1977) 2612.

10. L. Onsager, Phys. Rev. 65 (1944) 117.
    M.E. Fisher and R.J. Burford, Phys. Rev. 156 (1967) 583.

11. F. Wegner, J. Math. Phys. 12 (1971) 2259.

12. K. Osterwalder and E. Seiler, Ann. Phys. (NY) 110 (1978)
    440.

13. See for instance
    D.J. Gross, Lectures at the Les Houches Summer School, 1975,
    (North Holland P.C., ed. R. Balian).

14. Of course, this is not true for those dimensionful
    quantities which depend explicitly on momenta or co-ordi-
    nates like a scattering cross-section.

15. A. Hasenfratz and P. Hasenfratz, Phys. Lett. 93B (1980) 165.

16. Actually, no explicit proof, valid in every order, exists
    in the literature. At the one loop level it is easy to
    produce a general argument and the statement has been
    checked explicitly (see[15]). Although it is a common lore
    that the renormalized perturbation theory on the lattice
    is identical to that of the continuum prescriptions in
    every order, it would be nice to have an explicit demonstra-
    tion.

17. R. Dashen and D. Gross, Phys. Rev. D23 (1981) 2340.
    H. Kawai, R. Nakayama and K. Seo, Nuclear Phys. B189
    (1981) 40.
    A. Hasenfratz and P. Hasenfratz, Nucl. Phys. B193 (1981)
    210.

P. Weisz, Phys. Letters. 100B (1981) 331.

A. Gonzales-Arroyo and C.P. Korthals Altes, Nucl. Phys. B205 [FS5] (1982) 46.

H.S. Sharatchandra, H.J. Thun and P. Weisz, Nucl. Phys. B192 (1981) 205.

F. Karsch, Nucl. Phys. B205 [FS5] (1982) 285.

18. A partial list of references:

    J.B. Kogut, R.B. Pearson and J. Shigemitsu, Phys. Rev. Lett. 43 (1979) 484.

    J.B. Kogut and J. Shigemitsu, Phys. Rev. Lett. 45 (1980) 410.

    J.B. Kogut, R.B. Pearson and J. Shigemitsu, Phys. Lett. 98B (1981) 63.

    G. Münster, Phys. Lett. 95B (1980) 59.

    G. Münster, Nucl. Phys. B190 [FS3] (1981) 439, E.: B205 [FS5] (1982) 648.

    G. Münster and P. Weisz, Phys. Lett. 96B (1980) 119.

    K. Seo, Enrico Fermi Inst. preprint, EFI 82-10 (1982).

19. J. Smit, Nucl. Phys. B206 (1982) 309.

20. A. Hasenfratz, E. Hasenfratz and P. Hasenfratz Nucl. Phys. B181 (1981) 353.

    C. Itzykson, M. Peskin and J.B. Zuber, Phys. Lett. B95 (1980) 259.

    M. Lüscher, G. Münster and P. Weisz, Nucl. Phys. B180 (1980) 1.

21. K. Binder, in Phase Transitions and Critical Phenomena (eds. C. Domb and S. Green), Academic Press, New York (1976), Vol. 5B, Monte Carlo Methods in Statistical Physics, Springer Verlag (1979), (ed. K. Binder).

22. After the pioneering work of M. Creutz, L. Jacobs and C. Rebbi, Phys. Rev. Lett. 42 (1979) 1390, where the d=4, Z_2 gauge theory was investigated, and of M. Creutz, Phys. Rev. Lett. 43 (1979) 553, measuring for the first time the string tension in an SU(2) gauge theory, a large number of works has been published. For a detailed list of references, the reader is referred to the summary works in [2].

23. For an interesting attempt which works for Abelian gauge
    theories, see:
    T. Sterling and J. Greensite, Berkeley preprint, LBL-14769
    (1982).

24. R. Balian, J.-M. Drouffe and C. Itzykson, Phys. Rev. $\underline{D11}$
    (1975) 2104.

25. J.-M. Drouffe, Nucl. Phys. $\underline{B170}$ [FS1](1980) 211.
    E. Brezin and J.-M. Drouffe, Nucl. Phys. $\underline{B200}$ [FS4](1982)
    93.

26. V. Alessandrini, V. Hakim and A. Krzywicki, Nucl. Phys.
    $\underline{B215}$ [FS7] (1983) 109.
    V. Alessandrini, CERN preprint TH.3336 (1982).
    J. Greensite and B. Lautrup, Phys. Lett. $\underline{104B}$ (1981) 41.
    H. Flyvbjerg, B. Lautrup and J.B. Zuber, Phys. Lett. $\underline{110B}$
    (1982) 279.
    B. Lautrup, Niels Bohr Institute preprint NBI-HE-82-8 (1982).
    V.F. Müller and W. Rühl, Nucl. Phys. $\underline{B210}$ [FS6] (1982)
    289.

27. S. Elitzur, Phys. Rev. $\underline{D12}$ (1975) 3978.

28. K. Wilson, Lecture presented at the Abingdon Meeting on
    Lattice Gauge Theories (1981).

29. B. Berg, A. Billoire and C. Rebbi, Ann. Phys. $\underline{142}$ (1982)
    185.
    B. Berg and A. Billoire, Phys. Lett. $\underline{113B}$ (1982) 65.
    B. Berg, CERN Preprint TH.3327 (1982).
    M. Falcioni, E. Marinari, M.L. Paciello, G. Parisi,
    F. Rapuano, B. Taglienti and Zhang Yi-cheng, Phys. Lett.
    $\underline{110B}$ (1982) 295.
    K. Ishikawa, M. Teper and G. Schierholz, Phys. Lett. $\underline{110B}$
    (1982) 399,ibid $\underline{116B}$ (1982) 429, and DESY preprint 83-004
    (1983).

30. M. Creutz, Phys. Rev. Lett. $\underline{45}$ (1980) 313.

31. G. Bhanot and C. Rebbi, Nucl. Phys. $\underline{B180}$ [FS2] (1981) 469.

32. C.B. Lang and C. Rebbi, Phys. Lett. $\underline{115B}$ (1982) 137.

33. For recent reviews, see:
    Q. Shafi, in International Conference on High Energy Physics,

Lisbon (1981), ed. J. Dias de Deus.

L. Van Hove, in Quark Matter Formation and Heavy Ion Collisions, Bielefeld (1982), eds. M. Jacob and H. Satz.

34. N. Cabibbo and G. Parisi, Phys. Lett. $\underline{59B}$ (1975) 67.

J.C. Collins and M.J. Perry, Phys. Rev. Lett. $\underline{34}$ (1975) 1353.

P.D. Morley and M.B. Kislinger, Physics Reports $\underline{51C}$ (1979) 63.

M.B. Kislinger and P.D. Morley, Phys. Rev. $\underline{D13}$ (1976) 2771.

B.A. Freedman and L.D. McLerran, Phys. Rev. $\underline{D16}$ (1977) 1130, 1147, 1169.

J.I. Kapusta, Nucl. Phys. $\underline{B148}$ (1979) 461.

V. Baluni, Phys. Rev. $\underline{D17}$ (1978) 2092.

E.V. Shuryak, Physics Reports $\underline{61C}$ (1980) 71.

D.J. Gross, R.D. Pisarki and L.G. Yaffe, Rev. Mod. Phys. 53 (1981) 43.

35. A.M. Polyakov, Phys. Lett. $\underline{72B}$ (1978) 477.

L. Susskind, Phys. Rev. $\underline{D20}$ (1979) 2610.

J.B. Kogut, CLNS-402 (1978).

36. L.D. McLerran and B. Svetitsky, Phys. Lett. $\underline{98B}$ (1981) 195.

J. Kuti, J. Polónyi and K. Szlachányi, Phys. Lett. $\underline{98B}$ (1981) 199.

J. Engels, F. Karsch, I. Montvay and H. Satz, Nucl. Phys. $\underline{B205}$ [FS5] (1982) 545, Phys. Lett. $\underline{101B}$ (1981) 89.

37. K. Kajantie, C. Montonen and E. Pietarinen, Z. Phys. $\underline{C9}$ (1981) 253

I. Montvay and E. Pietarinen, Phys. Lett. $\underline{115B}$ (1982) 151.

38. J. Kogut, M. Stone, H.W. Wyld, W.R. Gibbs, J. Shigemitsu, S.H. Shenker and D.K. Sinclair, Phys. Rev. Lett $\underline{50}$ (1983) 393.

T. Celik, J. Engels and H. Satz, Bielefeld preprint, BI-TP83/04 (1983).

39. N.S. Manton, Phys. Lett. $\underline{96B}$ (1980) 328.

40. J.-M. Drouffe, Phys. Rev. $\underline{D18}$ (1978) 1174.

P. Menotti and E. Onofri, CERN preprint TH.3026 (1981).

41. J.B. Kogut and L. Susskind, Phys. Rev. $\underline{D11}$ (1975) 395.

42. For an example, see the third reference in 17.

43. C.B. Lang, C. Rebbi, P. Salomonson and B.-S. Skagerstam,

Phys. Lett. 101B (1981) 173, Phys. Rev. D26 (1982) 2028.

44. H.S. Sharatchandra and P.H. Weisz, DESY preprint 81-083 (1981).

45. R.V. Gavai, F. Karsch and H. Satz, Bielefeld preprint BI-TP 82/26 (1982).

46. G. Bhanot and R. Dashen, Phys. Lett. 113B (1982) 299.
    B. Grossman and S. Samuel, New York preprint RU 82/B/25 (1982)
    A. Gonzales Arroyo, C.P. Korthals Altes, J. Peiro and M. Perrottet, Phys. Lett. 116B (1982) 414.
    Yu.M. Makeenko, M.I. Polikarpov and A.V. Zhelonkin, Moscow preprint ITEP-21 (1983).

47. The string tension result is taken from:
    R.W.B. Ardill, M. Creutz and K.J.M. Moriarty, Brookhaven preprint BNL-32377 (1982).
    The mass gap is obtained by B. Berg and A. Billoire as given in:
    B. Berg, Lecture Notes presented at the Johns Hopkins Workshop, Florence (1982), CERN preprint TH.3327 (1982), and it is consistent with the results obtained by K. Ishikawa et al. in[29].The critical temperature is quoted from the second reference in [38] .

48. H. Hamber and G. Parisi, Phys. Rev. Lett. 47 (1981) 1792.
    E. Marinari, G. Parisi and C. Rebbi, Phys. Rev. Lett. 47 (1981) 1795.
    D.H. Weingarten, Phys. Lett. B109 (1982) 57.
    H. Hamber, E. Marinari, G. Parisi and C. Rebbi, Phys. Lett. B108 (1982) 314.
    A. Hasenfratz, P. Hasenfratz, Z. Kunszt and C.B. Lang, Phys. Lett. B110 (1982) 282 and Phys. Lett. B117 (1982) 81.
    F. Fucito, G. Martinelli, C. Omero, G. Parisi, R. Petronzio and F. Rapuano, Nucl. Phys. B210 (1982) 407.
    G. Martinelli, C. Omero, G. Parisi and R. Petronzio, Phys. Lett. 117B (1982) 434.
    G. Martinelli, G. Parisi, R. Petronzio and F. Rapuano, Phys. Lett. 116B (1982).
    D.H. Weingarten, Indiana University preprint IUHET-82 (1982).

C. Bernard, T. Draper and K. Olynyk, Phys. Rev. Lett. $\underline{49}$ (1982) 1076.

H. Hamber and G. Parisi, Brookhaven preprint BNL 31322 (1982).

49. P. Hasenfratz and I. Montvay, Phys. Rev. Lett. $\underline{50}$ (1983) 309.

C. Bernard, T. Draper and K. Olynyk, UCLA/82/TEP/22 (1982).

R. Gupta and A. Patel, CALTECH preprint CALT-68-966 (1982).

50. J. Kogut, M. Stone, H.W. Wyld, J. Shigemitsu, S.H. Shenker and D.K. Sinclair, Phys. Rev. Lett. $\underline{48}$ (1982) 1140.

See also the first reference in[38]and:

J. Engels, F. Karsch and H. Satz, Phys. Lett. $\underline{113B}$ (1982) 398.

J. Engels and F. Karsch, "The deconfinement transition for quenched SU(2) lattice QCD with Wilson fermions", CERN preprint TH.3481 (1982).

51. P. Hasenfratz and F. Karsch, CERN preprint TH.3530 (1983).

52. S.L. Adler, Brandeis University Summer Institute in Theoretical Physics, eds. S. Deser, M. Grisaru and H. Pendleton, MIT Press, Cambridge, MA, and London (1970).

53. S.L. Adler, Phys. Rev. $\underline{177}$ (1969) 2426.

J.S. Bell and R. Jackiw, Nuovo Cimento $\underline{60A}$ (1969) 47.

54. L.H. Karsten and J. Smit, Nucl. Phys. $\underline{B183}$ (1981) 103.

55. K.G. Wilson, in New Phenomena in Subnuclear Physics (Erice 1975), ed. A. Zichichi, Plenum Press, New York (1977).

56. A. Hasenfratz and P. Hasenfratz, Phys. Lett. $\underline{104B}$ (1981) 489.

C.B. Lang and H. Nicolai, Nucl. Phys. $\underline{B200}$ [FS4] (1982) 135.

I.O. Stamatescu. Phys. Rev. $\underline{D25}$ (1982) 1130.

57. On the lattice this problem is discussed in detail by B.E. Baaquie, Phys. Rev. $\underline{D16}$ (1977) 2612.

58. V. Baluni, Phys. Rev. $\underline{D17}$ (1978) 2092.

Acta Physica Austriaca, Suppl. XXV, 357–397 (1983)
© by Springer-Verlag 1983

RECENT DEVELOPMENTS IN 1/N EXPANSION FOR QCD
AND MATRIX MODELS[+]

by

A.A. SLAVNOV
Steklov Inst. Moscow

ABSTRACT

Development of a reliable calculational scheme for
low energy quantum chromodynamics (QCD) is at the moment one
of the most urgent problems for quantum field theory. Computer calculations and lattice space-time formulation of
the theory present a possible way to attack the problem
where the most active researches were carried out last years.
Another possibility is given by analytical approximation
methods dealing with the theory in the continuous space-
time.

In this lecture I shall describe some new developments
in the second direction, namely the perturbative expansion
with respect to the number of colours or so called N^{-1}
expansion.

The lecture is organized as follows. The first part contains a very elementary introduction to the method. Then
a brief review of different approaches to the problem is
given. Finally I describe in some more details the method
of singlet collective variables developed in my recent papers.

[+]Lectures given at the XXII. Internationale Universitätswochen
für Kernphysik,Schladming,Austria,February 23 - March 5,1983.

I. MOTIVATIONS AND GENERAL OUTLINE

1. Solution of practically any physical problem requires
some perturbation scheme. The scheme mostly used in quantum
field theory is the expansion with respect to the coupling
constant g. This method is known to work very well in QED
and weak interaction models. However, the situation is
different in the case of QCD. In this case the coupling
constant g is not an appropriate expansion parameter in the
low energy region.

Renormalization group tells us that the effective coupling
constant in QCD depends on the momentum in the following
way

$$\tilde{g}^2 = \frac{g^2}{1+c \ \ln\frac{p^2}{\mu^2}} \qquad , \qquad c > 0 \qquad\qquad (1.1)$$

where μ^2 is a normalization point. For large Euclidean
p^2, $\tilde{g}^2 \to 0$ and perturbation expansion is reliable. However
for small p^2, $\tilde{g}^2 \to \infty$ and perturbation theory makes no sense.

In fact in QCD the coupling constant g is not a free
parameter. This situation is common for the theories with
so-called dimensional transmutation [1,2], which at the
classical level have no natural mass scale. For example in
the two dimensional chiral model the physical mass of the
chiral field is expressed in terms of the coupling by

$$m = \mu\exp\{-\frac{4\pi}{g}\} \qquad\qquad (1.2)$$

where μ is a normalization point. It means that fixing
the physical value of the mass and the normalization point
we define uniquely the value of coupling constant g. In this
situation one hardly can expect that the expansion near the
point g = 0 does make sense.

Indeed if one tries to calculate the amplitude of some

physical process in QCD using the expansion with respect
to g, one encounters severe infrared divergencies. The infrared divergencies arise also in QED but there they may be
eliminated if one takes into account together with the given
process physically undistinguishable processes with the
soft photon emission. In QCD such cancellation is absent.
Contrary to QED the Bloch-Nordsiek type cross-section does
not exist in this case [3,4].

All these facts are quite expectable from the experimental point of view. It is well known that perturbation
theory works if the spectrum of the complete Hamiltonian
coincides with the spectrum of the free one. It is the case
in electrodynamics but it is far from true in QCD. The spectrum
of the free Hamiltonian contains the massless gluons and free
quarks which are not observed experimentally. Naturally the
expansion near this unphysical ground state is senseless.

To develop a sensible perturbation theory one should
look for the expansion which in the lowest order reproduces
a reasonable particle spectrum. The possible candidate for
such an expansion was pointed out originally by G.'t Hooft
[5,6], who proposed to develop the perturbation theory with
respect to the number of colours (N) near the point $N = \infty$.

The following arguments may be given in favour of N^{-1}
expansion.
a) N^{-1} expansion was proved to work successfully in some
two-and three-dimensional models. In two-and three-dimensional chiral models N^{-1} expansion reproduces correctly the
particle spectrum and allows to analyse the problem of phase
transitions [7,8,9]. In two dimensional QCD N^{-1} expansion
gives extremely promising results: the quarks are confined
and the physical spectrum consists of a discrete set of
singlet states [5,6,10].
b) in the framework of N^{-1} expansion the meson phenomenology
may be successfully explained.
c) Last but not least: N^{-1} is the only known dimensionless

parameter for QCD.

To conclude we note that N^{-1} expansion is of interest not only for QCD. In particular it seems to be useful for so-called matrix models which serve to investigate different aspects of QCD and also have independent applications to field theory and statistical mechanics.

2. To illustrate the general idea we shall consider the Yang-Mills theory based on the group SU(N)

$$L = \frac{1}{8} \, \text{Tr} \, F_{\mu\nu} F_{\mu\nu} + i\bar{\psi} \, \gamma_\mu \nabla_\mu \psi \tag{1.3}$$

where

$$F_{\mu\nu} = \partial_\nu A_\mu - \partial_\mu A_\nu + \frac{g}{\sqrt{N}} [A_\mu, A_\nu] \tag{1.4}$$

is the usual curvature tensor,

$$\nabla_\mu \psi^q = \partial_\mu \psi^q - \frac{g}{\sqrt{N}} \, A^q_{\mu b} \psi^b \tag{1.5}$$

is a covariant derivative. $A_\mu \equiv A^i_{\mu j}$ are NxN antihermitian matrices with zero trace. $\psi \equiv \psi_i$ belong to the fundamental representation of the SU(N).

The standard quantization procedure gives the following structure for the propagators

$$\overline{\psi_i \, \psi^j} = \delta_i{}^j S(x-y) \, , \tag{1.6}$$

$$\overline{A^i_{\mu j} A^k_{\nu l}} = (\delta_l{}^i \delta^k{}_j - N^{-1} \delta_j{}^i \delta_l{}^k) D_{\mu\nu}(x-y) \quad , \tag{1.7}$$

where the explicit form of $D_\mu(x)$ depends on the particular gauge chosen (in covariant gauges the ghost lines will also arise). The second term in brackets provides the condition $\text{Tr} A_\mu = 0$.

This term is not essential in the limit $N \to \infty$, and

to simplify the formulae we shall omit it and consider the propagator, corresponding to U(N) group

$$A^i_{\mu j}A^k_{\nu l} = \delta_1{}^i\delta_j{}^k D_{\mu\nu}(x-y) \ . \tag{1.8}$$

To analyse the N-dependence of amplitudes it is convenient to introduce the diagram technique indicating explicitely the flow of colour indices

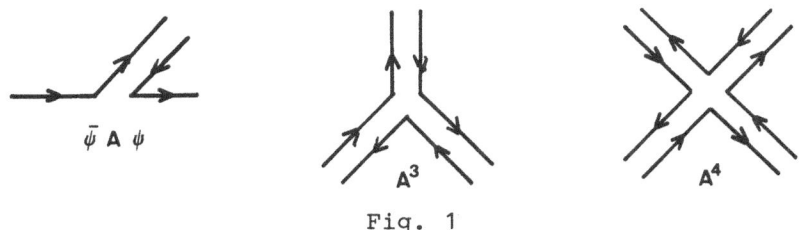

The interaction vertices will look as follows

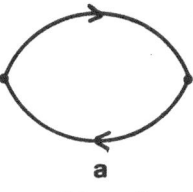

Fig. 1

Let us consider the Feynman diagrams corresponding to the quark-antiquark loop with radiation corrections. In the lowest order we have the diagram (a)

Fig. 2

Some higher order diagrams are shown at Fig. 3.

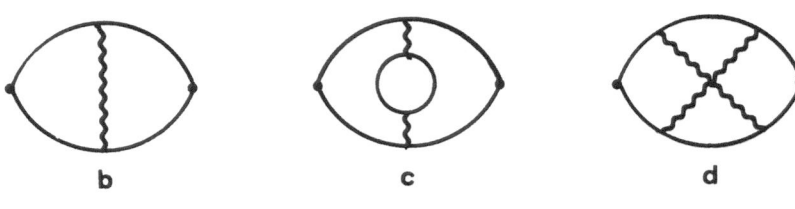

Fig. 3

In double line notations they look as follows

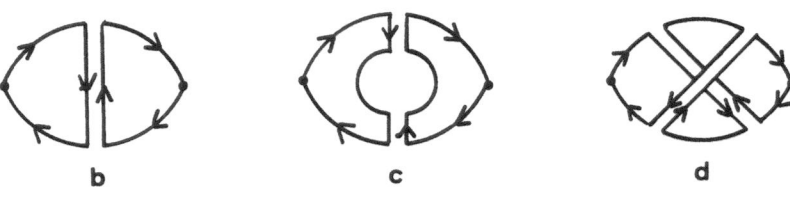

Fig. 4

Every closed loop formed by single lines corresponds to the summation over colour indices and therefore produces the factor N.

Having in mind that according to our choice of charge normalization the trilinear vertex contains the factor $N^{-1/2}$ and fourlinear vertex the factor N^{-1} we see that the diagram (b) has the same order in N as the diagram (a), the diagram (c) is supressed by the factor N^{-1}, and the diagram (d) is supressed by the factor N^{-2}. In this way it is obvious that all the diagrams may be classified according to the powers of N they produce.

Investigation of the diagrams (b,c,d) shows that they differ by their topological structure.

The diagram (b) can be drawn without selfintersection on the plane bounded by the external quark lines. Such diagrams are called planar diagrams. The diagrams (c) and (d) cannot be drawn on a plane without selfintersections. The diagram (d) may be drawn on the plane with a handle and the diagram (c) on the plane with a hole. This consideration shows that the order in N of a given diagram is related to it's topological structure, namely it is defined by the Euler characteristic of the manifold on which the diagram may be drawn without selfintersection. In the general case this statement may be proved as follows. Let us rescale the fields $A_\mu \to N^{1/2}A_\mu$, $\psi \to N^{1/2}\psi$. Then the Lagrangian (1.3) will contain N only as a common factor and therefore every propagator will contribute the factor N^{-1} and every vertex factor N. To any diagram we may put into correspondence a polyhedron with the number of vertices (V) equal to the number of vertices of the diagram, number of edges (L) equal to the number of propagators, and F borders. Every border corresponds to the closed cycle and produces the factor N. The overall factor N corresponding to the given diagram is equal therefore to N^{V-L+F}. Here the exponent is nothing but the Euler characteristic of an oriented surface. Any two-dimensional oriented surface is topologically equivalent to a sphere with some number of holes and handles. The Euler characteristic for such a sphere is equal to 2-2H-B, where H is a number of handles and B is a number of holes. The highest order in N have diagrams without holes and handles, i.e. planar diagrams. So the problem of calculation of the leading term in the limit N→∞ is reduced to the summation of planar diagrams.

3. Now we shall come back to the physical motivations of N^{-1} expansion. We shall show that in the framework of N^{-1} expansion meson phenomenology acquires a simple qualitative explanation if one accepts the hypothesis that in the limit N→∞ the colour states are confined. The facts stated below were discovered by several authors in different models and summarized by E. Witten [11] whose paper we shall follow here (see also [12]).

Let us show that assuming colour confinement N^{-1}

expansion predicts the existence of infinite set of stable
meson states. Consider the expectation value

$$G(p) = \int e^{-ipx} \langle J(x)J(-x) \rangle \qquad (1.9)$$

where $J(x)$ is a singlet current, e.g. $J(x) = \bar{\psi}^i(x)\psi_i(x)$.
In the limit $N \to \infty$ this matrix element is described by the
sum of planar diagrams of the type

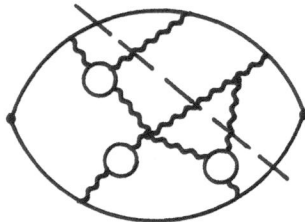

Fig. 5

Every possible cut of this diagram corresponds to some
intermediate state and all possible cuts of such diagrams
describe all possible states. One can easily see that only
one-meson states are possible in the limit $N \to \infty$. Indeed the
states with more than one quark-antiquark pair correspond
to the diagrams with holes and therefore are supressed.
The meson-gluonium states are also absent because to form
such states we need in the intermediate state the product
of several singlet factors, for example

$$\bar{\psi}_i\; A_j^i\; \psi^j\; A_k^l\; A_n^k\; A_l^n \;. \qquad (1.10)$$

One can verify that for planar diagrams such combination
is impossible (whereas it may be present in nonplanar diagrams). Therefore only irreducible singlet states containing
one quark-antiquark pair e.g. single meson states are
possible. The matrix elements (1.9) may be written as a
sum over these intermediate states

$$\langle J(p)J(-p) \rangle = \sum_n \frac{a_n^2}{p^2 - m_n^2} \;. \qquad (1.11)$$

Here m_n are the masses of mesons. The mesons in the limit $N \to \infty$ are stable because the singularities of the left hand side of eq. (1.11) may be situated only on the real axis. The number of mesons is infinite because for large p^2 the left hand side of eq. (1.11) has asymptotics $<J(p)J(-p)> \sim \ln p^2/\mu^2$ as follows from the asymptotic freedom. A finite number of poles cannot produce such a behaviour.

Analogous arguments show that any scattering amplitude of singlet states is dominated by the tree diagrams where the real mesons are exchanged. Such amplitudes correspond to the matrix elements of the type $<J_1,J_2..J_n>$, where J_i are singlet currents. By the same arguments as before an arbitrary cut of the corresponding diagrams contains only singlet meson states. The only possible singularities of the matrix elements under consideration are the sums of simple poles. That means that they are described by the tree diagrams with the lines corresponding to real mesons propagators. In other words Regge phenomenology is predicted.

A natural explanation is given in the frame-work of N^{-1} expansion to the fact of suppression of exotic states, Zweig rule and some other phenomenological facts. The absence of exotic states containing more than one quark-antiquark pair follows from the fact that the corresponding diagrams contain additional quark loops (holes) and therefore are supressed at least by the factor N^{-1}.

The Zweig rule is explained in the same way: The processes with virtual quark annihilation are described by the diagrams with extra quark loops and are supressed.

Although the arguments given above were not rigorous they seem rather convincing and indicate that QCD for $N \to \infty$ posesses reasonable properties from a phenomenological point of view (assuming that colour is confined).

II. HOW TO DEVELOP N^{-1} EXPANSION?

In the previous section it was shown that the limit $N \to \infty$ for QCD is described by the sum of planar diagrams. It is obvious however that it is completely hopeless to try to sum up planar diagrams directly because even to write explicitly a planar diagram of arbitrary order is a formidable problem. How can one try to solve the problem?

First of all due to the fact that planar diagrams are topologically distinguished one can hope that a closed equation may be written for the sum of planar diagrams which may be solved exactly or approximately.

Another possibility is given by the reduction of a problem to a more simple one. For example it was shown that a sum of planar diagrams can be calculated explicitly in zero-and one-dimensional models [13]. One can try to relate in some way a four-dimensional case to these lower dimensional models.

The third approach is the closest to the usual perturbation theory: one may reformulate a theory in such a way that N will enter only explicitly as a numerical factor. It is not the case in the original Lagrangian because there N enters implicitly via the number of field components. After such reformulation N^{-1} expansion may in principle be constructed in the same way as ordinary perturbation theory: firstly one separates the terms of the highest order in N and solves this reduced theory exactly and starting from this ground state calculates the higher corrections.

Let us discuss all these possibilities.

1. Consider the simplest matrix model described by the Lagrangian

$$L = \frac{1}{2}\text{Tr}(\partial_\mu \phi)^2 + \frac{g}{\sqrt{N}} \text{Tr}(\phi)^3 \qquad (2.1)$$

where $\phi = \phi^i_j$ is NxN matrix field. The discussion of the preceeding section applies equally to this model and the leading order in N is given by the sum of planar diagrams.

Starting from this Langrangian we can write equations for the Green's function

$$\Box_x <\phi^b_a(x) \prod_i \phi^{b_i}_{a_i}(y_i)> =$$

$$= \frac{g}{\sqrt{N}} <\phi^2(x)^b_a \prod_i \phi^{b_i}_{a_i}(y_i)> + \sum_j \delta(x-y_j)\delta^{b_j}_a \delta^{a_j}_b < \prod_{i \neq j} \phi^{b_i}_{a_i}(y_i)> . \quad (2.2)$$

This equation is valid in every order in g and in particular it is valid for planar diagrams

$$\Box_x G^{pl}(x,y_1,\ldots y_n) = \frac{g}{\sqrt{N}} G^{pl}(x,x,y_1,\ldots y_n) +$$

$$+ \sum_j \delta(x-y_j) G^{pl}(y_1,\ldots,y_{j-1},y_{j+1},\ldots y_n). \quad (2.3)$$

However this equation in general does not allow to separate the contribution of planar diagrams because it contains a planar Green's function with coinciding arguments. In general such Green's function is not described by a planar diagram. For example the diagram (a)

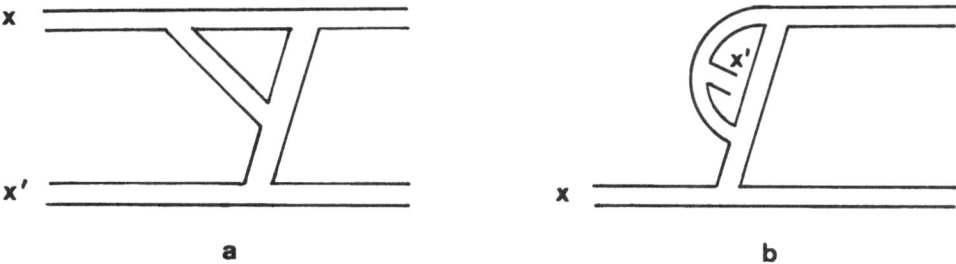

Fig. 6

preserves the planarity when x = x', but the diagram (b) loses this property. Therefore the eq. (2.3) in the general case is not a closed equation for planar diagrams.

A sufficient condition for a diagram to preserve the planarity when some arguments coincide is the possibility to place all external lines on a circle which does not intersect the internal lines. An example is shown in Fig. 7.

Fig. 7

The arguments of corresponding Green's functions can be ordered uniquely. Therefore one may expect that for such objects closed equations for planar diagrams do exist. The most important example is given by the Wilson loop.

Migdal and Makeenko succeeded to rewrite the Schwinger-Dyson equation for Yang-Mills theory as a variational derivative equation for the expectation value of Wilson loop products [14 - 17] (see also [18]):

$$\frac{1}{g^2 N} \partial_\mu^x \frac{\delta W(c)}{\delta \sigma_{\mu\nu}(x)} = \int_C dy_\nu \delta(x-y) [W(C_{xy}, C_{yx}) - N^{-2} W(c)] \qquad (2.4)$$

where

$$W(c_1 \ldots c_n) = \langle \phi(c_1) \ldots \phi(c_n) \rangle , \qquad (2.5)$$

$$\phi(c_n) = \frac{1}{N} \mathrm{Tr}\, P \exp \oint_{C_n} A_\mu(x) dx_\mu \qquad (2.6)$$

where P means the ordering along the contour C.

δW means the variation of the functional $W(c)$ when the contour C is changed by the infinitesimal addition $\delta C(x)$ at a point x

$\delta C(x)$ 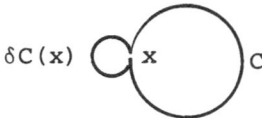 C

$\delta\sigma_{\mu\nu}$ is the area inside $\delta C(x)$

$$\delta\sigma_{\mu\nu}(x) = \oint_{\delta C(x)} y_\mu dy_\nu \quad . \tag{2.7}$$

The eq.(2.4) should be solved simultaneously with the Bianchi identity

$$\varepsilon_{\mu\nu\lambda\rho}\partial_\nu^x \frac{\delta W_n(c)}{\delta\sigma_{\mu\rho}(x)} = 0 \quad . \tag{2.8}$$

This equation reminds the eq.(2.3) derived earlier but contrary to the later it allows unique separation of planar diagrams. To see that we note that in the limit $N\to\infty$ the functional $W(c_1..c_n)$ is factorized

$$W(c_1,\ldots,c_n) = \prod_{i=1}^{n} W(C_i) + O(N^{-2}) \quad . \tag{2.9}$$

This property follows from the fact that disconnected graphs always contain more closed cycles than connected graphs of the same order and therefore produce the larger power of N. Having that in mind we can replace the first term in the r.h.s. of eq. (2.4) by the product and drop the second term. Finally we have the closed equation for $W(c)$ in the limit $N\to\infty$

$$\partial_\mu^x \; \frac{\delta W(C)}{\delta \sigma_{\mu\nu}(x)} = g^2 N \int_C dy_\nu \, \delta(x-y) W(C_{xy}) W(C_{yx}) \quad . \tag{2.10}$$

In principle the problem is solved. Unfortunately (2.10) is a complicated variational derivative equation in the contour space and no effective methods to solve such equations are known. It has been shown that it reproduces the ordinary perturbation theory in g, and that it is consistent with the area law for the functional W(c). One can show that for large smooth loops the asymptotic behaviour

$$W(c) \to e^{-\sigma S_{min}} \tag{2.11}$$

where S_{min} is the area of the minimal surface is compatible with the eq. (2.10). However other asymptotics are not excluded. The detailed investigation of the eq. (2.10) is still an open problem.

Analogous equations for scalar models were derived in [19].

2. Now we shall investigate the possibility to reduce a four-dimensional matrix model (e.g. QCD) to a zero-dimensional one. This possibility was discovered recently [20,21,22,23,24] and is now widely discussed. It was shown that to calculate the invariant Green's functions in a d-dimensional matrix model it is sufficient to calculate them in some effective zero-dimensional theory which is obtained by fixing ("quenching") the momenta, and after that to integrate over quenched momenta.

We illustrate this prescription by the simplest matrix model

$$S = \int d^d x \; \mathrm{Tr}[\tfrac{1}{2}(\partial_\mu \phi)^2 + \tfrac{1}{2} m^2 \phi^2 + \frac{g}{\sqrt{N}} \phi^3] \tag{2.12}$$

where $\phi_i{}^j$ is again NxN matrix.

The crucial observation is that for planar diagrams
there exists a one to one correspondence between momentum
flow and colour flow which allows to express the integrand
of any planar diagram in terms of momenta p_μ^i associated
with every colour index. To illustrate this statement let
us consider some vacuum planar diagram, for example Fig. 8a

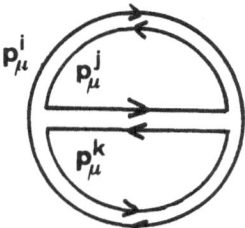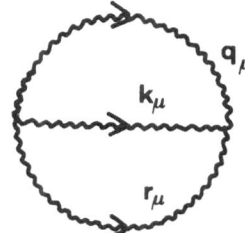

Fig. 8a

Here we associated with every colour line a momentum p_μ^i. Note
that the number of independent colour circles is equal to
the number of independent momenta, and therefore we may
put

$$p_\mu^i - p_\mu^j = q_\mu \, , \quad p_\mu^j - p_\mu^k = k_\mu \, , \quad p_\mu^k - p_\mu^i = r_\mu \quad . \tag{2.13}$$

This assignment is compatible with the momentum conservation
in every vertex because in every vertex the flow of colour
current is continuous. Indeed

$$q_\mu + k_\mu + r_\mu = p_\mu^i - p_\mu^j + p_\mu^j - p_\mu^k + p_\mu^k - p_\mu^i = 0 \quad . \tag{2.14}$$

Note that such assignment is possible only for planar diagrams. Consider some nonplanar diagram, e.g. Fig. 8b

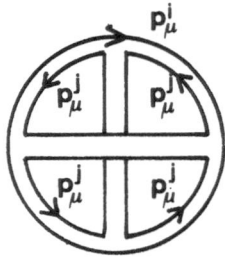
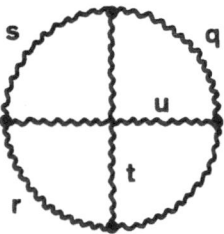

Fig. 8b

In this case there are only two independent colour circles
labelled by indices (i) and (j), and the number of independent momenta is 6.

For planar diagrams we can according to the eq. (2.13)
associate with every double line with indices (i,j) the
propagator

$$<\phi^i_j \; \phi^k_l> = \frac{\delta^i_1 \; \delta^k_j}{(p^\mu_i - p^\mu_j)^2 + m^2} \qquad . \qquad (2.15)$$

If we sum over all i and integrate over all p^i_μ we have N^2
identical contributions which reproduce up to a constant
factor the planar diagram under consideration. The numerical
factor arises due to the fact that the number of integrals
over p^i_μ is larger than the number of independent momenta
in the given diagram. To eliminate this factor we should
normalize all these extra integrals to 1.
This considerations may be summarized as follows. Let us
introduce a momentum cut off Λ, and consider the following
effective action

$$S(\phi, p_\mu) = (\frac{2\pi}{\Lambda})^d [\frac{1}{2} m^2 \phi^i_j \phi^j_i - \frac{1}{2}(p^i_\mu - p^j_\mu)^2 \phi^i_j \phi^j_i + \frac{g}{\sqrt{N}} \, \mathrm{Tr} \phi^3] \qquad (2.16)$$

where ϕ^i_j are constant matrices, d is a dimension of space-
time. This action obviously generates the propagators for

the fields $\phi^i_{\ j}$ of the form (2.15). Using this action one can calculate the integrand of any connected planar diagram. For the vacuum energy $F(p_\mu)$ we shall have

$$F(p_\mu) = \ln \int \exp\{-S(\phi,p_\mu)d\phi\} \ . \qquad (2.17)$$

To obtain the real vacuum energy per unit volume one must integrate $F(p_\mu)$ over p_μ with the following measure

$$\int dp_\mu \ = \ \int_{-\Lambda/2}^{+\Lambda/2} \prod_{i=1}^{N} \frac{d^d p_\mu^i}{\Lambda^d} (\frac{\Lambda}{2\pi})^d \ . \qquad (2.18)$$

Let us check this prescription for an arbitrary n-loop planar diagram. This diagram contains n+1 colour circles producing the factor N^{n+1}. It has 2(n-1) vertices producing the factor N^{1-n}.

So the overall power of N is N^2 as it should be. According to our choice of the action each propagator is proportional to $(\frac{\Lambda}{2\pi})^d$ and each vertex is proportional to $(\frac{2\pi}{\Lambda})^d$. That gives for a n-loop diagram a factor $(\frac{\Lambda}{2\pi})^{d(n-1)}$. The integration measure contributes the factor $\Lambda^{-Nd}(\frac{\Lambda}{2})^d$. Finally a n-loop diagram depends on n independent momenta and therefore (N-n-1) extra momentum integrations will produce the factor $\Lambda^{(N-n)d}$. Summarizing we see that in this way we recover the ordinary expression for the vacuum energy in the d-dimensional model with the ultraviolet cut off Λ. If the corresponding diagram is ultraviolet finite we may finally put $\Lambda \to \infty$. For the theories with logarithmic ultraviolet divergencies the same prescription holds if one introduces simultaneously necessary counterterms. Some more care is needed in the theory with quadratic and more severe divergencies. In this case to avoid non Lorentz invariant counterterms one must introduce firstly some covariant regularization procedure, e.g. higher derivatives method, then perform all the procedures described above, and finally remove the intermediate regularization.

So we succeeded to separate the problem of calculation of the vacuum energy into two parts. First one calculates the vaccum energy in the zero-dimensional theory described by the effective action (2.16) where p_μ^i are considered as fixed ("quenched") variables and then integrate over p_μ^i with the appropriate measure.

This prescription may be easily extended to calculate any invariant Green's function. Any such function may be calculated by adding to the action an invariant term pro-protional to the singlet source J, calculating the vacuum diagrams according to the prescription given above, and finally differentiating the answer with respect to J at J=0. In this way we have for example for the two-point function

$$G(x) = <\text{Tr}\phi(x)\phi(0)> \ , \tag{2.19}$$

$$G(x,p_\mu) = \frac{1}{Z}\int d\phi \ \exp\{-S(\phi,p_\mu) \sum_{i,j} \phi_{ij}^2 \ e^{i(p_i-p_j)x}\} \ , \tag{2.20}$$

$$Z = \int d\phi \ \exp\{-S(\phi,p_\mu)\} \ .$$

To recover the real planar Green's function we must again integrate $G(x,p_\mu)$ over the quenched variables p_μ.

The prescription described above reduces the d-dimensional quantum field theory problem to the zero-dimensional one. Unfortunately this zero-dimensional problem is still too complicated to be solved exactly. The method of summation of zero-dimensional planar diagrams developed in the paper [13] uses havily the invariance of the action under the group U(N) (or SU(N), SO(N)) and is not applicable to the effective action (2.16) which does not posess this invariance due to the presence of the factor $(p_\mu^i - p_\mu^j)^2$.

The price we paid for the reduction of the problem to the zero-dimensional one appears to be rather high: the theory lost it's invariance properties. Nevertheless the quenched momentum prescription is a considerable step in understanding of the structure of planar theory. In particular it clarifies the relation between this theory and statistical mechanics models. The problem of summation of planar diagrams in the quenched momentum prescription is equivalent in fact to the investigation of disordered systems. The constant matrices $\phi^i_{\ j}$ are distributed according to the Gibbs distribution, and the momenta p^i_μ are stochastic ("quenched") variables, whose distribution does not depend of the fluctuations of the variables ϕ^i_j. The very important example of such a system in statistical mechanics is given by spin glass [25].

This parallel allows to apply the methods developed in the theory of disordered systems to the planar theory. It shows at the same time that in spite of apparent simplification due to reduction to the zero-dimensional problem no simple solution should be expected because the theory of disordered system is known to be extremely complicated.

One of the methods borrowed from the disordered system theory which have been used for the investigation of large N limit is the so called replica method [26]. In this method to calculate the vacuum energy according to the quenched momentum prescription

$$\frac{F}{N^2} = \int \frac{d\mu(p)}{N^2} \ln\{\int \exp\{-(\frac{2\pi}{\Lambda})^d \ [\sum_{i,j} (p^i_\mu - p^j_\mu)^2 \phi^+_{ij} \phi_{ij}$$

$$+ m^2 \mathrm{Tr}\phi^+\phi \ - \frac{g^2}{N} \mathrm{Tr}(\phi^+\phi)^2] d\phi^+ d\phi\} \tag{2.21}$$

one represents the ln which enters into the integrand (2.21) according to the eq.

$$\ln x = \lim_{n\to 0} \frac{1}{n}(x^n - 1) \quad . \tag{2.22}$$

Using the fact that the n-th power of an integral may be represented as a product of n identical integrals over fields ϕ^α_{ij}, "replicas", one can rewrite (2.21) as follows

$$F = \lim_{n \to 0} \int d\mu(p) \{ \prod_{\alpha=1}^{n} d\phi^{+\alpha} d\phi^{\alpha} \exp\{(\frac{2\pi}{\Lambda})^d [-\sum_{i,j} (p_i^\mu - p_j^\mu)^2 \phi_{ij}^{+\alpha} \phi_{ji}^{\alpha}$$

$$- m^2 \sum_{\alpha} \text{Tr}\phi^{+\alpha}\phi^{\alpha} - \frac{g^2}{N} \sum_{\alpha} \text{Tr}(\phi^{+\alpha}\phi^{\alpha})^2]\} - 1\} n^{-1} \quad . \tag{2.23}$$

This representation allows to perform explicit integration over momenta p_i^μ. In this case it is more convinient to use a smooth factorizable cut off e.g.

$$d\mu(p) = (\frac{\Lambda}{2\sqrt{\pi}})^d \prod_{i=1}^{N} \frac{d^d p_i}{(\Lambda\sqrt{\pi})^d} e^{-\frac{p_i^2}{\Lambda^2}} \quad . \tag{2.24}$$

The remaining integral over ϕ^α describes some zero-dimensional model

$$F = \lim_{n \to 0} \frac{1}{n}[Z(n) - 1] \quad ,$$

$$Z(n) = \int \prod_{\alpha} d\phi^{+\alpha} d\phi^{\alpha} \exp\{-S_{eff}(\phi)\} \tag{2.25}$$

where $S_{eff}(\phi)$ is a nonpolynomial function of ϕ whose explicit form may be found from the eq. (2.23). Due to the nonpolynomial form of S_{eff} the integral (2.25) can not be calculated exactly. To estimate it the Bogoliubov-Feynman variational principle [27] is used. According to this principle $S_{eff}(\phi)$ is approximated by the quadratic form

$$S_o(\phi) = \sum_{\alpha,\beta} \phi_{i,j}^{+\alpha} J_{ij,kl}^{\alpha\beta} \phi_{kl}^{\beta} \tag{2.26}$$

where the kernel J is determined from the maximum condition of the functional F(J)

$$F(J) = \ln Z_o - <S_{eff} - S_o>_o \quad , \tag{2.27}$$

$$Z_o = \int \exp\{-S_o(\phi)\} \prod_\alpha d\phi^{+\alpha} d\phi^\alpha \quad , \tag{2.28}$$

$$\langle A \rangle_o = \frac{1}{Z_o} \int A \exp\{-S_o(\phi)\} \prod_\alpha d\phi^{+\alpha} d\phi^\alpha \quad . \tag{2.29}$$

The functional $F(J)$ is known to estimate $\ln Z(n)$ from below

$$\ln Z(n) \geq \max F(J) \tag{2.30}$$

and therefore finding the J which maximizes $F(J)$ we get the best estimation of $\ln Z(n)$ in the considered class of functions.

The analysis performed in [26] indicates that in the framework of this method the new phase which is analogous to the spin-glass phase is predicted for sufficiently small coupling constants g, and for low dimensions d.

This phase is characterized by non-zero value of the parameter q

$$q = \sum_{i,j} \langle \phi_{ij}^+(x) \rangle \langle \phi_{ji}(x) \rangle \tag{2.31}$$

whereas the expectation value $\langle \mathrm{Tr}\phi_{ij} \rangle$ is equal to zero.

Analogous technique may be applied also to Yang-Mills theory. The main drawback of this approach is a lack of a reliable method to control the accuracy of the approximation. Due to this fact it is hard to say whether the observed properties really characterize the theory or they are artefacts caused by the approximation.

It must be clear from the preceeding discussion that the quenched momentum prescription is valid for any planar theory, in particular for Yang-Mills theory in the limit $N \to \infty$. However, no systematic method to estimate higher orders is known in this approach. The diagrams with quark loops may be included into this scheme (see [24]), but in the general case the problem is open.

At present it is not clear how far one can go in this way to the analytic solution of the problem. At any rate the quenched momentum prescription helps in the computer calculations of the large N-limit.

To conclude this discussion we mention that this prescription obtained originally by direct inspection of planar diagrams may be derived in an elegant way using the stochastic quantization method [28].

3. Now we shall sketch the main idea of the third approach to the investigation of large N-limit, the singlet collective variables method. We use for illustration the "vector" or "spin" models, for example the model with the Lagrangian

$$
L = \frac{1}{2}\partial_\mu \phi^i \partial_\mu \phi^i - \frac{g}{N}(\phi^i\phi^i)^2 - \frac{m^2}{2}\phi^i\phi^i \quad . \tag{2.32}
$$

Here ϕ^i is the N-component field.

The Euclidian Green's function generating functional may be written in the following form

$$
Z = \int \exp\{ \ (\frac{1}{2}\phi^i \Box \phi^i - \frac{1}{2}m^2\phi^i\phi^i - (\phi^i\phi^i)\sigma + \frac{g^2}{4g}N + s.t.)dx\}d\sigma d\phi \tag{2.33}
$$

where the auxiliary field $\sigma(x)$ is introduced to make the integral Gaussian in ϕ; s.t. means source terms, which we omit in the following. The interaction in the eq. (2.33) has the form of a product of singlet current $(\phi^i\phi^i)$ by the singlet field σ. Therefore performing explicitely the Gaussian integration over ϕ^i we eliminate all nonsinglet variables and get the effective action which depends only on singlet variables. Namely

$$
Z = \int \exp\{\frac{\int\sigma^2 N dx}{4g} - N \ \text{Tr} \ \ln\{-\Box + m^2 + 2\sigma \}\}d\sigma \quad . \tag{2.34}
$$

The dependence of the effective action on N now is explicit, and to calculate Z in the limit N→∞ one can use the steepest descent method. The stationary point of the exponent is defined by

$$\frac{\sigma(x)}{2g} - 2(-\square+m^2+2\sigma)^{-1}_{xx} = 0 \quad . \tag{2.35}$$

Being interested only in translationally invariant solutions we may put σ_{st}=const., and rewrite the eq. (2.35) in the form

$$\frac{\sigma_{st}}{2g} - \frac{2}{(2\pi)^d} \int\frac{d^dp}{(p^2+m^2+2\sigma_{st})} = 0 \quad . \tag{2.36}$$

For d>1 this equation is singular and subtractions corresponding to the wave function and charge renormalization should be done.

The stationary solution being found the N^{-1} expansion is constructed in a standard way. One expands the exponent in eq. (2.34) near the point $\sigma = \sigma_{st}$, and rescaling the fields $\sigma \to N^{-1/2}\sigma$ leads to the following effective action

$$Z = \int\exp\{\int\sigma(x)[\frac{1}{4g}\delta(x-y)+K_2(x,y)]\sigma(y)\,dxdy - \frac{1}{\sqrt{N}}\int K_3(x,y,z)$$

$$\times \sigma(x)\sigma(y)\sigma(z)\,dxdydz + \frac{1}{N}\int K_4(x,y,z,u)\sigma(x)\sigma(y)\sigma(z)\sigma(u)\,dxdydzdu$$

$$+ \dots\}d\sigma \quad . \tag{2.37}$$

Here $K_2(x,y)$, $K_3(x,y,z)$,... denote the kernels generated by the expansion of the second term in the exponent (2.34). The quadratic term defines the propagator of σ-fields, and the higher order terms define the interaction vertices. (In fact one must be somewhat more careful about the boundary conditions when integrating over ϕ, but we ignore here these complications).

This method allows to construct in a simple way the N^{-1} expansion for vector models. It can not be used however directly in the matrix models because contrary to the vector models the interaction is not a product of the singlet currents in this case. In the following section we shall show to handle this problem in matrix models.

III. SINGLET COLLECTIVE VARIABLES METHOD

1. Let us consider the models described by the Lagrangian

$$L = K_1(\phi) + K_2(\psi) + V(\phi) + \tilde{g}\bar{\psi}^a \Gamma \phi^b_a \psi_b \ . \tag{3.1}$$

Here $\phi^q_b(x)$ are NxN matrix fields which may have also additional Lorentz or some other indices which shall not be indicated explicitely. $\bar{\psi}^a$, ψ^a are N-component vectors which also may have additional Lorentz or other indices. The Lagrangian (3.1) is invariant with respect to some "colour" group: SU(N), U(N) or SO(N). Different groups correspond to the different choice of matrices ϕ^i_j, in particular in SU(N) case tr ϕ^i_j = 0. The quadratic forms $K_1(\phi)$ and $K_2(\psi)$ define the propagators of the fields ϕ,ψ. We assume that these forms are nondegenerate and the propagators have the following structure

$$\overline{\bar{\psi}^a(x)\psi_b}(y) = \delta^a_b S(x-y) \ , \tag{3.2}$$

$$\overline{\phi^a_b(x)\psi^c_d}(y) = \delta^a_d \delta^c_b D(x-y) \ . \tag{3.3}$$

For simplicity we wrote the propagators for unconstrained matrices ϕ^a_b. In the case of SU(N) group the propagators differ by terms $\sim N^{-1}$ which are nonessential in the limit $N\to\infty$, and may be easily taken into account in the higher orders.

The interaction U(ϕ)is a finite or infinite series

$$V(\phi) = \tilde{g}_3 \ \text{Tr} \ \phi^3 + \tilde{g}_4 \ \text{Tr} \ \phi^4 + \dots \ . \tag{3.4}$$

It is clear that QCD (after the gauge fixing) belongs to this class. In that case the ψ^a are quark fields and ϕ^a_b are gluon-fields. If the constants $\tilde{g}$, $\tilde{g}_3$, $\tilde{g}_4$ ... are normalized as follows

$$\tilde{g} = \frac{g}{\sqrt{N}} , \qquad \tilde{g}_3 = \frac{g_3}{\sqrt{N}} , \qquad \tilde{g}_4 = \frac{g_4}{\sqrt{N}}, \ldots \qquad (3.5)$$

the leading order in N corresponds to the sum of all planar diagrams.

Our strategy will be analogous to the strategy sketched at the end of the preceeding section. We shall show that the Green's function generating functional corresponding to the Lagrangian (3.1) may be rewritten in the following equivalent form [29,30].

$$Z(J) = \int \exp\{i \int L(\psi,\phi) dx + \ldots\} d\phi d\psi = \int \exp\{i\, S_{eff}(\zeta_i)\} d\zeta_i \qquad (3.6)$$

where the fields ζ_j are singlets with respect to the colour group. (The constant normalization factor $Z(O)$ will be systematically omitted).

We start with the relatively simple case $V(\phi) = O$. In this case the interaction in (3.1) may be easily rewritten in terms of the interaction of singlet currents. To illustrate the idea consider the simplest diagram shown at Fig. 9a.

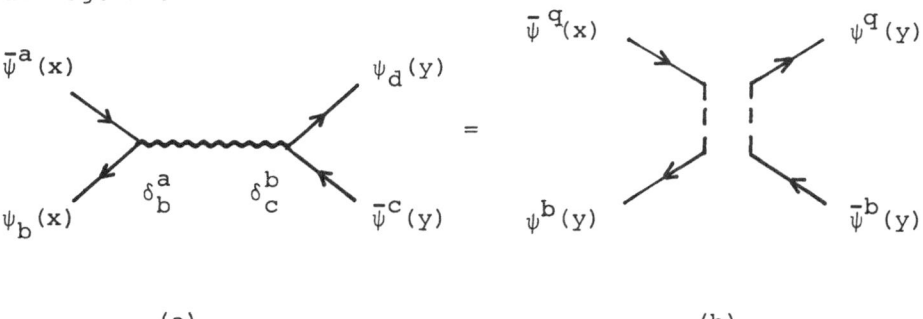

(a)   (b)

Fig. 9

Due to the structure of the gluon propagator this diagram may be presented in the equivalent form in Fig. 9b where the interaction vertices are formed by singlet bilocal currents. More precisely due to the fact that in the case $V(\phi) \equiv 0$ the integral over ϕ is Gaussian we may perform it explicitly with the result

$$Z = \int \exp\{i \int K_2(\psi) dx - \frac{ig^2}{2N} \int (\bar{\psi}^a(x) \psi_b(x)) D(x-y)$$

$$\times (\bar{\psi}^b(y) \psi_a(x)) dxdy + \ldots\} d\bar{\psi} d\psi \quad . \tag{3.7}$$

The effective interaction in eq. (3.7) can be written as the interaction of singlet bilocal currents mediated by the exchange of a singlet bilocal field $\xi(x,y)$:

$$Z = \int \exp\{i \int K_2(\psi) dx + i \int [\bar{\psi}^a(x) \psi_a(y) \xi(y,x)$$

$$+ \frac{N}{2g^2 D(x-y)} \xi(x,y) \xi(y,x) + \ldots] dxdy\} d\xi d\bar{\psi} d\psi \quad . \tag{3.8}$$

The next step is completely analogous to the case of vector models. The integral over $\bar{\psi}, \psi$ is calculated explicitly giving the effective action depending only on singlet variables ξ.

$$Z = \int \exp\{N \text{ Tr } \ln[\frac{\delta^2 K_2}{\delta \bar{\psi}(x) \delta \psi(y)} + \xi(y,x)]$$

$$+ \frac{iN}{2g^2} \int \frac{1}{D(x-y)} \xi(x,y) \xi(y,x) dxdy + \ldots\} d\xi \quad . \tag{3.9}$$

The dependence on N is explicit and the derivation of N^{-1} expansion is straight-forward. N is a common factor multiplying the N-independent action. Therefore N^{-1} expansion is completely analogous to the usual quasi-classical expansion over $\hbar$. One should find the stationary point of the exponent defined by the eq.

$$\frac{i}{g^2 D(x-y)} \xi(x,y) + [\frac{\delta^2 K}{\delta\bar\psi(x)\delta\psi(y)} + \xi(x,y)]^{-1} = 0 \qquad (3.10)$$

and expand the effective action near this point. The quadratic part will define the propagators of the fields σ and the higher order terms define the interaction vertices.

Any model of this type may be considered in the same way. The problem of the construction of the N^{-1} expansion is reduced therefore to the solution of the integral equation (3.10). In particular the two dimensional QCD in $A_0=0$ gauge falls into this class. In this gauge the gluon self interaction vanishes and the only remaining vertices have the form

$$\frac{g}{\sqrt{N}} \bar\psi^a \gamma^\mu A^b_{\mu a} \psi^b \quad . \qquad (3.11)$$

Using the procedure described above one obtains in this case the G.'t Hooft's equations [5,6], (see [31,11]). One can construct in this way the quark Green's functions gluon self-interaction [32]. The quark Green's function has no one-particle pole which is replaced by the branch point. It does not mean however that this model demonstrates quark confinement. In fact the situation reminds the situation in quantum electrodynamics where the vacuum degeneracy also leads to the replacement of the electron pole by the branch point. Of course the model neglecting the gluon self-interaction is too crude to be taken seriously. A possible way to improve it is to modify in some way the gluon propagator assuming that it is the main effect of gluon self-interaction. If the modified propagator has a singularity stronger than κ^{-4} at $\kappa^2=0$, then one can show [32] that confining solutions for the quark propagator in the limit N are possible.

Now we shall analyse the general case $U(\phi)\neq 0$. To save the place we perform the explicit calculations for the vertex

Tr ϕ^3. Other vertices are considered analogously.

The main observation is that the ϕ^n interaction can be linearised. Introducing N-component ghost fields one can reduce "gluon-gluon" interaction $Tr(\phi^n)$ to the ghost-gluon interaction having the same structure as the quark-gluon interaction considered above.

There are several equivalent ways to achieve such linearization all of them using the well known property of a triangle matrix (more precisely it's infinite dimensional analogue): the determinant of a triangle matrix is equal to the product of the diagonal elements. For example

$$\exp\{i\int \frac{g}{\sqrt{N}}\ Tr\ \phi^3 dx\} \equiv \int\ \prod_{i,j=1}^{3}\ d\bar{\chi}_i d\chi_j\ \exp\{i\int[\frac{g^{1/3}}{N^{1/6}}\ [\bar{\chi}_i^{-a}\phi_a^b\chi_{2b} + \bar{\chi}_2^{-b}\phi_b^c\chi_{3c}$$

$$+ \bar{\chi}_3^{-c}\phi_c^d\chi_{1d}] + i\bar{\chi}_1^{-a}\hat{\Theta}^{-1}\chi_{1a} + \bar{\chi}_2^{-a}\chi_{2a} + \bar{\chi}_3^{-a}\chi_{3a}]dx\}\ , \qquad (3.12)$$

$\hat{\Theta}^{-1}$ denotes the inverse of some operator which posesses properties of Θ-function at least in one of the arguments. That means that the propagator of the field χ_1, which will be denoted symbolically as $\hat{\Theta}(x-y)$ has one of the following forms:

$$\prod_{\alpha=0}^{3}\Theta(x_\alpha-y_\alpha),\ \text{or}\ D_{ret}(x-y)\ \text{etc.}$$

We define the propagator $\hat{\Theta}(x-y)$ in such a way that $\hat{\Theta}(0)=const\neq0\neq\infty$. One may put $\hat{\Theta}(0)=1$. The intermediate ultraviolet regularization is also assumed so that the propagators of the fields $\phi(x)$ are nonsingular at $x=0$.

The validity of the representation (3.10) is checked by the direct calculation. Integration firstly over $\bar{\chi}_2,\chi_2,\bar{\chi}_3,\chi_3$ and then over $\bar{\chi}_1,\chi_1$ we get

$$\int d\bar{\chi}_1 d\chi_1\ \exp\{i\int[\frac{g}{\sqrt{N}}\ \bar{\chi}_1^{-a}\phi_a^b\phi_b^c\phi_c^d\chi_{1d}+i\chi_1^{a}\hat{\Theta}^{-1}\chi_{1a}]dx\} =$$

$$= \exp\{\text{Tr } \ln[\delta^{ab}\hat{\Theta}^{-1}] + \text{Tr } \ln[\delta^{ab} + \hat{\Theta} \frac{ig}{\sqrt{N}}\phi^a_b\phi^b_c\phi^c_d] =$$

$$= \exp\{i \frac{g}{\sqrt{N}}\int \text{Tr } \phi^3(x)dx\} \, . \qquad (3.13)$$

where the nonessential constant factor $\text{Tr } \ln\hat{\Theta}^{-1}$ was omitted. The last equality follows from the well-known theorem of nonlinear analysis which generalizes the triangle matrix property mentioned above. In the language of Feynman diagrams it means that all closed loops formed by the χ_1-propagators are equal to zero because they correspond to the products of Θ-functions

$$\Theta(x_1 - x_2)\Theta(x_2 - x_3) \ldots \Theta(x_4 - x_1) \, . \qquad (3.14)$$

The only term which is different from zero is the "tadpole" diagram, which is proportional to $\hat{\Theta}(0)$. This term reproduces the left hand side of eq. (3.12).

So we succeeded to write the gluon-gluon interaction in the same form as the quark-gluon interaction. Now we can apply the same technique as before. Integrating over the fields ϕ we shall obtain the effective interaction of singlet bilocal currents formed by the ghost fields χ_i. Then in complete analogy with the eq. (3.8) we introduce singlet bi-local fields $\xi_j(x,y)$ and finally integrate over the ghost fields χ_i: As a result we obtain the effective action which depends only on singlet variables ξ_j, and contains N only as a numerical factor. The Green's function generating functional will have a form

$$Z = \int\exp\{NK(\xi_j) - \frac{N^{1/3}}{g^{2/3}} C^{ik}\int\xi_i(x,y)\xi_K(y,x)dxdy + \ldots\} \, . \qquad (3.15)$$

Here $K(\xi_j)$ is the functional arising after the integration over the ghost fields χ_i.

Formally the functional (3.15) is analogous to the functional (3.9). However the N dependence in (3.15) is more complicated. Here there are terms of different order in N, and some terms proportional to N are multiplied by the Θ-function. That does not allow to use directly the steepest descent method. In particular if one tries naively to define the stationary point from the condition $\frac{\delta K(\xi)}{\delta \xi}=0$ treating the term proportional to $N^{-1/3}$ as a small correction, one finds that the quadratic form generated by the expansion of K near the stationary point is degenerate as a general rule.

Nevertheless in some interesting cases we succeeded to solve this problem exactly or approximately and to construct an explicit N^{-1} expansion. Some examples will be given below.

2. Exact summation of planar diagrams up to now have been performed only for a limited class of simplified models. One of such models is a matrix Bloch-Nordsieck type model [30]

$$L = \frac{1}{2}\partial_\mu \phi_b^a \partial_\mu \phi_a^b + i\bar{\psi}_b^a u_\mu \partial_\mu \psi_a^b - m\bar{\psi}_b^a \psi_a^b + \frac{g}{\sqrt{N}}\bar{\psi}_b^a \phi_c^b \psi_a^c \quad . \tag{3.16}$$

Here ψ_b^a are complex NxN matrixes, ϕ_b^a is a Hermitian NxN matrix; u_μ is a constant vector, $u_\mu^2 = 1$. Due to the special choice of the kinetic term the propagator of ψ-fields has only one pole and closed loops formed by the ψ-lines are absent. Such model is known to describe heavy nonrelativistic particles interacting with the relativistic field ϕ.

We shall calculate the Green's function of the ψ-field

$$G(x,y) = \langle \bar{\psi}_b^a(x)\psi_a^b(y)\rangle = \frac{\delta Z}{\delta J(x,y)}\Big|_{J=0} \quad ,$$

$$Z = \int \exp\{i\int L(x)dx + \int \bar{\psi}_b^a(x)\psi_a^b(y)J(x,y)dxdy\}d\bar{\psi}d\psi d\phi \quad . \tag{3.17}$$

First of all we perform the Gaussian integration over $\bar{\psi},\psi$

$$G(x,y) = N\int Tr[\delta^{ab}(i\partial_\mu u_\mu - m) + \frac{g}{\sqrt{N}}\phi^{ab}]^{-1}_{xy}$$

$$\times \exp\{i\int \frac{1}{2} Tr \partial_\mu\phi\partial_\mu\phi dx\}d\phi \quad . \tag{3.18}$$

Due to the nonrelativistic nature of ψ-propagators the determinant arising after this integration is equal to one.

The integral (3.18) may be written with the help of the ghost fields χ^a

$$G(x,y) = N\delta Z/\delta J(x,y)|_{J=0} \quad ,$$

$$Z = \int \exp\{i\int[\frac{1}{2} Tr(\partial_\mu\phi\partial_\mu\phi) + \bar{\chi}^a(x)(iu_\mu\partial_\mu - m + J(x,y))\delta^b_a$$

$$+ \frac{g}{\sqrt{N}}\phi^b_a]\chi_b(x)]dx\}d\bar{\chi}d\chi d\phi \quad . \tag{3.19}$$

The effective action in (3.19) is transformed according to the prescription described above. Integrating over ϕ^a_b and introducing a singlet bilocal field $\xi(x,y)$ one gets

$$Z = \int \exp\{i\int[\bar{\chi}^a(iu_\mu \frac{\partial}{\partial\chi_\mu} - m)\chi_a dx + \int[(\bar{\chi}^a(x)\chi_a(y))\xi(y,x)$$

$$- \frac{1}{2}\frac{N}{g^2}\frac{\xi(x,y)\xi(y,x)}{D(x-y)} + (\bar{\chi}^a(x)\chi_a(y))J(x,y)dxdy\}d\bar{\chi}d\chi d\xi \quad . \tag{3.20}$$

Here $D(x-y)$ is the Green's function of the field ϕ.

Finally integrating over χ and differentiating Z over J we have

$$G(x,y) = N^2\int M^{-1}(x,y)\exp\{N Tr \ln M(\xi)$$

$$- \frac{1}{2}\frac{N}{g^2}\int \frac{\xi(x,y)\xi(y,x)}{D(x-y)} dxdy\}d\xi \quad , \tag{3.21}$$

$$M(\xi) = (iu_\mu\frac{\partial}{\partial x^\mu} - m)\delta(x-y) + \xi(y,x) \quad . \tag{3.22}$$

The steepest descent method may be applied. In the leading

order in N

$$N^{-2}G(x,y) = M^{-1}(y,x)\big|_{\xi=\xi_{st}} \quad . \tag{3.23}$$

Where the stationary point is defined by the eq.

$$M^{-1}(y,x) - \frac{i}{g^2} \frac{\xi(x,y)}{D(x-y)} = 0 \quad . \tag{3.24}$$

Combining this eq. with the definition of the operator M we obtain the equation for the Green's function G which may be conveniently written in the momentum space as follows

$$\hat{G}^{-1}(p) - (up+m) - \frac{g^2}{(2\pi)^4}\int \frac{1}{(p-k)^2} \hat{G}(k)d^4k = 0 \quad . \tag{3.25}$$

Iterations of this eq. in g reproduces all the planar diagrams which are in this case "rainbow" type diagrams. Of course to get finite results one must make necessary subtractions. The eq. (3.25) may be solved exactly. To do that note that $\hat{G}(p)$ depends only on the variable (up). Going over to the Euclidean region and applying to the both sides of eq. (3.25) D'Alambert operator $\square$ we obtain the differential equation

$$\frac{d^2\hat{G}^{-1}}{dz^2} = \alpha G; \qquad z = (up); \qquad \alpha = -\frac{g^2}{4\pi^2} \quad . \tag{3.26}$$

Here we used the following property of the four-dimensional massless Green's function

$$\square_p \frac{1}{(p-k)^2} = - 4\pi^2\delta(p-k) \quad . \tag{3.27}$$

The general solution of eq. (3.27) is defined by the implicit function

$$z = \int_0^{\hat{G}^{-1}} [(\hat{G}^{-1})'(\mu) + 2\alpha \ln \frac{y}{\hat{G}^{-1}(\mu)}]^{-1/2}dy + C \quad . \tag{3.28}$$

Here C and μ are arbitrary constants which are related to
arbitrary subtraction constants. The Green's function $\overset{\curvearrowright}{G}(p)$
has a singularity at up=C, and

$$\lim_{up \to C} (up-C)\overset{\curvearrowright}{G}(up) = 0 \quad . \tag{3.29}$$

One particle pole is absent due to the vacuum degeneracy with
respect to the soft φ-quanta. Analogously the sum of rainbow
diagrams may be calculated in the theories with relativistic
fields ψ. However in these theories rainbow diagrams do not
form the complete set of planar diagrams.

3. The problem of analytic summation of all planar diagrams
in realistic models seems to be too complicated to be solved
exactly. Some approximation methods should be developed.
Unfortunately in the planar theory there is no small parameter
exept for the usual coupling constant g, and it is hard to
control the accuracy of the approximation.

In this section we shall describe a method of approximate
summation of planar diagrams in the matrix ϕ^4 models for
arbitrary dimension of space-time which allows at least
qualitatively to estimate the accuracy of approximation [33].
Being applied to the zero-and one-dimensional models it
gives the results in a good agreement with the paper [13].

The model is described by the Euclidean action

$$S = \int d^d x \, Tr\{\partial_\mu \bar{\phi} \partial_\mu \phi + m^2 \bar{\phi}\phi + \frac{2g}{N} \bar{\phi}\phi\bar{\phi}\phi\} \quad , \tag{3.30}$$

ϕ_a^b is a complex NxN matrix. The case of Hermitian matrices
is treated analogously and to the leading order the result
differs only by known numerical factors.

We shall sum up explicitly some subset of all planar
diagrams generated by (3.30) which includes all types of
planar diagrams. Although the choice of this subset seems
somewhat artifical we may estimate how representative our

set is by comparing our result for the zero-dimensional case with the exact result known from [13]. In the zero-dimensional case the expectation values calculated with the help of (3.30) give just the number of corresponding planar diagrams calculated with the weight depending on the coupling constant g. Therefore if the approximate result is close to the exact one that means that all essential diagrams are taken into account.

To formulate our approximation we shall write the Green's function generating functional for the model (3.30) in the form

$$Z = \int \exp\{ \int d^d x \ \mathrm{Tr}[\bar{\phi}\Box\phi - m^2\bar{\phi}\phi - (\tfrac{8g}{N})^{1/2}\bar{\phi}\phi\lambda + \lambda\lambda] + i\int d^d x\, d^d y\, J(x,y)$$

$$\times \mathrm{Tr}[\bar{\phi}(x)\phi(y)]\}d\lambda d\bar{\phi}d\phi \quad . \tag{3.31}$$

Here the auxiliary field λ_b^a with the local propagator was introduced to make the action quadratic in ϕ. Our approximation scheme is the expansion of the integral (3.31) in the limit $N\to\infty$ over the number of loops of the fields ϕ. In other words integrating over ϕ

$$Z = \int \exp\{ \int d^d x \ \tfrac{N}{8g}\ \mathrm{Tr}(\lambda\lambda) - N\ \mathrm{Tr}\ \ln[\delta^{ab}(m^2 - \Box - J) + \lambda_b^a]\}d\lambda \tag{3.32}$$

we may put the parameter α in front of the second term in the exponent and expand the integral over α, putting in the final answer $\alpha=1$.

Every fixed order in α includes an infinite set of planar diagrams. Calculating the different orders in α and comparing the results with the exact results for zero-dimensional case we shall estimate the accuracy of approximation.

So we begin with the calculation of the integral (3.32) in zero-dimensional case. We shall study the vacuum energy so we put $J=0$ and $m^2=1$. According to the general scheme described above we rewrite this integral in terms of singlet variables and then apply the steepest descent method.

To do that we represent the Tr ln in the exponent (3.32) as an integral over the auxiliary variable s

$$Z = \int \exp\{\frac{N}{8g} \, \mathrm{Tr}\,(\lambda\lambda) - \alpha N \int_0^1 ds \, (\delta_b^a + s\lambda_b^a)^{-1} \lambda_a^b\} d\lambda \quad . \tag{3.33}$$

This interaction may be linearized with the help of the ghost fields χ^α as follows

$$Z = \int \exp\{\frac{N}{8g} \, \mathrm{Tr}\,(\lambda\lambda) + \int \bar{\chi}^q(s) [(\delta_a^b + \lambda_a^b s) \delta(s-t) + \alpha N \lambda_a^b \theta(s-t) \theta(s) \theta(t-s)]$$

$$\times \chi_b(t) ds dt\} d\bar{\chi} d\chi d\lambda \quad . \tag{3.34}$$

Here it is assumed that local measures formally proportional to $\delta(0)$ are trivial as it takes place for example in dimensional regularization (this assumption is not essential but simplifies the calculations). As usual we define $\theta(0)$ to be equal to 1. The validity of eq. (3.34) is checked directly in the same way as we have done in the general case.

Integrating (3.33) over λ and introducing the singlet bilocal variables ξ we get the following final result

$$Z = \int \exp\{\frac{N}{8g} \int \xi(s,t) \xi(t,s) ds dt - N \, \mathrm{Tr}\, \ln\{\delta(s-t)$$

$$+ \int du [u\delta(u-s) + \alpha N \theta(u-s) \theta(1-u) \theta(u)] \xi(u,t)\}\} d\xi \quad . \tag{3.35}$$

To the lowest order in α one has

$$Z = \int \exp\{\frac{N}{8g} \int \xi(s,t) \xi(t,s) ds dt - N \, \mathrm{Tr}\, \ln\{\delta(s-t) + t\xi(t,s)\}\}$$

$$\times \{1 - \alpha N^2 \int \theta(u-t) \theta(1-u) \theta(u) \xi(u,s) [\delta(s-t) + s\xi(s,t)]^{-1}\} du ds dt\} d\xi \quad .$$

The value of the integral in the limit $N\to\infty$ is determined by the stationary point of the exponent

$$s[\delta(s-t)+s\xi(s,t)]^{-1} - \frac{1}{4g}\,\xi(s,t) = 0 \quad . \tag{3.36}$$

The solution of this eq. is

$$\xi_{st}(s,t) = \delta(s-t)\,\frac{f(s)}{s}; \qquad f(s) = \frac{-1+\sqrt{1+16gs^2}}{2} \quad . \tag{3.37}$$

Substituting this solution into the order α term we obtain the following vacuum energy

$$E^1 = -\ln Z^{(1)} = \frac{1}{16g}\,(\sqrt{1+16g}-1)+\ln(\frac{1+\sqrt{1+16g}}{2}) - \frac{1}{2} \quad . \tag{3.38}$$

This expression should be compared with the exact sum of the planar diagrams given in [13].

$$E_o = \frac{1}{12}(a^2-1)(9-a^2)-\ln a^2 \quad ,$$

$$a^2 = \frac{1}{24g}\,\sqrt{1+48g} - 1 \quad . \tag{3.39}$$

For small g the difference between (3.38) and (3.39) is $\sim 50\%$. It had to be expected because for small g the ordinary perturbation works and our one loop approximation includes exactly one half of the lowest order diagrams. For $16g=10^{-2}$, $E_o^1 \sim 0.00125$. whereas $E_o \sim 0.00248$. However with the increasing of g the agreement improves considerably. For $16g=10^3$, 10^4, 10^5 we have $E_o^1=2.32$; 3.43; 4.57 and $E_o=2.60$; 3.76; 4.68 respectively. Comparing the eqs. (3.38) and (3.39) one sees that they give the same asymptotic behaviour for E_o

$$E_o \xrightarrow[g\to\infty]{} \frac{1}{2}\,\ln(g) \quad . \tag{3.40}$$

For large g our first approximation gives the exact answer for the sum of planar diagrams. Therefore in this limit all essential diagrams are taken into account already at the one loop level, and one can hope that this is true in higher dimensions as well.

As an additional test we shall calculate the vacuum energy

for one-dimensional model which describes unharmonic oscillators.

For non zero-dimensional case the representation of the generating functional analogous to (3.34) looks as follows

$$Z = \int \exp\{\frac{N}{8g}\int d^d x \mathrm{Tr}\,(\lambda(x)\lambda(x)) + \bar{\chi}^b(s,x)\,[\delta(s-t)\,[\delta_b^a K_\tau + s\lambda_b^a(x)]$$

$$+ \alpha N \lambda_b^a(x)\,\theta(s-t)\,\theta(1-s)\,\theta(s)\,\delta(x-y)]\chi_a(y,t) - N^2\,\mathrm{Tr}\,\ln K_\tau\}d\bar{\chi}d\chi d\lambda;$$

$$K_J = m - \square - J. \tag{3.41}$$

In this case the ghost fields χ^α depend not only on auxiliary variable s but also on the space time-point s.

Transforming this integral in complete analogy with the zero dimensional case we get

$$Z = \int \exp\{\frac{N}{8g}\int \xi(s,t,x)\,\xi(t,s,x)\,dsdtdx - N^2 \mathrm{Tr}\,\ln K_J - N\,\mathrm{Tr}\,\ln\,[\delta(s-t)K_J$$

$$+ \int du[u\delta(u-s) + \alpha N\theta(u-s)\,\theta(1-u)\,\theta(u)]\xi(u,t,x)]\}d\xi \ .$$

$$\tag{3.42}$$

The stationary point of the exponent at $\alpha=0$ is defined by the equation

$$s[(m^2-\square)\delta(s-t) + s\xi(s,t,x)]_{xx}^{-1} - \frac{1}{4g}\xi(s,t,x) = 0 \ . \tag{3.43}$$

Translationally invariant solutions of this equation have a form

$$\xi_{st}(s,t,x) = \delta(s-t)f(s) \tag{3.44}$$

where the function f(s) satisfies

$$s\int\frac{d^d p}{(2\pi)^d}\,\frac{1}{p^2+m^2+sf(s)} - \frac{1}{4g}f(s) = 0 \ . \tag{3.45}$$

Integrating over p for such dimensions d when the integral (3.45) exists we get the algebraic equation

$$s \frac{\pi^{d/2}}{(2\pi)^d} (m^2+sf(s))^{\frac{d}{2}-1} \Gamma(1-\frac{d}{2}) - \frac{1}{4g} f(s) = 0 \quad .$$ (3.46)

In one dimensional case this equation reduces to

$$(m^2+sf) f^2 = 4g^2 s^2$$ (3.47)

As we expect that the first approximation in α gives the correct answer for large g, we shall solve this eq. in this limit. For large g the mass term may be neglected and

$$f = (2g)^{2/3} s^{1/3} \quad .$$ (3.48)

To calculate the vacuum energy in this approximation we substitute the solution (3.48) into the order α term in the eq. (3.42) and put $J = 0$.

$$\frac{E_o^1}{VN^2} = \int_0^1 \frac{1}{4gs} (2g)^{4/3} s^{2/3} ds = 3(2)^{-5/3} g^{1/3} \quad .$$ (3.49)

In the case of Hermitian fields ϕ_b^a the corresponding value is

$$\frac{E_{H_o}^1}{VN^2} = \frac{1}{2} \frac{E_o^1 (2g)}{VN^2} = 3 \cdot 2^{-7/3} g^{1/3} \simeq 0.595 g^{1/3} \quad .$$ (3.50)

This is to be compared with the asymptotic estimate given in [13]

$$\frac{E_{rev}}{VN^2} \simeq 0.5899 g^{1/3} \quad .$$ (3.51)

Again for large g already our first approximation takes into account all essential diagrams.

Up to now we considered only one loop approximation which seems to work very well for large g. For small g the higher order terms must be taken into account. We calculated the two loop correction to the zero-dimensional vacuum energy. The results are presented below. The second line gives the sum of one- and two-loop approximations and the third line gives the exact sum of planar diagrams.

16g	0,01	0,1	1	10	100
E^1+E^2	0.00248	0.0239	0.182	0.748	1.7
E	0.00248	0.0237	0.179	0.696	1.56

One sees that the two-loop approximation practically saturates the sum of planar diagrams in the broad interval of the values of g. The higher orders do not change this result for sufficiently small g. The problem of the convergence of our expansion for large g is questionable. As has been shown by G't Hooft [36] the sum of planar diagrams in the theories satisfying certain reasonable requirements converges within the finite radius of convergence.

The approach described above leaves many question to be answered. At present we know no deep reasons why it works so well at large g. The most urgent problem is the extension of this approach to other theories, in particular to QCD. The direct extension to QCD is prevented by the presence of the A^3 vertices which requires a special treatment. These problems are now under investigation.

REFERENCES

1. S. Coleman, E. Weinberg, Phys. Rev. D7 (1973) 1888.

2. D. Gross, A. Neveu, Phys. Rev. D10 (1974) 3235.

3. L.E. Carneiro et al., Nucl. Phys. B183 (1981) 445.

4. A. Andraŝi et al., Nucl. Phys. B182 (1981) 104.

5. G.'tHooft, Nucl. Phys. B72 (1974) 461.

6. G.'t Hooft, Nucl. Phys. B75 (1974) 461.

7. S. Coleman, R. Jackiw, H.Politzer, Phys. Rev. D10 (1974) 242.

8. M.Kobayashi, T. Kugo, Progr. Theor. Phys. 54 (1975) 1537.

9. I.Ya. Arefieva, Ann. Phys. 117 (1979) 393.

10. C. Callan, W. Coote, D. Gross, Phys. Rev. D13 (1976) 1649.

11. E. Witten, Nucl. Phys. B160 (1974) 57.

12. I.Ya. Arefieva, A.A. Slavnov, Lectures in the XIV. International School of Young Scientists, Dubna 1980.

13. E. Brezin, C. Itzykson, G. Parisi, J-B. Zuber, Com.Math. Phys. 59 (1978) 35.

14. Yu.M. Makeenko, A.A. Migdal, Phys. Let. 88B,(1979) 135

15. T.L. Gervais, A. Neveu, Nucl. Phys. B153 (1979) 445.

16. A.A. Migdal, Ann. of Phys. 126 (1980) 279.

17. Yu.M. Makeenko, A.A. Migdal, Yad. Phys. 32 (1980) 848.

18. A.M. Polyakov, Nucl. Phys. B164 (1980) 171.

19. I.Ya. Arefieva, Phys.Lett. 104B, (1981) 453.

20. T. Eguchi, H. Kavai, Phys. Rev.Lett. 48 (1982) 1063.

21. G. Bhanot, U. Heller, H. Neuberger, Phys.Lett. 113B (1982) 47; 115B (1982) 237.

22. G. Parisi, Phys.Lett. 112B (1982) 463.

23. G. Parisi, Zhang-Yi-Cheng, Phys.Lett. 114B (1982) 319.

24. D. Gross, Y. Kitazawa, Princeton University Preprint PRE2564. (1982)

25. S. Edwards, P. Anderson, T. Ph. F-5 (1975) 965.

26. I.Ya. Arefieva, Theor. Math. Phys. 54 (1983) 154.

27. S.V. Tyablicov, Methods of Quantum Theory of Magnetism, Moscow, Nauka, 1965.

28. D. Greensite, B. Halpern, Berkeley Preprint B-PTH-82/14.

29. A.A. Slavnov, Phys. Lett. 112B (1982) 154.

30. A.A. Slavnov, Theor. Math. Phys. $\underline{51}$ (1982) 307.

31. D. Ebert, V.N. Pervushin, Theor. Math. Phys. $\underline{36}$ (1978) 313.

32. A.A. Slavnov, Theor. Math. Phys. $\underline{54}$ (1983) 46.

33. A.A. Slavnov, Theor. Math. Phys. in print.

34. M. Aizenman, Com. Math. Phys. $\underline{86}$ (1982) 151.

35. J. Fröhlich, Nucl. Phys. $\underline{B200}$ (1982) 281.

36. G.'t Hooft, Utrecht University Preprints (1982).

Acta Physica Austriaca, Suppl. XXV, 399–487 (1983)
© by Springer-Verlag 1983

RECENT RESULT FROM THE $\bar{\text{p}}$p COLLIDER[+]

by

Horst D. WAHL
Inst. f. Hochenergiephysik, Wien
and
CERN

1. INTRODUCTION

The suggestion to convert the CERN SPS into a proton-anti-proton collider[1] was motivated by the expectation that the centre-of-mass energy of 540 GeV (an order of magnitude higher than that at existing accelerators) would make it possible to search for the intermediate vector bosons (IVB) $W^{\pm}$, Z^{0} predicted by the standard model of electroweak interactions[2].

Apart from this main motivation, it was clearly also considered to be of interest to study general features of $\bar{\text{p}}$p collisions in this new energy domain and possibly discover some new hitherto unexpected phenomena.

These lectures give a short review of the status of the collider and the experiments installed in its experimental areas, as well as a summary of physics results obtained so far. The main emphasis is put on the recent observation of events whose signature agrees with that expected for production and decay of IVB's via the reaction

_____

[+] Lectures given at the XXII. Internationale Universitätswochen für Kernphysik,Schladming,Austria,February 23-March 5,1983.

$$\bar{p}p \rightarrow W^{\pm} + x$$
$$\hookrightarrow e^{\pm} + \nu \tag{1}$$

In Sect. 2 we give a short description of the collider and its performance up to now, Sect. 3 contains information about the experiments installed at the collider and their main physics goals. In Sect. 4 we present a summary of results on general features of $\bar{p}p$ collisions at $\sqrt{s}$ = 540 GeV, and the observation of jets is discussed in Sect. 5. Sect. 6 is concerned with the signature of events of the type (1) and how to recognize them in an experiment, and Sect. 7 describes the search for such events. An outlook on future developments is presented in Sect. 8, and conclusions are given in Sect. 9.

Previous summaries of the CERN $\bar{p}p$ collider physics programme can be found, e.g. in refs.[3,4].

Being a member of UA1, I am somehow biased in favour of this experiment, which will therefore be treated in more detail than the others. I hope that the reader will forgive me.

2. THE CERN $\bar{p}p$ COLLIDER (S$\bar{p}p$S)

2.1 Why a $\bar{p}p$ Collider?

When in the early seventies indirect experimental evidence supporting the electroweak theories began to accumulate, it appeared more and more derivable to submit these theories to a more direct test, namely to verify the existence of the gauge bosons $W^{\pm}$ and Z^{0} with the properties predicted by the standard model[5]. Since the maximum energy reached by existing accelerators ($\sqrt{s}$ = 63 GeV at the CERN ISR) was not sufficiently high to produce these particles of masses of $\sim$100 GeV, a new machine was necessary to perform this test which was felt to be crucial for further progress in our understanding of particle physics. The most economic way to achieve higher energies was to modify an existing large fixed target machine (SPS, FNAL) such that it could run as a collider in which two beams of particles circulate in opposite direction and undergo head-on collisions. In this case, the centre-of-mass energy $\sqrt{s} \sim 2E$, while in the fixed target mode $\sqrt{s} \sim \sqrt{2mE}$ (where E = beam energy in the laboratory). Since there is only one ring of magnets, the two beams must consist of particles of the same mass but opposite charge. Furthermore, due to the high energy loss through synchrotron radiation of electrons (energy loss proportional to γ^4/R), the existing rings are not large enough to reach the required energy E with electrons, leaving only a $\bar{p}p$ collider as a viable solution. The energy that can be reached is then limited by the maximum guide field compatible with the limits on power dissipation in d.c. mode, which is 270 GeV per beam in the CERN SPS, thus yielding a centre-of-mass energy of 540 GeV.

2.2 Beam Cooling and the Antiproton Accumulator

The disadvantage of colliders is that even extremely intense beams are very tenuous compared with targets made of solid or even liquid material. Therefore it is difficult to reach a high luminosity L(L= the quantity which when multiplied by a cross section, gives the counting rate for that process).

If it is difficult to make a dense proton beam in a collider, it is even more difficult to produce a high intensity antiproton beam, given the fact that the typical relative yield in the production of antiprotons is $\lesssim 10^{-6}$ $\bar{p}$'s per proton. With the number of protons available from the proton synchrotron ($\sim 10^{13}$ protons per pulse, i.e. every 2.4 seconds) it will take many hours to get $\sim 10^{11}$ antiprotons, the number deemed necessary to get reasonable counting rates for the rare processes that one wants to study. Therefore it is necessary to have an intermediate storage for the antiprotons (the accumulator) before filling them into the collider, so that accumulation of antiprotons can proceed while the collider is operating.

If one wants to have as high a yield of antiprotons as possible, one has to accept them in a large angular and momentum range from the production process, a condition which is incompatible with the limited acceptance in position, angle and momentum bite of the SPS. This problem was solved by the invention of "beam cooling", which allows effective compression of the phase space occupied by the particles in the beam. It consists in the reduction of deviations of the particles from the "wanted trajectory" in phase space, i.e. a damping of individual motion as opposed to collective motion; thus the term "cooling" is appropriate. It was the invention and successful demonstration of beam cooling that paved the way to obtaining sufficiently dense antiproton beams, thus making the construction of a p̄p collider feasible.

Two techniques of beam cooling have been developed: "electron cooling" and "stochastic cooling".

Electron cooling uses well-ordered electron beams to cool disordered proton or antiproton beams. It was invented by G. Budkov in 1966 at Novosibirsk[6] and demonstrated on protons in 1974[7].

Stochastic cooling uses correction signals derived from beam fluctuations in a feedback loop to correct for displacements, thereby compressing the phase space (see ref. 8 for more information). It was invented by S. van der Meer at CERN[9] and demonstrated at CERN in the ISR on a proton beam in 1974[10].

After the suggestion to convert the SPS into a p̄p collider, the "Initial Cooling Experiment" (ICE) was set up to study both methods of beam cooling. It demonstrated the full validity of the stochastic cooling principle[8] and led to the authorization of the p̄p project based on stochastic cooling.

Accumulator operation:

During accumulation (stacking), protons are accelerated in the PS (see fig.1) to 26 GeV, extracted and directed on to a target which is followed by a focussing horn. The particle which are produced are spead over a wide range of angles and momenta, and contain a small fraction of anti-protons. The maximum antiproton yield occurs at a momentum of 3.5 GeV/c, corresponding to production at rest in the centre-of-mass system. The antiprotons are guided into the accumulator where they are precooled and added to the stack of p̄'s already in the ring. By cooling in the antiproton accumulator (AA), the product of phase space and momentum bite occupied by the antiprotons is reduced by a factor $\sim 10^9$ in order to match the acceptance of the SPS. Typical stacking rater are $\sim 5 \times 10^9$ p̄/hour. More details can found in the design reports [11,12] and in refs. 3.

2.3 Collider Operation and Performance

The lay-out of the collider complex is shown in fig.1. When a sufficiently dense stack of antiprotons has been accumulated in the AA, filling of the SPS can start. First, protons are accelerated in the PS to 26 GeV and a number of proton bunches (in general 3) are injected (through transfer line TT10) into the SPS ring and left coasting (circulating) in clockwise direction.

Then the densest part of the p̄ stack is extracted from the AA and (via transfer line TTL2) injected into the PS, where it circulates in the opposite sense to that previously used by the protons, so that no change of field direction is necessary. After acceleration from 3.5 GeV to 26 GeV, the p̄'s are extracted from the PS and (via transfer line TT70) injected into the SPS, where they circulate counterclockwise

404

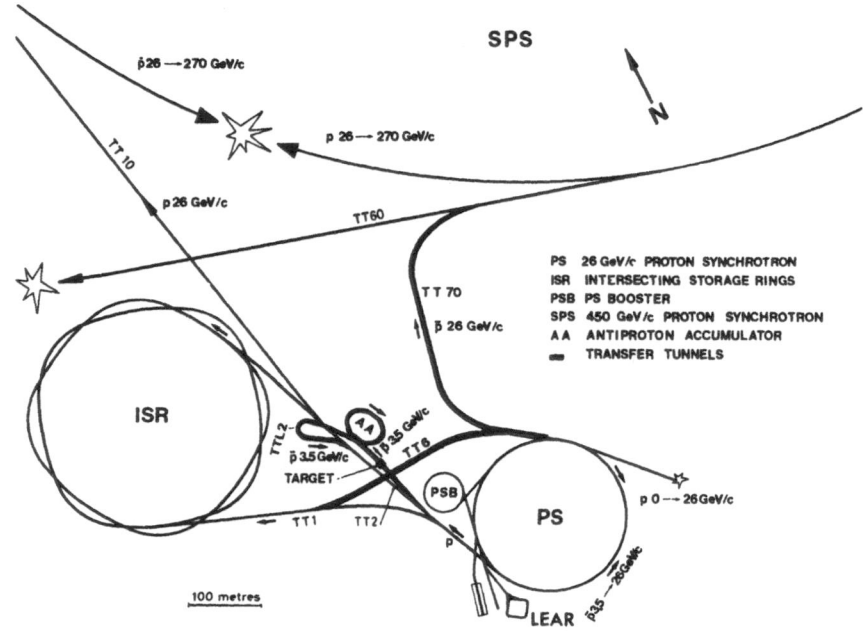

SPS

p̄ 26 → 270 GeV/c

p 26 → 270 GeV/c

TT 10

p 26 GeV/c

TT60

TT 70

p̄ 26 GeV/c

PS   26 GeV/c PROTON SYNCHROTRON
ISR INTERSECTING STORAGE RINGS
PSB PS BOOSTER
SPS 450 GeV/c PROTON SYNCHROTRON
AA   ANTIPROTON ACCUMULATOR
━   TRANSFER TUNNELS

ISR

AA

TTL2

p̄ 3.5 GeV/c   p̄ 3.5 GeV/c

TARGET

TT6

PSB

PS

p 0 → 26 GeV/c

TT1   TT2

p

LEAR

p̄ 3.5 → 26 GeV/c

100 metres

N

Fig. 1(a): Over-all site lay-out of the p̄p project.

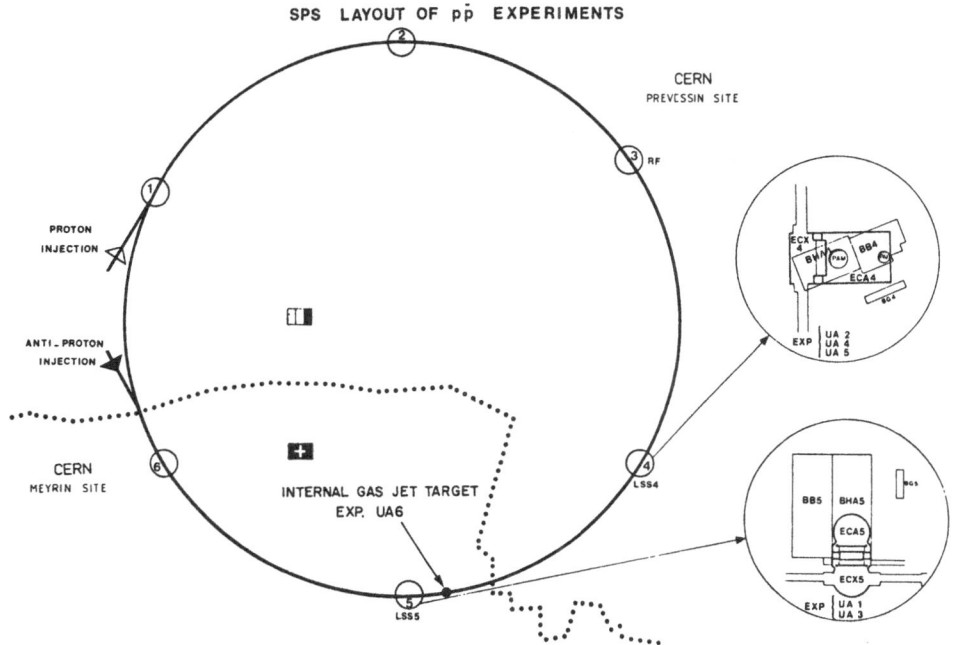

Fig. 1(b): Lay-out of the SPS ring as p̄p collider. The long
straight sections WS1 to LSS6 are denoted by
small circles.

(i.e. opposite to the protons). In general, this sequence of extraction from AA, acceleration in PS and injection into SPS is repeated three times in order to get three antiproton bunches into the collider.

The two beams of $\bar{p}$'s and p's are then accelerated by RF travelling-wave structures to 270 GeV. After this, the bunches circulate for many hours, as long as the beam quality and luminosity is considered good enough and until there is again a sufficiently dense $\bar{p}$ stack available in the AA.

The bunches have a length of several tens of centimeters ($\sim$2ns) and their transverse diameter is of the order of a millimetre. By powering a set of quadrupoles ("low-β insertion" to reduce the beta-function of the machine),the transverse dimensions of the beams can be squeezed locally and the luminosity thus increased. (More information can be found in refs. 11,12).

In the now standard operation mode with three p and three $\bar{p}$ bunches, the bunches cross each other every 7.7 μs at six points around the ring centred in the six long straight sections LSS1-LSS6. There are typically $\sim10^{11}$ protons and $\sim10^{10}$ antiprotons per corresponding bunch.

Collider performance:

The most important dates in the history of the CERN $\bar{p}$p project are shown in table 1, and table 2 gives a summary of the running conditions during the $\bar{p}$p collider periods in 1981 and 1982. Even though appreciable progress has been made since the first collider period[13], the maximum luminosity achieved in 1982 (5 $\times$ 10^{28}cm^{-2}s^{-1}) is still a factor of 20 lower than the design luminosity of 10^{30}cm^{-2}s^{-1}. A number of improvements have been envisaged by the PS/AA/SFS operations crew in order to increase the luminosity and the over-all performance of the collider[14] :

PS: increase intensity (protons on target per pulse from 10^{13} to 1.3 10^{13})

AA: . increase AA acceptance (i.e. bring it from 70% to 100%

1966	Electron cooling proposed at Novosibirsk.
1968	Stochastic cooling proposed at CERN.
1974	Electron cooling first observed at Novosibirsk. Stochastic cooling first observed at CERN ISR.
1976	CERN $p\bar{p}$ project suggested. ICE started. First $p\bar{p}$ design report (using electron cooling).
1978 (June)	Second $p\bar{p}$ design report (using stochastic cooling). Authorization of project. Start AA construction. Approve experiments UA1, UA2, UA3, UA4, UA5.
1980	Approve LEAR. $\bar{p}$ into AA and to PS, accelerate to 26 GeV.
1981 (April) (July,Aug.) (Oct.,Nov.) (Dec.)	$p\bar{p}$ collisions in ISR. $p\bar{p}$ collisions in SPS observed by UA1. Machine development; $L \sim 10^{25} \text{cm}^{-2} \text{s}^{-1}$; UA1,UA3,UA4,UA5. first $p\bar{p}$ collider period: max.lum. = $5 \times 10^{27} \text{cm}^{-2} \text{s}^{-1}$ UA1,UA2,UA3,UA4
1982 (Sept.) (Oct.-Dec.)	machine development, high β running: UA1,UA4,UA5 second $p\bar{p}$ collider period max. lum. = $5 \times 10^{28} \text{cm}^{-2} \text{s}^{-1}$ UA1,UA2,UA3,UA4

Table 1: Important dates in the history of the CERN $p\bar{p}$ project.

	Dec. 1981	Oct-Dec. 1982
luminosity L($cm^{-2}s^{-1}$) typical	3×10^{26}	2×10^{28}
best	5×10^{27}	5×10^{28}
beta-values $\beta_H \times \beta_V$(m) typical	7×3.5	2×1
lowest	2×1	
luminos. lifetime (hours)	~ 16	~ 16
best transfer efficiency, % (AA → PS → SPS)	25	60
total collision time (hours)	~ 100	~ 700
integrated luminosity, $\int$Ldt in nb^{-1}	~ 0.1	~ 25

Table 2: Summary of running conditions in the $\bar{p}p$ collider periods of 1981 and 1982.

of nominal acceptance)

. improve cooling by installation of additional pick-ups

SPS: . improve transfer efficiency

. lower β (go grom $\beta_H \times \beta_V = 2 \times 1 \ m^2$ to 0.75×1.5 or $0.5 \times 1.0 \ m^2$)

. increase luminosity lifetime (by reducing noise in the RF system, better control of tune, gain of experience)

All these measures together are expected to yield on over-all improvement by a factor of 2 to 5. Thus it appears realistic to assume that luminosities of a few $10^{29} cm^{-2} s^{-1}$ are within reach, but it seems improbable that the full design luminosity will be achieved in the near future. Important increases in performance however, could be possible with a major improvement programme for the collider (see Sect. 8.4)

3. THE UA EXPERIMENTS

3.1 Main Physics Goals

Apart from the hope to prove or disprove the existence of the IVB's with the properties expected from the standard model[2], which was the main motivation for the construction of the collider, there is always a lot of phenomena to be studied whenever a new collider comes into operation; and a number of articles have appeared formulating the physics expectations[15-19].

Five collider experiments UAn, n= 1,...,5 (UA for "underground area") were approved in 1978 (see table 3).Their main physics goals as formulated in their proposals [20-24] are to study the following subjects:

experiment	participating institutions	references
UA1	Aachen - Annecy (LAPP) - Birmingham - CERN - Helsinki- Queen Mary College, London - Paris (Collège de France) - Riverside - Roma - Rutherford Appleton Laboratory - Saclay (CEN) - Vienna	20 25-36, 57,58
UA2	Bern - CERN - Copenhagen (NBI) - Orsay - Pavia - Saclay (CEN)	21, 37-43
UA3	Annecy (LAPP) - CERN	22,44
UA4	Amsterdam (NIKHEF)- CERN Genova - Napoli - Pisa	23,45-47, 56
UA5	Bonn - Bruxelles - Cambridge - CERN - Stockholm	24 48-54, 68
UA6	CERN - Lausanne - Michigan - Rockefeller	55

Table 3: The experiments at the $\bar{p}p$ collider.

elastic scattering
total cross section
          (UA1, UA4)

study of general features of $\bar{p}p$ collisions,
e.g. particle production as a function of
particle kind, rapidity, transverse momentum;   (UA1,UA4,UA5)
multiplicity distributions

monopole production   (UA3)
Centauros   (UA1,UA5)

high p_T phenomena, jets
new flavours and quarks
new $q\bar{q}$ states
Drell-Yan continuum   (UA1, UA2)
IVB's
Higgs Bosons

Of course, this list is not exhaustive and nobody will be
disappointed if we find something which had not been foreseen.

3.2 Experimental Areas

    Two experimental areas, ECX4 and ECX5, are available
for $\bar{p}p$ physics at the SPS. They are located at the two conse-
cutive long straight sections LSS4 and LSS5 separated by
1/6 of the SPS ring. In each experimental area a garage position, called ECA4 and ECA5 respectively, is provided in
which the experimental set-up can be withdrawn during fixed-
target operation of the SPS.

    ECX4 is in one of the deepest parts of the accelerator
(about 60 m). Two access shafts, 8m and 5m in diameter,
connect the underground area to auxiliary buildings on the
ground level. The floor surface ($40 \times 20m^2$) of the hall is pre-
pared in such a way as to allow heavy experimental equipment

to be moved on air cushions. The beam height is 5.5 m above floor level. A surface for electronic barracks (about 250 m^2) is provided on four levels around the personnel access pit. The area ECX4 houses the experiments UA2, UA4, UA5.

At ECX5 the SPS ring is at its shallowest point (22m) below ground. The experimental area consists of two cylindrical shafts, of 20 m diameter, with a connecting chamber running between the two. One of the shafts (pointing to the inside of the SPS ring) is used for assembly and parking of the UA1 experiment. The experimental equipment is mounted on a single platform which can be moved on rails into the second hall, straddling the SPS tunnel. The beam height above floor level in this hall is 6.3 m. A local electronics facility, MEC (mobile electronic chariot), moving with the experimental apparatus, is available in ECA5. About 250 m^2 of floor space for electronics racks and computers is, in addition, provided in the auxiliary building above the assembly shaft. The area ECX5 houses experiments UA1 and UA3.

Recently, an additional small experimental area has been constructed in the medium straight section adjacent to LSS5 and equipped to house experiment UA6.

3.3 The Experiments and Their Detectors

3.3.1 UA1

The general-purpose detector of UA1[20,25-27] was designed to have an almost complete coverage of the whole solid angle down to a polar angle $\theta = 0.2^\circ$ with both track detection and calorimetry. Its central part ($5^\circ < \theta < 175^\circ$) consists of a "Russian box" like structure of track detectors and calorimetry (see fig. 2). Starting from the interaction point, a particle encounters first a track detector (central detector), then an electromagnetic calorimeter("gondolas" and "bouchons"), followed by the magnet yoke = hadron calorimeter

Fig. 2: The UA 1 detector

(a) Over-all view of the central part

(b) Cut through the central part by a plane
     perpendicular to the beam acis.

(c) Cut through the central part by a vertical
     plane containing the beam axis. The main
     components of the detector are indicated.

(d) Schematic view of the over-all detector
     lay-out, including the forward detector.
     (The Roman pots shown at 25 m from the
     interaction point have meanwhile been
     moved to 43 m from the interaction point).

(e) Detail of the elastic scattering detector.

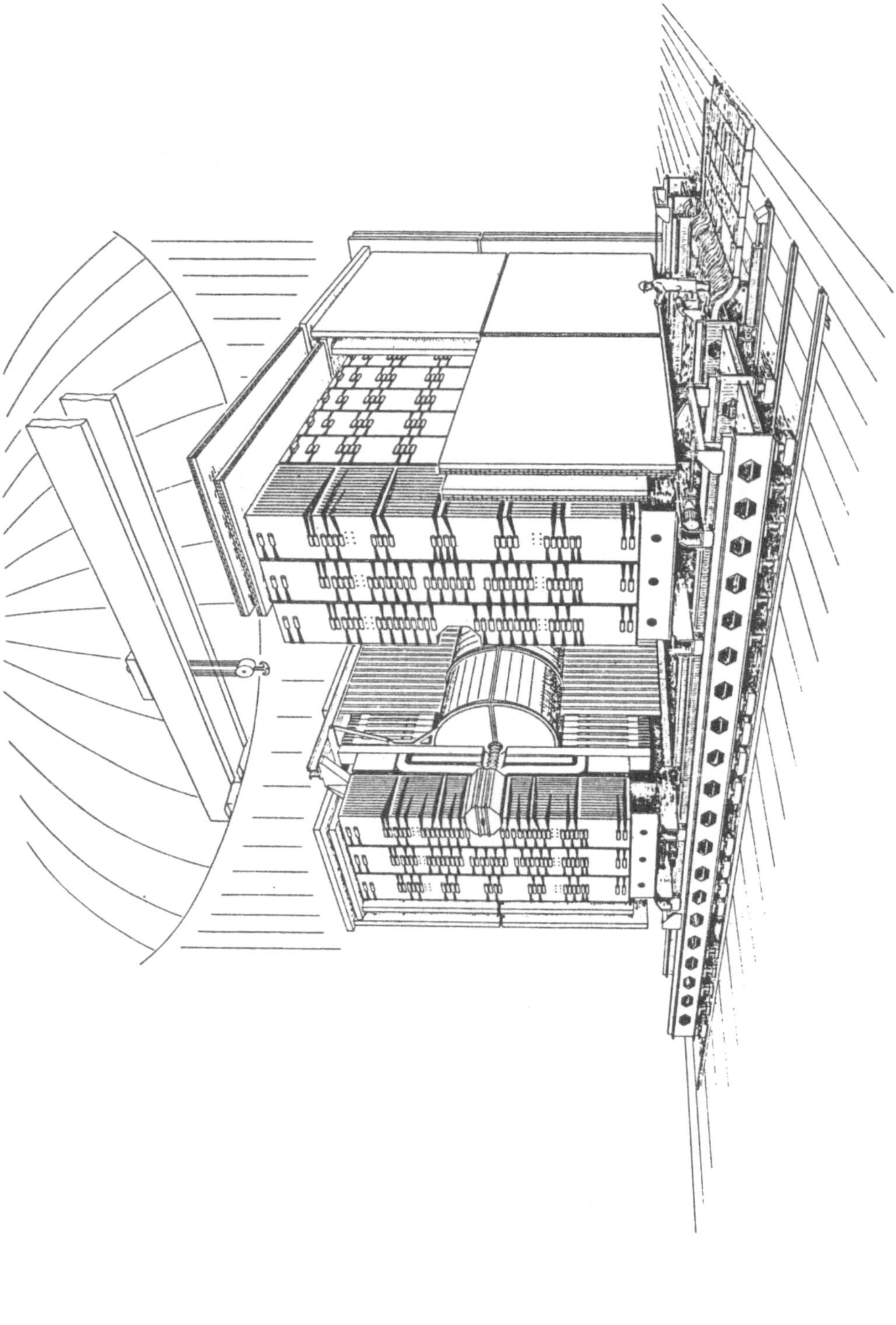

Fig. 2a

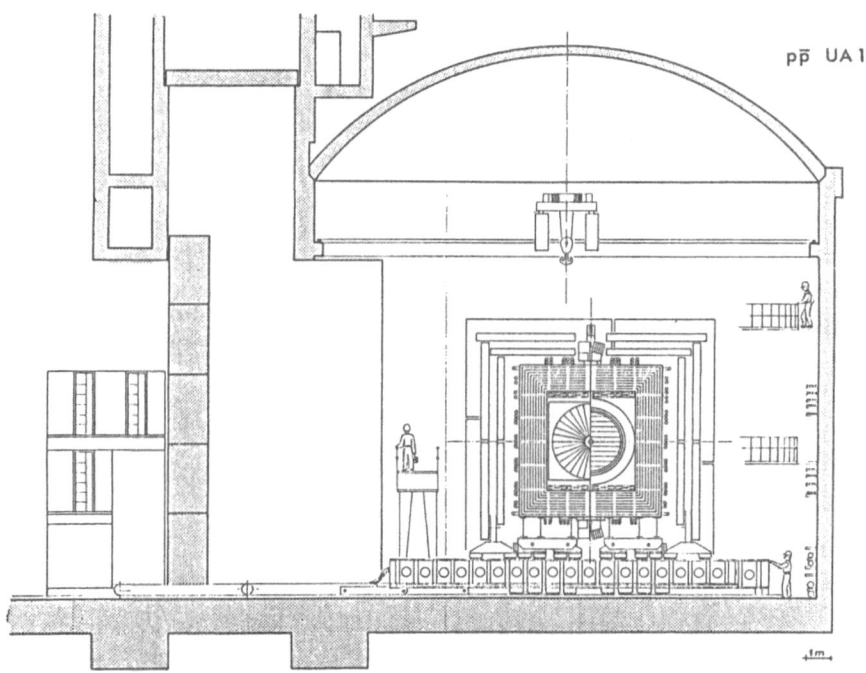

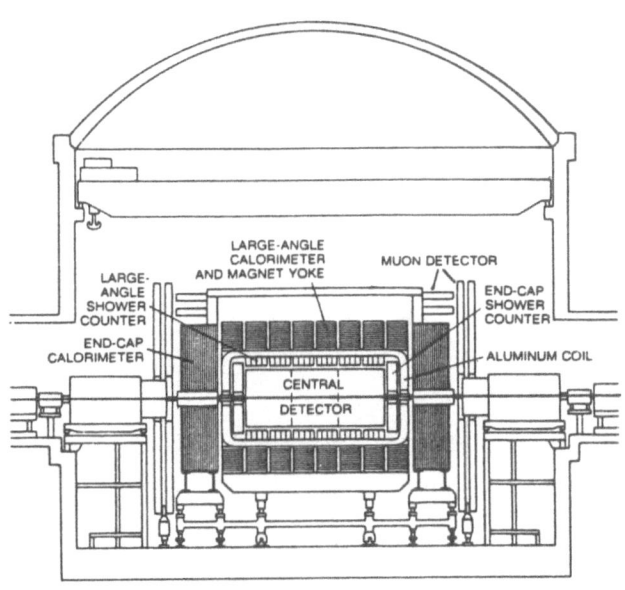

Fig. 2b and 2c

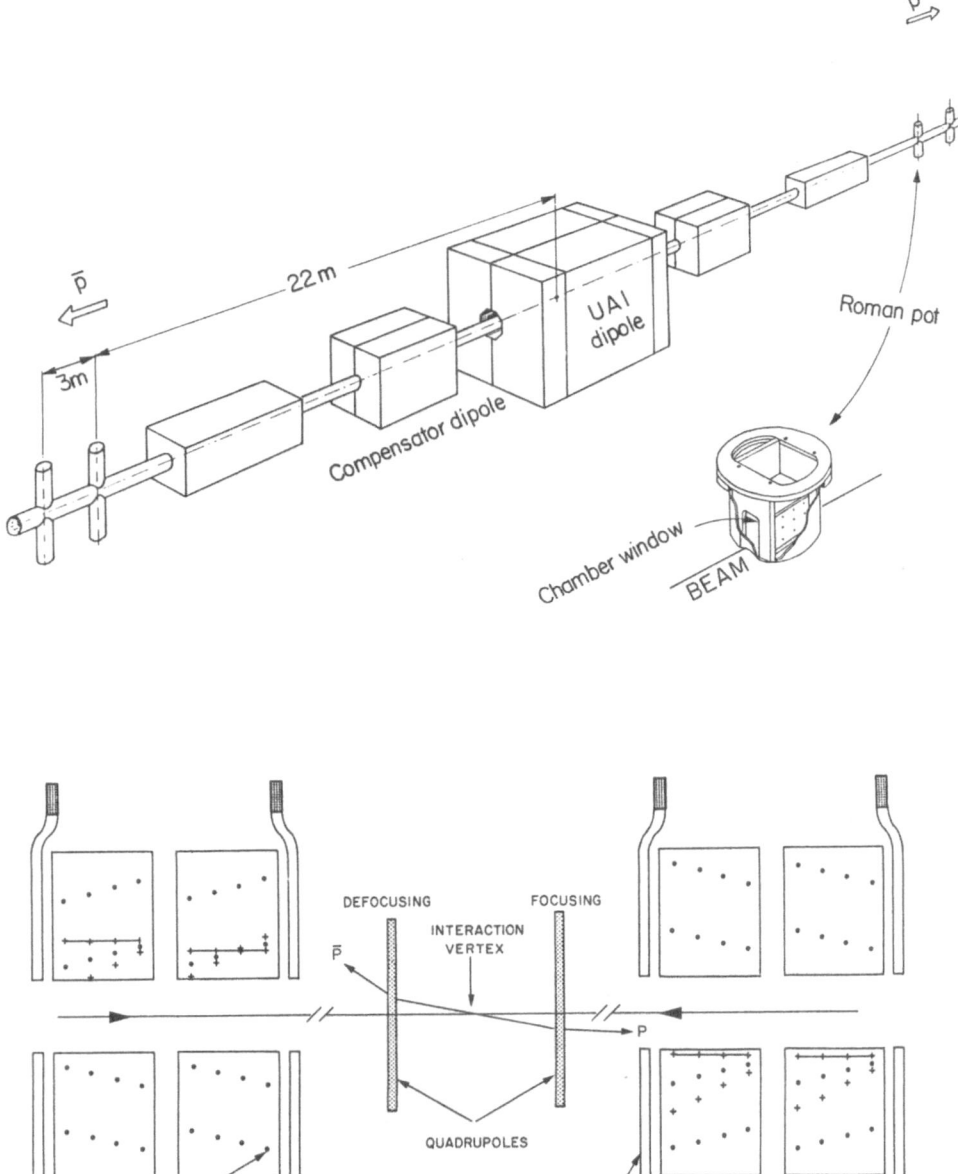

Fig. 2d and 2e

("C" and "I"), and finally the μ-detector.

The detector is extended into the forward region ($0.2°<$ $<θ<5°$) by a calorimetrized compensator magnet ("Calcom") and small angle calorimeters ("very forward calorimeter"), as well as track detection (drift chamber). Furthermore, Roman pots with drift chambers at ∿25 m from the intersection allow measurements of elastic scattering angles of a few mrad.

The spectrometer magnet is a dipole magnet which produces a very uniform field of maximum value 0.7 T over a volume of $7 × 3.5 × 3.5$ m^3. The field direction is horizontal, perpendicular to the beam direction. The magnet contains the central detector and the central electromagnetic (EM) calorimeters (gondolas and bouchons).

The central detector consists of six semicylindrical drift-chambers which together cover a cylindrical volume of diameter 2.3 m and length 5.8 m. The image read-out of the 6110 sense wires yields space points at centimetre intervals on the detected tracks, giving a bubble-chamber like picture of the collision. All the wires (sense and field wires) are parallel to the magnetic field (i.e. horizontal). They are arranged in horizontal (forward) and vertical (centre) planes so that the drift direction is approximately vertical in the four forward chambers and approximately horizontal in the two centre chambers. This way, the drift direction is approximately perpendicular to the particle trajectories, thus optimizing the measurement of sagittae for high momentum tracks. On the average, ∿100 points per track are recorded. The spatial resolution is ∿290 μm in the drifttime coordinate and ∿2% of the wire length for the charge division coordinate. This results in a typical relative momentum resolution of ∿20% for a 1m long track of p=40 GeV in the plane normal to the magnetic field (better for longer tracks, worse for other directions). The ionization loss can be measured to ∿10% (truncated mean of the lowest 60%). Thus it is possible to identify narrow high - energy particle bundles (e^+e^- pairs or pencil jets)

Calorimeter		Angular coverage θ (°)	Thickness		Cell size		Sampling step	Segmentation in depth	Reso-lution
			No. rad. lengths	No. abs. lengths	Δθ (°)	Δφ (°)			
Barrel	e.m.: gondolas	25–155	26.6/sinθ	1.1/sinθ	5	180	1.2 mm Pb / 1.5 mm scint.	3.3/6.6/10.1/6.6 X_0	$0.15/\sqrt{E}$
	hadr.: C's		–	5.0/sinθ	15	18	50 mm Fe / 10 mm scint.	2.5/2.5 λ	$0.8/\sqrt{E}$
End-caps	e.m.: bouchons	5–25 and 155–175	27/cosθ	1.1/cosθ	20	11	4 mm Pb / 6 mm scint.	4/7/9/7 X_0	$0.12/\sqrt{E_T}$
	hadr.: I's		–	7.1/cosθ	5	10	50 mm Fe / 10 mm scint.	3.5/3.5 λ	$0.8/\sqrt{E}$
Calcom	e.m.	0.7–5 and 175–179.3	30	1.2	4	45	3 mm Pb / 3 mm scint.	4 × 7.5 X_0	$0.15/\sqrt{E}$
	hadr.		–	10.2	4	45	40 mm Fe / 8 mm scint.	6 × 1.7 λ	$0.8/\sqrt{E}$
Very forward	e.m.	0.2–0.7 and 179.3–179.8	24.5	1.0	0.5	90	3 mm Pb / 6 mm scint.	5.7/5.3/5.8/7.7 X_0	$0.15/\sqrt{E}$
	hadr.		–	5.7	0.5	90	40 mm Fe / 10 mm scint.	5 × 1.25 λ	$0.8/\sqrt{E}$

Table 4a: General characteristics of the central detector

which cannot be resolved by using the space coordinates.

Calibration is done with cosmic rays, X-rays, laser beams and with an electronic testpulse system.

The main characteristics of the central detector are summarized in table 4a; more details can be found in ref.25.

The large angle EM calorimeter surrounds the central detector, covering the polar angle region $25° < \theta < 155°$. It consists of two semicylindrical half shells, one on either side of the beam axis, with an inner radius of 1.36 m. In each half shell there are 24 modules ("gondolas"), each of which measures 4m around its circumference and 22.5 cm in the beam direction. The light produced in each of the four segments in depth is seen by wavelength shifter plates on each side of the gondola, which in turn are connected via light guides to four photomultipliers (PM's), located outside the magnet. A comparison of the pulseheights of these four PM's in each segment allows a measurement of azimuthal angle ϕ and position along the beam direction, x, for localized energy depositions, with resolutions $\sigma(\phi) \approx 0.3E^{-1/2}$, $\sigma(x) \approx 6.3E^{-1/2}$ (with E in GeV, ϕ in rad, x in cm). The energy of an electron is measured with a resolution of $\sigma(E) \sim 0.15 \ E^{1/2}$ (E in GeV).

Calibration is done by measurements in testbeams and with a Co^{60} source (7 Ci). The stability of the gain of the PM's is monitored with a laser system, which in turn is monitored by silicon photodiodes (More information can be found in [32,35] and the technical notes quoted there; see also table 4b)

The endcap EM calorimeter ("bouchons") covers the end faces of the central detector ($5°<\theta<25°$). It consists of 2 × 32 radial sectors ("petals"), segmented four times in depth and positioned at 3m from the interaction point. The attenuation length of the scintillator has been chosen to match the variation of $\sin\theta$ over its angular range, so that the sig-

Type	Drift chamber with charge division readout of the second coordinate
Gas mixture	Argon (40%) + ethane (60%)
Drift field and gap length	1.5 kV/cm, 18cm
Drift velocity	5.3 cm/μs
Drift angle	23° at $|\vec{B}| = 0.7$ T
Anode plane arrangement:	
a) Distance between sense wires	10 mm
b) Wire length	80 cm min., 220 cm max.
c) Sense wire charac.	35 m Ni-Cr stretched at 80 g
d) Field wire charac.	100 μm gold-plated Cu-Be stretched at 200 g
Cathode plane structure:	
a) Distance between wires	5 mm
b) Wire characteristics	150 μm gold-plated Cu-Be stretched at 200 g
Total number of wires	22800
Total number of sense wires	6110

Table 4b: Angular coverage, segmentation and resolution of
the calorimeter.

nal size is proportional to the transverse energy $E_T = E \cdot \sin\theta$
for every cell. The shower position is measured by a position
detector (proportional tubes) with a precision of ±2 mm. (see
talbe 4b and refs.32,36).

The central hadron calorimeter (large angle - C's and end-
caps - I's), surrounding the gondolas and bouchons, is built
from scintillator plates inserted into the luminated yoke
of the magnet. (see table 4b and refs.26,35).

The μ-detector surrounds the magnet yoke with two layers
of drift tube chambers, covering the polar angle range
$5^O < \theta < 175^O$ and full azimuth, with a total sensitive area
of ∿500 m². Every chamber (4 × 6m²) consists of 2 × 2
staggered planes of drift tubes (made of extruded aluminium
and therefore light and self-supporting). The two layers
of chambers are separated from each other (distance between
corresponding planes = 62 cm). Tracks are measured in two
orthogonal projections with four wire planes per projection.
The angular resolution is ∿1 mrad. More details are given
in ref.27.

3.3.2 UA2

The UA2 detector[21,37-43] , shown in fig.3, was designed
mainly with the aim to observe electroweak boson decays
and to study final states containing high transverse momentum hadron jets. It is composed of fine segmented calorimeter cells, arranged in a tower structure pointing to the
interaction region, allowing electron identification in
the full angular acceptance $(20^O < \theta < 160^O)$. The forward
regions $(20^O < \theta < 40^O, 140^O < \theta < 160^O)$ are equipped
with two magnetic spectrometers, each consisting of a toroidal
field magnet followed by nine drift chamber planes. Preshower counters in front of the calorimeter serve to improve
the space resolution for shower localization. The central
preshower counter is a cylindrical proportional chamber
with helicoidol cathode strips and the forward one consists

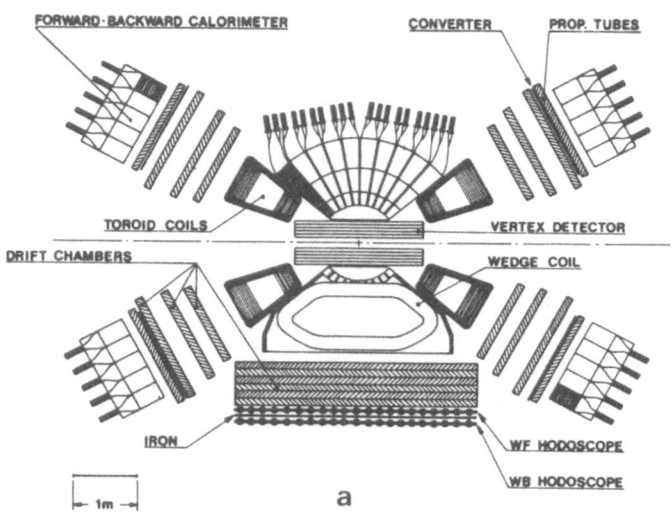

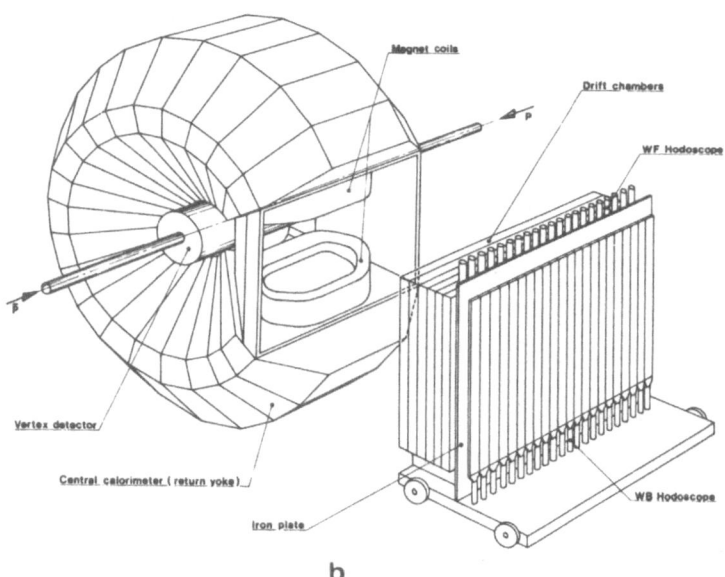

Fig.3: The UA2 detector
   (a) Schematic cross section in the horizontal plane
       containing the beam
   (b) Isometric view of the detector
   (The wedge detector has been removed after the 1982 run).

of proportional tube planes. Each is preceded by a converter (∿1.5 radiation lengths).

A set of cylindrical wire chambers densely packed around the beam in the collision region provides measurements of the position of the vertex and the directions of the charged particles produced in the collision. It is made of four proportional chambers with helicoidal cathode strips and of two drift chambers.

Up to the end of 1982, a wedge ($\Delta\phi = 60^{\circ}$) was left open in the central region to house a large angle magnetic spectrometer instrumented with drift chambers, scintillator hodoscopes and a lead-glass away. It has been used to provide information about inclusive particle production around 90°[38,41], as well as for a search for fractionally charged particles[40]. It has now been closed to lead to an azimuthally symmetric detector configuration, better suited to the study of IVB's and jets.

The main detector parameters and listed in table 5.

3.3.3 UA3

Experiment UA3[22,44] is a search for magnetic monopoles. It consists of plastic sheets, at present Kapton of thickness 125 μm. These are placed inside and outside the vacuum tube in the intersection region, and around the half cylinders of the UA1 central detector image chamber, as shown in fig.4. They make use of the expected high ionization of monopoles

$$(\frac{dE}{dx})_{mono} / (\frac{dE}{dx})_{min.ion.} = \beta^2(g^2/e^2) ,$$

where β is the velocity of monopoles, g their pole strength, and e the usual unit of electric charge. The factor (g^2/e^2) has a value ∿5×10^3 for Dirac monopoles. Some theories predict a pole strength a factor of 12 higher than this.

The radiation damage caused by this massive ionization

Table 5: Main parameters of the UA2 detector

I. SOLID ANGLE COVERAGE

    Central region. Electron-hadron calorimetry.

$$\Delta y \simeq 2, \ \Delta\theta \simeq 100^{O}, \ \Delta\phi \simeq 300^{O}$$

    Forward-backward regions. Electron calorimetry and magnetic
spectroscopy.

$$\Delta y \simeq 1.5, \ \Delta\theta \simeq 35^{O}, \ \Delta\phi \simeq 82\% \ of \ 360^{O}$$

    Wedge region. Electron calorimetry and magnetic spectroscopy.

$$\Delta y \simeq 1.3, \ \Delta\theta \simeq 68^{O}, \ \Delta\phi \simeq 28^{O}$$

II. CENTRAL CALORIMETER

    200 cells, each covering $\Delta\theta \times \Delta\phi = 10^{O} \times 15^{O}$.
    Longitudinal segmentation

            17 r.l. (lead-scintillator) + 2 × 2 abs. l.
            (iron-scintillator)

    Electron calorimetry 26 × 3.5 mm lead plates
                            27 × 4 mm NE 104 scintillator plates
    Hadron calorimetry (18 + 22) × 15 mm iron plates
                            (18 + 22) ×5 mm scintillator plates
                            (PMMA, 10% naphtalene, 1% PBD, 0.01%
                            POPOP)
    Light guides: 2 mm lucite, 80 mg/l BBQ
    Photobus: 7 per cell, XP2012 for electron calorimetry and
           XP2008 for hadron calorimetry

III. FORWARD DETECTORS

    24 identical sectors each covering $\Delta\theta \times \Delta\phi = 17.5^{O} \times 25^{O}$
    Field integral 0.38 Tm.
    9 drift chambers per sector
        - wire orientation with respect to field: $-7^{O}, 0^{O}, +7^{O}$.
        - drift cell width ±5 cm
        - field shaping wires every 5 mm
        - total number of signal wires 2304.

preshower counter
- preconverter 1.4 r.l. lead + iron
- 4 tube planes (brass) 20 mm O.D., 0.3 mm thick.
- tube orientation with respect to field: 0°, 0°, 77°, 77°.
- anode 30 micron gold plated tungsten

forward calorimeters
- 10 cells per sector, each covering $\Delta\theta \times \Delta\phi \simeq 4^{\circ} \times 15^{\circ}$
- cell transverse sizes 27×33 to 27×60 cm^2
- longitudinal segmentation: $33 \times$ (4 mm lead + 4 mm Altustipe 10105) + $8 \times$ (4 mm lead + 4 mm Altustipe 10105) = 24 + 6 r.l.
- light guides and phototubes as in central calorimeter.

IV. VERTEX DETECTOR
- five proportional chambers with cathode strip read-out, one of which is located behind a 1.5 r.l. tungsten converter.
  number of strips 480, 480, 528, 672 and 480.
  number of wires 288, 384, 576, 864 and 576.
  chamber radii 100, 124, 236, 315 and 355 mm.
  chamber lengths 104, 110, 150, 178 and 80 cm.
  wire pitch 2.2, 2.0, 2.6, 2.3 and 3.9 mm.
  strip angle $\tan \alpha = \pm 0.9$, 1.3, 1.0, 1.0, 1.0.
  half gap 4 mm
  strip pitch $\simeq$ 4 mm.
- two drift chambers of 24 azimuthal cells each, 6 sense wires/cell (charge division, multihit capability), sense wire lengths 1520 and 1785 mm
- 24 scintillator plates.

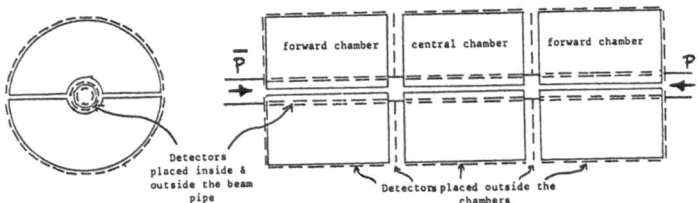

Fig.4: The UA3 detector lay-out. The position of the
Kapton foils around the beam pipe and the UA1 central
detector is indicated.

in Kapton sheets is made visible by chemical etching. The
sheets are completely insensitive to normal minimum-ioni-
zing electrically charged particles, the threshold being
a factor $\sim 2 \times 10^3$ higher. Possible background tracks might
come from heavy nuclear fragments knocked out of the vacuum
pipe material, but the range for these is typically only
10 μm in Kapton. Monopoles would be accelerated in the direction of the dipole magnet (maximum strength 0.7 T) which is
part of the UA1 apparatus.

No monopoles have been seen so far, and upper limits
for their production cross section have been published[44].

3.3.4 UA4

Experiment UA4[23,45-47] is designed to
measure the $\bar{p}p$ total cross section and elastic scattering.
The apparatus consists of an "inelastic detector" (needed
for the measurement of the total interaction rate), shown
in fig.5a and an "elastic detector", whose lay-out is shown

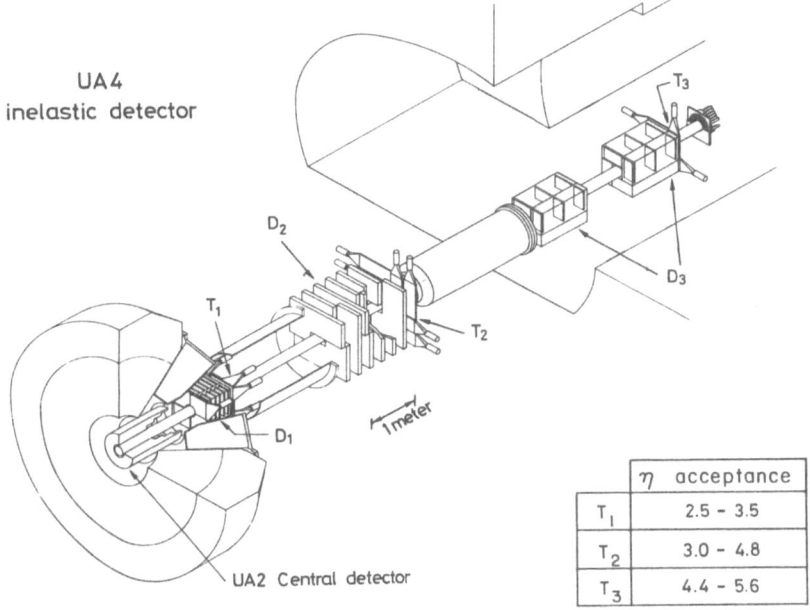

	η acceptance
T_1	2.5 - 3.5
T_2	3.0 - 4.8
T_3	4.4 - 5.6

Fig.5 : The UA4 detector

(a) lay-out of one arm of the inelastic detector; the assembly is symmetric with respect to the crossing region. The central detector of experiment UA2 is also sketched. The η acceptance of the trigger counters T_1, T_2 and T_3 is indicated.

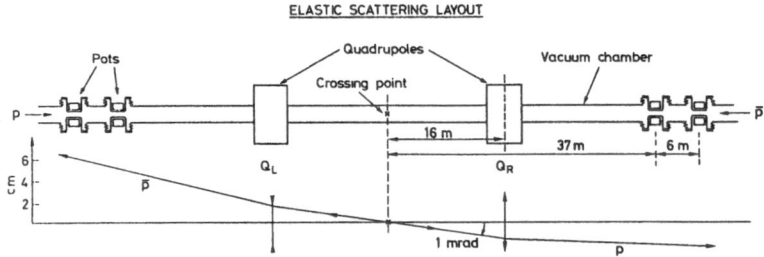

Fig.5b: Sketch of the elastic scattering set-up (Note that meanwhile additional Roman pots have been added at 37 m and 43 m from the interaction point).

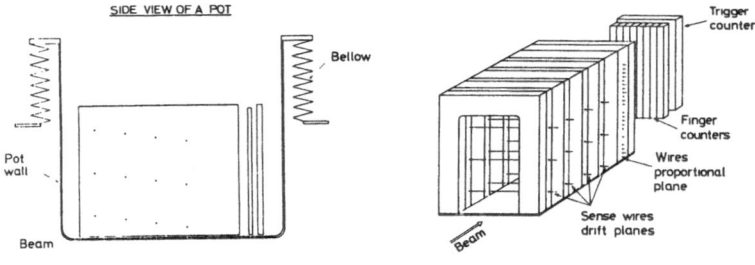

Fig. 5c: A detail of the detector inside the pot, together with a perspective view.

in fig.5b. Four pairs of detectors are placed above and below the beams, about 40 m away from the interaction point. Each detector consists of four small drift planes with vertical drift directions, a multiwire proportional chamber (MWPC), and some scintillation counters as shown in fig.5c, and is placed inside a "Roman pot", a thin vessel with vacuum bellows which enables the detector to be withdrawn during beam injection and then brought close to the circulating beams when conditions are sufficiently quiet. An up/down or down/up timing coincidence is required between the two arms, and the two tracks must extrapolate collinearly at the vertex. The long lever arms, which are required in order to observe scattering of around 1 mrad, cause the detectors to be placed beyond the first set of quadrupole magnets, and hence an accurate knowledge of their transfer matrices is required. This is also needed for Monte Carlo acceptance calculations. The detectors must be moved to within a few centimetres of the beam centre. When low-beta schemes are in use to squeeze the beams at the interaction point for high luminosity, the beams will diverge strongly and have a significant size at the position of the Roman pots. Hence low-beta operation tends to be incompatible with good conditions for elastic scattering.

For the run at the end of 1982, the UA4 beam has added additional Roman pots located at 22 and 25 m from the interaction point, thereby extending their angular coverage and making the detector more flexible to adapt to variations in the operation scheme of the machine[47].

3.3.5 UA5

The main aim of experiment UA5[24,48-53] was to play the role of a survey experiment providing fast information about the general features of $\bar{p}p$ collisions at the collider. This requirement was met by using a detector which was already well understood at the beginning of the collider

430

operation since it had already been used in a run-in experiment at the ISR[54].

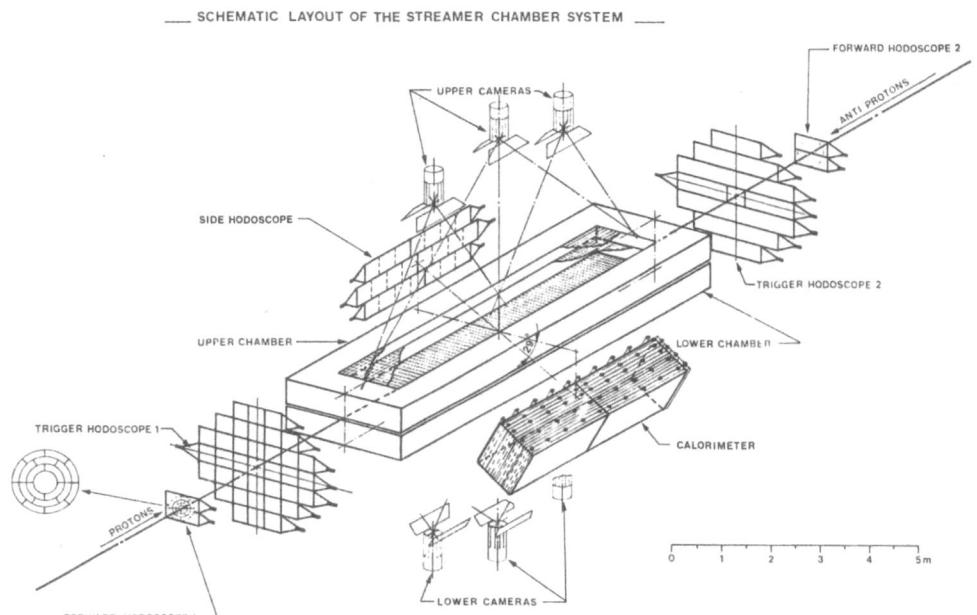

SCHEMATIC LAYOUT OF THE STREAMER CHAMBER SYSTEM

Fig.6: The UA5 detector

A schematic lay-out of the streamer chamber system is shown, as well as the calorimeter (being constructed) which will be used to trigger on high transverse energy events.

The detector, shown in fig.6, consists of two large streamer chambers ($6.0 \times 1.25 \times 0.5 \ m^3$), placed 4.5 cm above and below the SPS beam pipe. Each chamber is viewed by

three stereoscopic cameras: two main cameras, one on each
end, viewing slightly more than half of the chamber at a
demagnification of 50, and a central supplementary camera
viewing the whole chamber at a demagnification of 80. Image
intensifier tubes of gain $\approx$2000 are used with each camera;
this allows to run the streamer chambers with small streamers
($\approx$5 mm) and thus to obtain the necessary two-track resolution
of 2-3 mm. There is no magnetic field, so the observed tracks
are straight lines spreading out radially from the interaction point. The geometrical acceptance for charged particles
is $\sim$95% for $|\eta| <$ 3, falling to zero at $|\eta| \sim$5.

3.3.6 UA6

The experiment   UA6[55] is not a collider experiment
in the strict sense of the word, since it is not intended to
study $\bar{p}p$ interactions at $\sqrt{s}$ = 540 GeV by observing head-on
collisions between the p and $\bar{p}$ bunches circulating in the
ring. Instead, it is designed to investigate certain features
of $\bar{p}p$ interactions at $\sqrt{s}$ =22.5 GeV by observing collisions of
the p and $\bar{p}$ beams with a fixed target. For this purpose, a hydrogen cluster jet target is installed in a medium straight
section of the ring, where p and $\bar{p}$ bunches do not arrive
simultaneously. This target will allow running in parallel
with the collider experiments, with a $\bar{p}p$ and pp luminosity
of 10^{31}cm^{-2}s^{-1}.

The apparatus (shown in fig.7) will consist of a double-
arm spectrometer with high resolution wire chambers and EM calorimeters, The purpose of the experiment is to measure
the production of large mass electron pairs and high $p_T\pi^o$
and single photons in pp and $\bar{p}p$ interactions, as well as the
measurement of Λ and $\bar{\Lambda}$ polarization at large angles in $\bar{p}p$
interactions. Furthermore, a set of solid state counters at

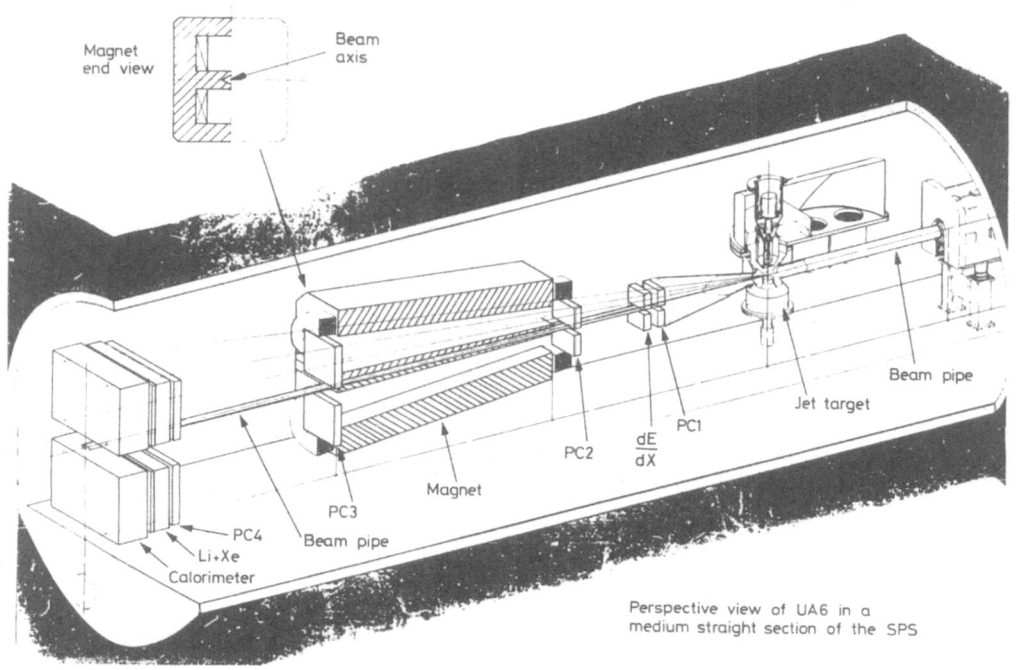

Fig.7: The UA6 detector - perspective view

90°(lab.) will be used to measure low-t elastic and inelastic cross sections.

4. GENERAL FEATURES OF $\bar{p}p$ COLLISIONS AT $\sqrt{s}$ = 540 GEV

4.1 Elastic and Total Cross Section

The angular dependence of elastic scattering has been
measured by UA1 and UA4 in various t regions (see table 6
for a summary of the available data). It is found that both
in the low ($|t| \lesssim$ 0.15 GeV^2) and medium t range (0.15$\lesssim|t|\lesssim$0.5)

-t range, GeV^2	number of events	slope b, GeV^{-2}	expt.	ref.	remark
.05 to .19	1480	17.2±1.0	UA4	45	
.14 to .26	787	13.3±1.5	UA1	31	
.21 to .45	7050	13.7±0.3	UA1	56,57	
.04 to .18	708	17.1±1.0	UA1	56,57	
.21 to .50	∿7000	13.7±0.3	UA4	47	preliminary
.03 to .19	840	17.6±1.0	UA4	47	preliminary
.45 to 1.5	∿40000	-	UA4	47	being analyzed

Table 6: Summary of elastic scattering data from the collider.

the data are reasonably well described by an exponential, exp(bt), but with different values in the two regions:

$$\text{low } t : b \sim 17 \text{ GeV}^{-2}$$
$$\text{medium } t : b \sim 13.6 \text{ GeV}^{-2},$$

i.e. the slope seems to change around $|t| \sim 0.15$ GeV^2. Unfortunately, the different t regions are in general covered in different data taking runs with different machine conditions (β-values). Therefore it is at present not possible to state whether there is a "kink" or a smooth transition between the two t-regions.

As can be seen from fig.8b and 8c, the values for the slopes found at the collider agree very well with extrapolations from measurements at lower energies; the diffraction peak continues to shrink.

By the optical theorem, the total cross section is related to the elastic cross section:

434

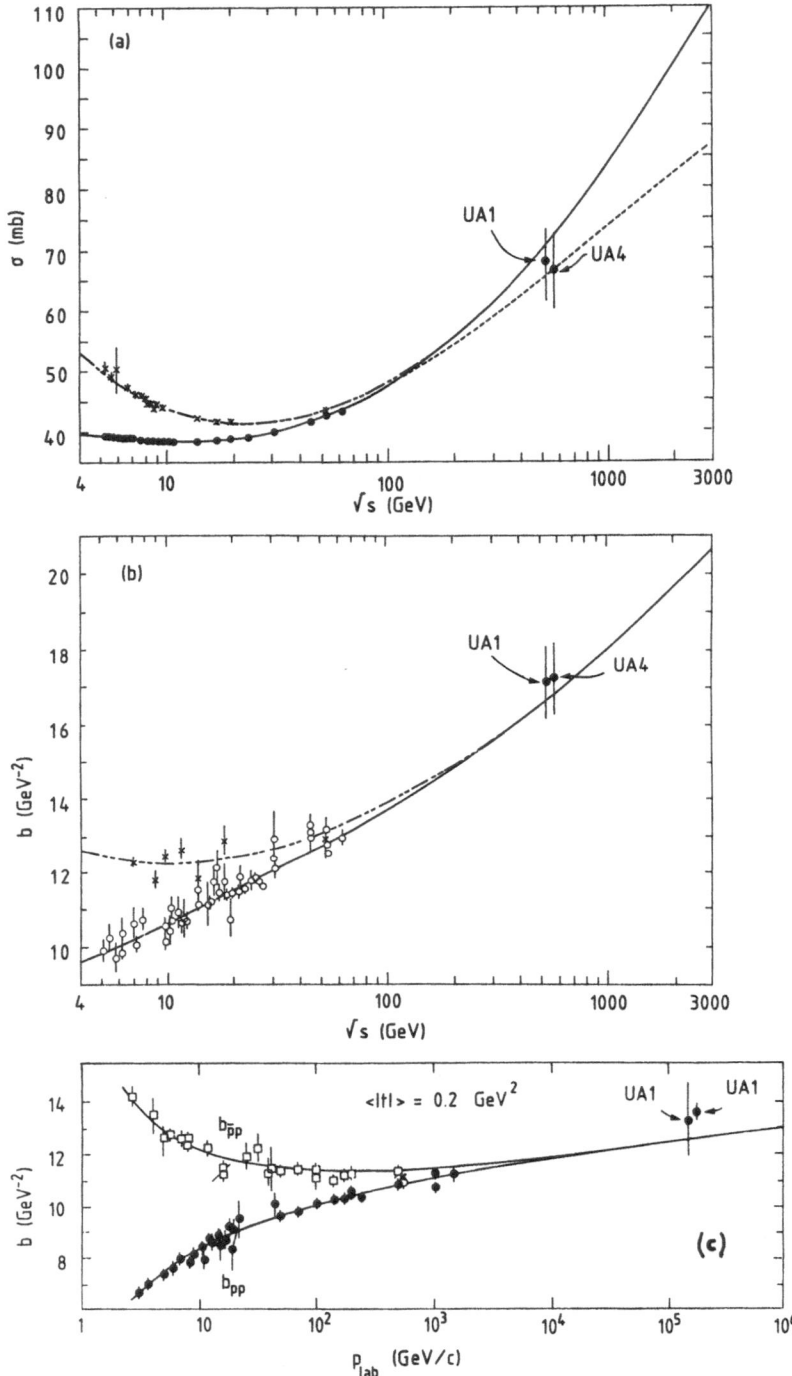

Fig. 8

Fig.8: Energy dependence of total and elastic cross section, and comparison with extrapolations based on fits to data at lower energy (figure from ref.56).

(a) total cross section (below 100 GeV: pp: full dats, p̄p crosses); collider data from UA4 [46] and UA1[58] The curves are fits to data at lower energy[59] and two possible extrapolations to √s>100 GeV (same curve for pp and p̄p for √s>100 GeV): the upper curve (solid line) corresponds to a $(\ln s)^2$ behaviour, the lower curve (dashed line) has a lns behaviour for √s>100 GeV.

(b) Slope b of dσ/dt at low t (pp: open circles, p̄p: crosses). The curves are the fits[59] corresponding to the $(\ln s)^2$ solution for σ_{tot}:

(c) Slope b of dσ/dt at $|t| \sim 0.2$ GeV^2, compared with extrapolation from low energy data[60].

$$\sigma_{tot}^2 = \frac{16\pi(\hbar c)^2}{1+\rho^2} \left. \frac{d\sigma_{el}}{dt} \right|_{t=0} \tag{2}$$

where ρ is the ratio of the real to the imaginary part of the forward elastic scattering amplitude.

Furthermore, σ_{tot} and σ_{el} are related to counting rates by the luminosity L:

$$\sigma_{tot} = L (N_{in} + N_{el}) , \qquad \frac{d\sigma_{el}}{dt} = L \frac{dN_{el}}{dt} , \tag{3}$$

where N_{in}, N_{el} are counting rates for inelastic and elastic events. Inserting (3) into (2) yields

$$\sigma_{tot} = \frac{16\pi(\hbar c)^2}{1+\rho^2} \frac{1}{N_{el}+N_{in}} \left. \frac{d\sigma_{el}}{dt} \right|_{t=0} . \tag{4}$$

If ρ is known, σ_{tot} can be determined by a simultaneous measurement of low t elastic scattering and either

(a) the total counting rate (without knowledge of the
    luminosity)
or (b) the luminosity (without knowledge of the total coun-
    ting rate).
Both UA4[46,56] and UA1[57,58] have determined the total
cross section from their data on elastic scattering. The
assumptions are:

    . $\rho = 0$ (note that for collider energies, $\rho < 0.2$ is
      expected from dispersion relation calculations[59,61]),

    . no change of slope for $|t| < 0.04$.

UA4 used an inelastic rate measurement (i.e. method (a)),
while UA1 used the luminosity determined from a measurement
of beam profile and beam currents (see[58] for more details).
The results, shown in table 7, are compatible with extrapolations from data at lower energies (see fig.8a), both
with lns and $(lns)^2$ dependence.

experiment	σ_{tot}, mb	σ_{el}/σ_{tot}	b/σ_{tot}, GeV^{-2} mb^{-1}	reference
UA4	66±7	0.20±.02	0.26±.03	46
UA1	67.6±5.9	0.21±.02	0.25±.03	58
UA4	71 ± 7	0.21±.02	0.25±.03	56

Table 7: Total cross section and related quantities, as
         measured by UA1 and UA4.

Table 7 also gives the results obtained for σ_{el}/σ_{tot}
and $b(t=0)/\sigma_{tot}$. At the collider, one finds

$$\frac{\sigma_{el}}{\sigma_{tot}} \sim 0.21 \pm .02, \qquad \frac{b}{\sigma_{tot}} \sim 0.25 \pm .03 \ GeV^{-2} \ mb^{-1},$$

while at the ISR these quantities were found to be ~ 0.18
and $0.30 \ GeV^{-2} mb^{-1}$ respectively, independent of energy over
the ISR energy range[62].

It has been pointed out[63] that if the Froissart
bound is qualitatively saturated, i.e.

$$\frac{\sigma_{tot}}{(lns)^2} \rightarrow const \neq 0 \ ,$$

then the asymptotic behaviour for σ_{el} and the slopes at
t=0 and t>0 is:

$$\frac{\sigma_{el}}{\sigma_{tot}} \rightarrow const \neq 0,$$

$$b(s,t = 0) \quad \sim \ c \cdot (lns)^2$$

$$b(s,t > 0) \quad \sim \frac{c'}{\sqrt{|t|}} \cdot lns \ ,$$

i.e. there is a non-uniform behaviour of the slope parameter
at high energies. The data appear to be compatible with this
behaviour, but more precise measurements of σ_{tot}, σ_{el}/σ_{tot}
and the slope are needed. A more detailed discussion can be
found in [64].

4.2 Particle Production

Rapidity distributions (UA1 [34], UA5[49]

The distribution of charged particles as a function
of pseudorapidity $\eta = -ln \tan \theta/2$ is shown in fig.9a (the
slight discrepancies between the UA1 and UA5 data are presumably due to differences in the trigger, i.e. different
fractions of diffractive events are excluded in the two experiments). For comparison, a curve representing ISR data[65]
is also shown. The distribution obtained at $\sqrt{s}$ = 540 GeV is
narrower than one would expect from a simple extrapolation
of ISR data: the width of the central plateau has grown by

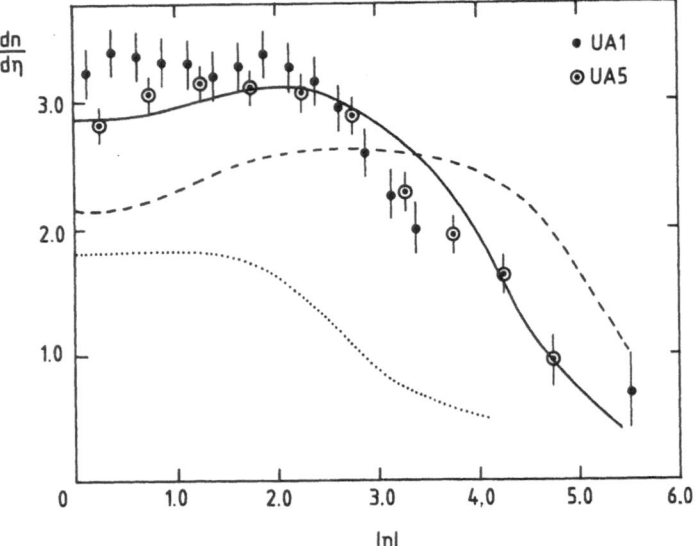

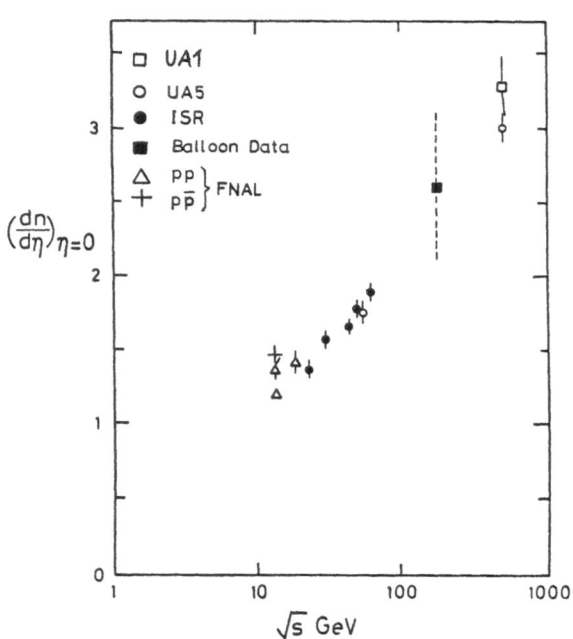

Fig. 9

Fig.9: Charged particle density dn/dη

    (a) dn/dη vs pseudorapidity η, as measured by UA1[34]
       and UA5[49]. The dotted curve represents ISR data
       at $\sqrt{s}$ = 53 GeV[54], the solid and dashed lines are
       results of Monte Carlo calculations for $\sqrt{s}$ = 540 GeV
       [49] with $<p_T>$ = 350 MeV (dashed line) and $<p_T>$ = 500 MeV
       (solid line).

    (b) $dn/d\eta|_{\eta=0}$ vs energy for $\bar{p}p$ and pp data (fig. from[34])

---

only ∿2 units of rapidity in going from $\sqrt{s}$ = 53 GeV to
$\sqrt{s}$ = 540 GeV, whereas the separation of the two beam particles
has grown by 4.6 units. This is interpreted as a kinematical
effect, due to the growth of the average transverse momentum
between ISR and collider energies.

    The central particle density, $\frac{dn}{d\eta}|_{\eta=0}$ is measured as
3.3 ± 0.2 by UA1 and 3.0 ± 0.1 by UA5, i.e. it rises by
∿70% from the value at the highest ISR energy $\sqrt{s}$ = 63 GeV
[66].

    The energy dependence of this quantity (fig.9b) is
compatible with a lns behaviour. The dependence of the rapidity distribution on event multiplicity shows similar
features to those observed at the ISR[66]; the central particle density grows linearly with the event multiplicity.

Multiplicity distributions (UA1[34], UA5[49,52].

    The charged multiplicity distribution has been
measured by UA1 for $|\eta|$ < 3.5 and by UA5 for $|\eta|$<5. It obeys
approximate KNO-scaling[67], but closer inspection shows
derivations at the high multiplicity tail of the distribution
[52](there are more high multiplicity events than expected
from KNO scaling).

    The total mean charged multiplicity is found (UA5) to
be $<n_{ch}>$ = 28.9±0.4 for non-single diffractive events;
correcting for the exclusion of single diffractive events

brings this number down to 26.5 ± 1.0[68]. This is compatible with the value ∿25 obtained from extrapolating the best fit [66] to ISR and FNAL data ($<n> \simeq 0.88 + 0.44$ lns + + 0.118 (lns)^2).

Transverse momentum distributions (UA1[29,36], US2[38,41])

Fig.10 compares inclusive transverse momentum distributions for charged and neutral particles at $\sqrt{s}$ = 540 GeV with these obtained at the ISR. There is an increase of about 3 orders of magnitude in cross section at p_T = 10 GeV/c. Both the ISR and collider data are in good agreement with QCD predictions[69].

The spectrum measured by UA1 is found to be well described by the empirical formula

$$E \frac{d^3\sigma}{d^3p} = A \cdot (\frac{p_o}{p_T + p_o})^n \text{ with } p_o \simeq 1.3 \text{ GeV/c,} \quad n \sim 9 .$$

The mean transverse momentum is $<p_T>$ = 424 MeV/c, i.e. ∿20% higher than at the ISR.

As shown in fig.11, the p_T spectrum is multiplicity dependent: it becomes flatter for higher multiplicity, in qualitative agreement with observation in cosmic ray experiments[70]. In fig.11b the mean transverse momentum is plotted as a function of the central particle density; $<p_T>$ increases from ∿350 to ∿470 MeV/c and saturates above a density of ∿10 charged particles per unit of rapidity. This saturation may be due to kinematics, but could also be interpreted[71] as evidence for a phase transition of hadronic matter (deconfinement phase transition from hadron gas to quark-gluon plasma).

Particle composition (UA2[38,41], UA5[52])

The typical event at $\sqrt{s}$ = 540 GeV has ∿43 primary stable particles produced in $|\eta|$ < 5, of which 62% are charged, 12% are kaons and ∿9% baryons and antibaryons. At

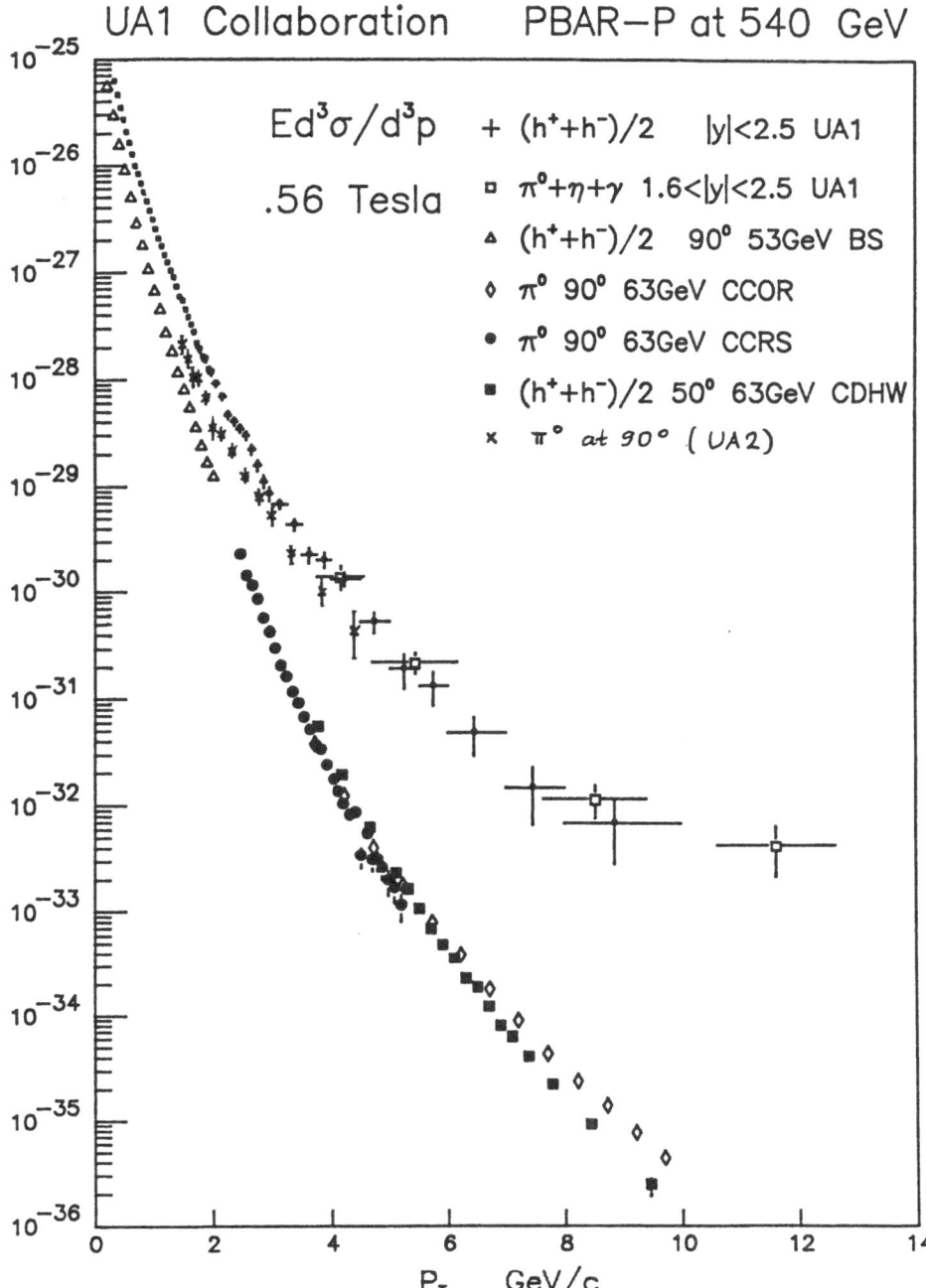

Fig.10: Invariant cross section $\dfrac{Ed^3\sigma}{d^3p}$ vs p_T for charged and
neutral particles (UA1, refs.29,36) and π^0(UA2,ref.38).
Also shown are data from various ISR experiments(see
[29] for references to ISR data).

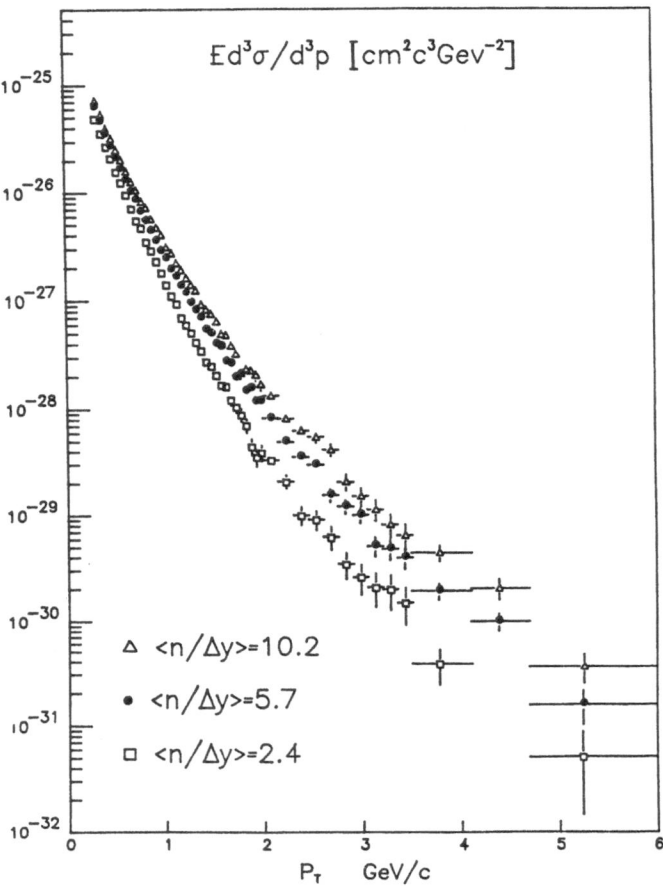

Fig.11: Multiplicity dependence of p_T spectra (fig.from[29])
(a) Invariant cross section as function of transverse
momentum for three bands of charged track multiplicity
in the rapidity interval $|y|<2.5$ (normalised to the
full cross section of $p_T=0$).

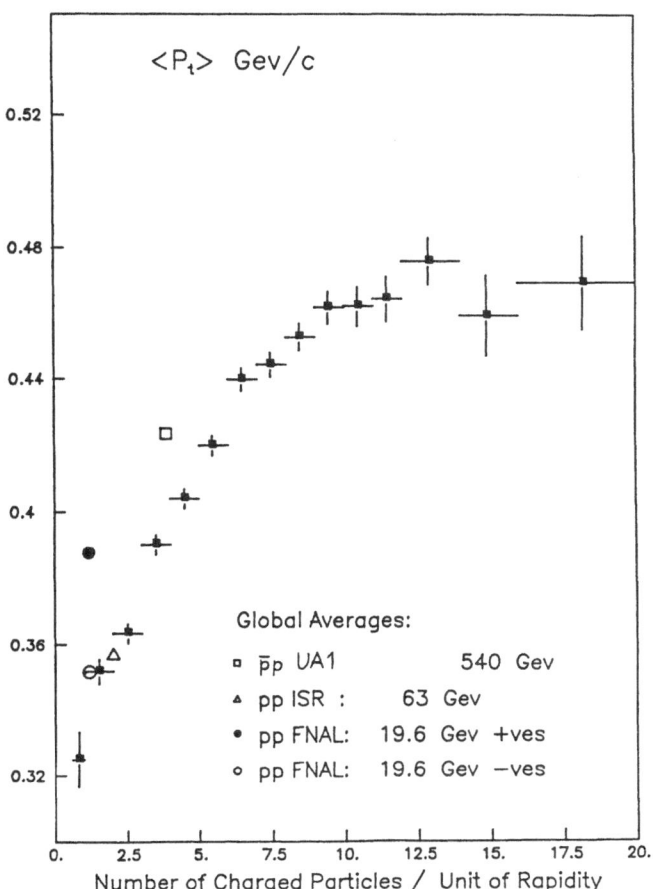

Fig.11: Multiplicity dependence of p_T spectra (fig. from[29]).
(b) The mean transverse momentum of charged hadrons as a function of charged track multiplicity in the rapidity interval $|y|<2.5$ (full squares: UA1 data, $\bar{p}p$ at 540 GeV). Points without error bars give the global average of p_T as a function of the mean number of charged particles at FNAL, ISR and collider energies (see[29] for references)

$\sqrt{s}$ = 53 GeV, there are ∿9% K's and ∿5% baryons, i.e. the
ratio of non–π stable particles to π's has increased appre-
ciably from the ISR to the collider. However, the proportion
of strongly decaying resonances does not seem to have
changed very much (inferred from the observed similarity
of correlation phenomena at ISR and collider [52]).

5. JETS

5.1 Importance and Difficulty of Jet Studies in Hadronic Interactions

The suggestions that hard scattering of hadron constituents
(partons) should manifold itself in hadron collisions by the
presence of high p_T jets originating from the frequentation
of the scattered partons [72], has stimulated intense experimental effort[73].

The interest in this subject is motivated by the fact
that hadron hadron collisions offer the only "laboratory" for
studying the interaction between the basic constituents of
hadrons (quarks and gluons). Unfortunately, jets in hadron
collisions stick out much less clearly than in e^+e^- interactions
[74], since due to the more complicated initial state, they
carry only a fraction of the energy available in the collision:
jets from the hard scattering process are accompanied by
"soft" (low p_T) hadrons originating from the fragmentation
of the spectator partons and from gluon bremsstrahlung [75].
Furthermore, since the parton-parton c.m.s. is different from
the over-all c.m.s. of the collisions the two jets are in
general not collinear.

Therefore, in order to be able to identify jets and
thus get a clear signature of the constituent structure of
hadrons, one has to select particular, rare types of events
(see[76] for a discussion of recent jet studies).

One of the possibilities is to trigger on a single

high p_T particle and to study the event structure, exploiting the fact that a particle with high p_T tags events in
which a hard scattering occured. Investigation of such
events of studying correlations of particles among each
other and with the trigger particle, both on the trigger
side (the hemisphere containing the trigger particle) and the
"away" side (hemisphere opposite in azimuth to the trigger)
have established evidence for jet-structure in hadron collisions and provided valuable information about the underlying
process [30,73,77].

Triggering on a high p_T particle however, selects pre-
ferably those events in which one of the jets fragments in
such a way that one particle (the trigger) gets most of the
jet energy, i.e. the jet yield is strongly suppressed due
to the fast fall-off of the fragmentation function.

5.2 Jet Search by Calorimetric Methods

A more efficient (and less biased) way to find jets
is to use calorimetric measurements to look for localized
energy concentrations above the soft background. To prove
the existence of jets it is important to have an unbiased
trigger so as to make sure that it is not the event selection
that generates the jet structure. The least biased selection
appears to be a so-called "large E_T trigger", which requires that a large amount of transverse energy be observed
in a large solid angle, without additional a priori condition on how the energy is distributed. If it is found that
a large fraction of this energy is concentrated in a small
fraction of the solid angle, then one may conclude that jets
have been seen.

In order to find jets, a cluster algorithm is applied
to look for local energy concentrations. The cluster algorithm contains an operational definition of a jet in terms
of energy deposition in the calorimeter cells (i.e. a

prescription about the condition under which a new cluster
has to be started, how to combine cells to a cluster, when
assignment of cells to the cluster is to be stopped, etc.).
This involves a certain arbitrariness which may affect
quantitative statements made about the jets. In order to correct
for this, Monte Carlo programs are used in which jets are
generated and fragment into particles which are then tracked
through the detector, and the showers in the calorimeter
are simulated. Then the jet-finding algorithm is applied
to these "events", and one compares the jets found with those
generated. This way the efficiency of the jet finding algorithm can be determined. This procedure is also used to
optimize the algorithm in order to reach a high efficiency
without generating artefact jets.

The approach of studying the ΣE_T dependence and looking
for jets in high ΣE_T events has been adopted by a number of experiments (NA5[78] at the CERN SPS, E557[79] and E609[80]
at FNAL, R 807 at the CERN ISR[81] and UA1[35] and UA2[39]
at the $\bar{p}p$ collider).

It turned out that triggering on high ΣE_T in a large
solid angle is rather ineffective in terms of jet finding
at lower energies (SPS/FNAL and ISR). Contrary to the expectation of the naive parton model [72], the high ΣE_T events
do not abound with jets or high p_T particles. It appears
that to get high ΣE_T, it is "easier" to add more particles
with low p_T than to give more p_T to the particles. In fact,
all observations can be described by models with multi-par-
ticle production in cylindrical phase space, with KNO-type
multiplicity distributions. This is now believed to be understood in the framework of QCD as being due to multi-gluon
bremsstrahlung[75]. Monte Carlo calculations based on a
combination of soft background and a hard scattering component[82] show that the contribution to the ΣE_T spectrum from
the hard (jet) component becomes comparable with that from
the soft multiparticle background only when ΣE_T is bigger
than a minimum value $E_o(\Delta R)$. The "cross-over" energy $E_o(\Delta R)$

depends on $\Delta R = \Delta\eta\Delta\phi$, the region in phase space over which the sum is taken:

$$E_0(\Delta R) \simeq 16\Delta\eta \, \frac{\Delta\phi}{2\pi} \qquad (E_0 \text{ in GeV}),$$

but it is rather insensitive to the c.m. energy and details of the model. Since the largest ΣE_T value accessible to the experiments at SPS and FNAL[78-80] is approximately 20 GeV, it is not surprising that they did not find evidence for the jet dominance expected from the parton model.

5.3 Jets Seen in Collider Experiments

At the ISR[81] and much more so at the $\bar{p}p$ collider, sufficiently large ΣE_T values can be reached. In the runs of 1981 and 1982 both UA1 and UA2 have accumulated high ΣE_T data, and even with the very modest sensitivity of the 1981 data sample (UA1 : $22\mu b^{-1}$, UA2: $79 \ \mu b^{-1}$) values beyond ~ 100 GeV were reached. The angular region over which the sum is taken is different for the two experiments:

$$\text{UA1:} \ \Sigma_1 \ \equiv \Sigma E_T(\Delta R_1), \Delta R_1 \ : \ |\eta| < 1.5 \quad ,$$
$$\Delta\phi \ = 2\pi$$

$$\text{UA2:} \ \Sigma_2 \equiv \Sigma E_T(\Delta R_2), \Delta R_2 \quad : \ |\eta| < 1.0$$
$$\Delta\phi = \frac{5\pi}{6} \quad .$$

Performing a search for localized concentrations of energy (see [35,39] for details), both experiments find:

(a) For sufficiently high $\Sigma E_T (\Sigma_1 \gtrsim 40$ GeV, $\Sigma_2 \gtrsim 25$ GeV), the "jettyness" of events begins to increase with ΣE_T, and eventually (for $\Sigma_1 \gtrsim 100$ GeV, $\Sigma_2 \gtrsim 60$ GeV) all events have a large fraction (>2/3) of their total transverse energy concentrated in a few (most of the time two) clusters.

(b) In the case of two-cluster events, the transverse energies of the two clusters are approximately equal, and

(c) the two clusters are back to back in azimuth.

These observations are consistent with the interpretation of these clusters as being jets originating from hard scattering processes. It is also worth mentioning that the values of ΣE_T above which evidence for jets begins to show up, corresponds to the values expected from the calculations of[82].

Having found jets, one can proceed to determine the cross section for jet production and the dijet mass distribution. Both are shown in fig.12, together with predictions of hardscattering models based on QCD[83,84]. It can be seen that for the E_T dependence there is good agreement between the experiments and with the predictions, while there are discrepancies as far as the absolute normalization is concerned: The UA1 data are slightly but systematically above and the UA2 data below the theoretical curves, and there is a discrepancy of a factor of ~ 5 between the two data sets.

It should be noted however, that the errors shown are statistical only. There are additional systematic errors due to the uncertainty in the total integrated luminosity and, more important, the uncertainties in the transverse energy scale coming from the absolute calibration, the correction for the different calorimeter response to neutral and charged particles, effects of the magnetic field and the jet definition. Due to the steep drop of the cross section with E_T, a 10% change in the energy scale will change $d\sigma/dE_T$ by a factor of ~ 2.

Preliminary results[85,86] based on the analysis of the 1982 data (with much higher statistics) indicate that the discrepancy between the two experiments has appreciably diminished, i.e. the cross sections seem to be compatible with each other. In these new data, the jet cross section is measured out to ~ 100 GeV (UA1 has found 2035 jets with $E_T(jet) > 35$ GeV in $|\eta| < 1.0$) and the multijet (mainly di-jet) invariant mass up to ~ 300 GeV, i.e. the collider is already

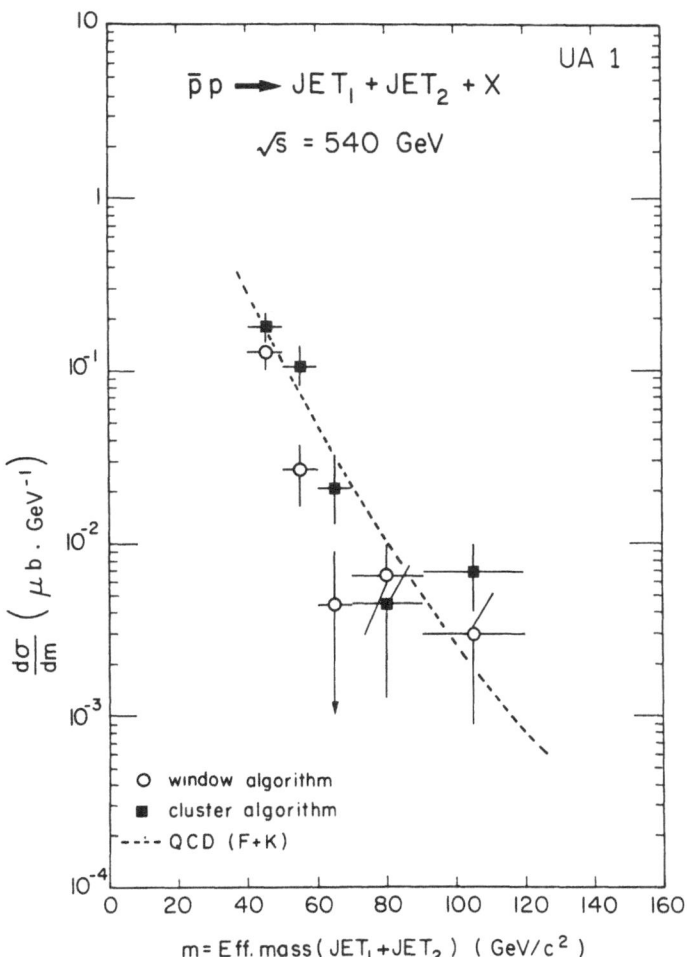

Fig. 12a: Comparison of jet cross section data of UA1[35] and UA2[39] with QCD predictions . "H+J" = calculation of[83] , "F+K" = calculation of ref.[84]. Dijet effection mass distribution, for jets with |η| < 1.4 and E_T(jet) > 15 GeV.

450

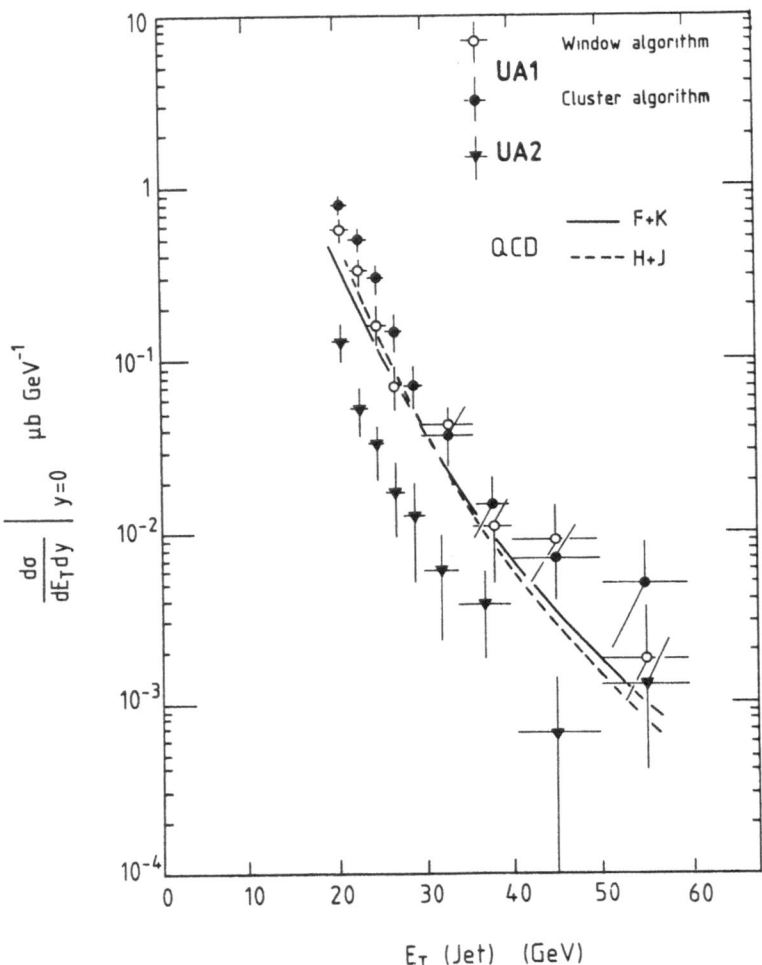

Fig. 12b: Comparison of jet cross section data of UA1[35]
and UA2[39] with QCD predictions. "H+K" = calculation of[83], "F+K" = calculation of ref.[84].
Transverse energy distribution of jets.

now probing constituent interactions far above the regime to be covered by LEP.

As an example, some of the UA1 events with $\Sigma_1 > 140$ GeV are shown in fig.13. The jet structure is obvious.

Jet fragmentation into charged particles can also be studied by using the information from the UA1 central detector in addition to the calorimeteric measurements which define the jet. Preliminary results[87] indicate that the fragmentation functions look similar to the quark fragmentation functions measured in e^+e^- interactions at PETRA[74,88].

5.4 Concluding Remarks About Jet Studies

The collider experiments UA1 and UA2 have provided clear and clean evidence for the existence of jets. Their findings are qualitatively and quantitatively in agreement with expectations from QCD, but more work has to be done (and is being done) on the understanding of systematic effects, both due to detector properties and the jet finding methods.

These studies are not only interesting in itself, but understanding of jets is also a prerequisite for other exciting investigations (IVB's, heavy flavour production, etc.)

6. PROPERTIES OF IVB's

Since there exists an abundant literature on the subject, only some of the main points are mentioned here, and the reader is referred to the reviews of collider physics[15,16,17] and is particular to the recent review article on the physics of IVB's by J. Ellis et al.[89], where also previous references can be found.

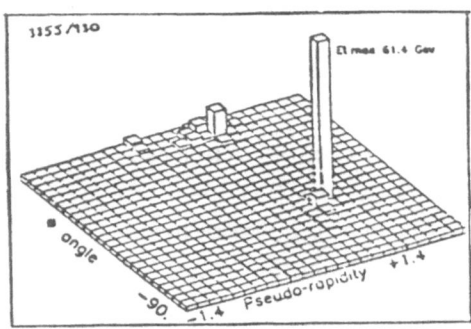
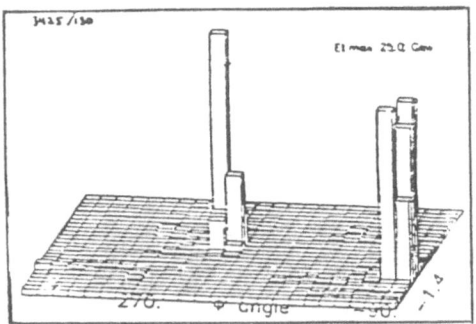
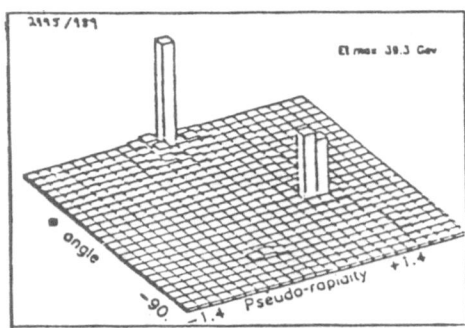
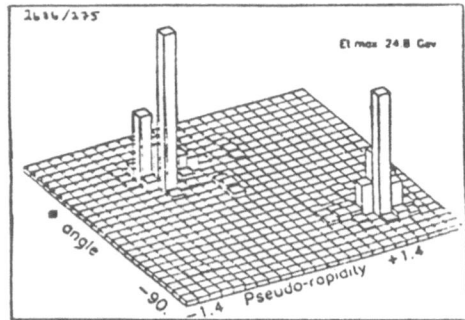
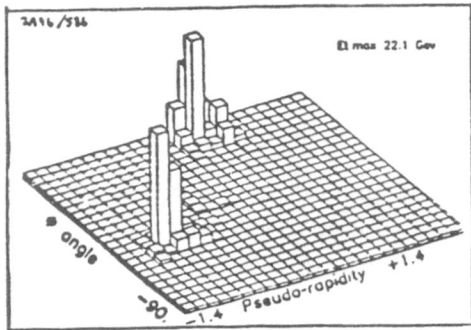
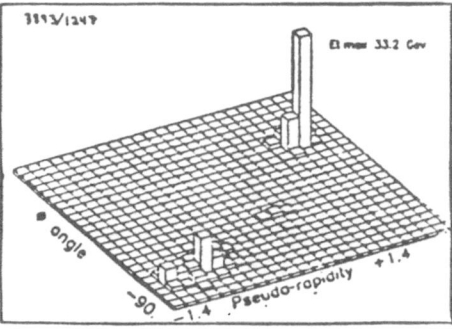

Fig. 13: "Lego plot" of some of the UA1 events with $\sum_{1} > 140$ GeV. For every bin in pseudorapidity η and azimuth ϕ, the total transverse energy E_T in that bin is shown.

6.1 Masses

In SU(2) ⊗ U(1) models of electroweak interactions [90,91] the SU(2) and U(1) coupling constants g and g' are related to measurable quantities as follows:

$$\tan \theta_W = g'/g \quad,$$
$$G_F/\sqrt{2} = g^2/M_W^2 \quad,$$
$$e = g \sin\theta_W = gg'/(g^2 + g'^2)^{1/2} \quad,$$

where θ_W is the electroweak mixing angle ("Weinberg angle"), G_F the Fermi constant ($G_F = 1.1663 \times 10^{-5}$ GeV^{-2}), and e is the positron charge. In the Born approximation (at the tree graph level) the mass of the W boson is

$$M_W = (\frac{\pi\alpha}{\sqrt{2}\cdot G_F})^{1/2}/\sin\theta_W \approx \frac{37.3 \text{ GeV}}{\sin\theta_W}$$

and the Z mass is

$$M_Z = M_W/(\sqrt{\bar{\rho}} \cdot \cos\theta_W) \quad.$$

The ρ parameter describes the relative strength of neutral and charged current effects and depends on the isospin structure of the Higgs fields; $\rho = 1$ if all Higgs bosons are in doublets and singlets. In the standard model, all left-handed fermions are in doublets, all right-handed fermions in singlets and there is only one Higgs doublet (i.e. $\rho = 1$).

If radiative corrections to the Born approximation (tree graph amplitudes) are taken into account, some of the relations given above are modified, depending on the renormalization scheme chosen [90]. Radiative corrections up to one loop contributions have been calculated [92], and for the current value of the mixing angle $\sin^2\theta_W = 0.215 \pm 0.014$ the present predictions are [93,94]:

$$M_W = 83.0 \, ^{+3.1}_{-2.8} \quad, \qquad M_Z = 93.8 \, ^{+2.5}_{-2.4}$$

6.2 Decay Widths and Branching Ratios

The widths of the principal decay modes of W and Z can be
deduced from the standard model Lagrangian [89,94,95]. The
leptonic widths are found to be ($\ell = e, \mu, \tau$)

$$\Gamma(W^- \to \ell\bar{\nu}) = G_F M_W^3/(6\pi\sqrt{2}) \approx 250 \text{ MeV} ,$$

$$\Gamma(Z^0 \to \ell^+\ell^-) = G_F M_Z^3[1+(1-4\sin^2\theta_W)^2]/(24\pi\sqrt{2}) \approx 90 \text{ MeV} .$$

For the total width one obtains N_G = number of generations):

$$\Gamma(W \to \text{all}) \sim \Gamma(Z^0 \to \text{all}) \sim N_G \text{GeV} \sim 3 \text{ GeV} ,$$

and the leptonic branching ratios are

$$\Gamma(W^- \to \ell\bar{\nu})/\Gamma(W \to \text{all}) \sim 1/4N_G \sim 8\% ,$$

$$\Gamma(Z^0 \to \ell^+\ell^-)/\Gamma(Z^0 \to \text{all}) \sim 1/11N_G \sim 3\% .$$

6.3 Production in p̄p Collisions

To calculate IVB production in hadronic collisions
one starts with the Drell-Yan model[96,97], which was
originally designed to describe lepton pair production via
a virtual photon. The hadronic cross section for production
of a heavy photon is obtained as an incoherent sum of all
possible fermion subprocesses weighted by the respective
probability distributions of the incoming constituents (structure functions).

The same physical picture can be applied to IVB production
but different elementary fusion subprocesses at the
constituent level have to be considered [15,16,17,89,98,99]:

$$u\bar{u}, d\bar{d}, s\bar{s}, \ldots \quad \to \quad Z^0 ,$$
$$u\bar{d}, u\bar{s}, \ldots \quad \to \quad W^+ ,$$
$$d\bar{u}, s\bar{u}, \ldots \quad \to \quad W^- .$$

In the naive version of the model in which quarks and antiquarks are treated as free objects inside the hadron and the structure functions scale, the cross section depends only on $\tau = M^2/s$. Taking into account strong interaction effects (QCD) leads to departures from this simple behaviour: additional diagrams involving gluons have to be considered [100], the structure functions have a logarithmic dependence on the energy scale Q^2 of the constituent process ($Q^2 \simeq \hat{s} \sim M_{W,Z}^2$), and the IVB gets a transverse kick much larger than expected from the Fermi motion of partons within the hadrons ($<Q_T> \sim 5$ to 10 GeV). Predicted cross sections[99] for W and Z^0 production in pp an $\bar{p}p$ collisions, including scale breaking effects are shown in fig.14. Using the expected IVB masses given in sect. 6.1($\sqrt{\tau} = 0.154$ for W and 0.173 for Z^0), the cross sections for IVB production at the CERN $\bar{p}p$ collider ($\sqrt{s} = 540$ GeV) are predicted to be

$$\sigma(\bar{p}p \to W^+) = \sigma(\bar{p}p \to W^-) \sim 1.8 \quad nb \quad ,$$

$$\sigma(\bar{p}p \to Z^0) \qquad\qquad \sim 0.85 \ nb \ .$$

Note that virtual gluon exchange (non-leading terms) give <u>additional</u> corrections to the parton model calculation which induce a correction factor K of order of 2 to the Drell-Yan cross section[101] (not included in above estimates). Enhancements of this order have in fact been observed in dilepton production at SPS and ISR energies (see B. Cox[97] for a summary of measured K values). The actual size of this enhancement factor for IVB production is rather uncertain, but it is expected that the shape of the distributions will not be modified drastically[102].

6.4 Signature of W and Z Events in $\bar{p}p$ Collisions

Hadronic decay modes

The dominant decay modes of the IVB's are those into $q\bar{q}'$ pairs which will show up as hadronic jets of high trans-

456

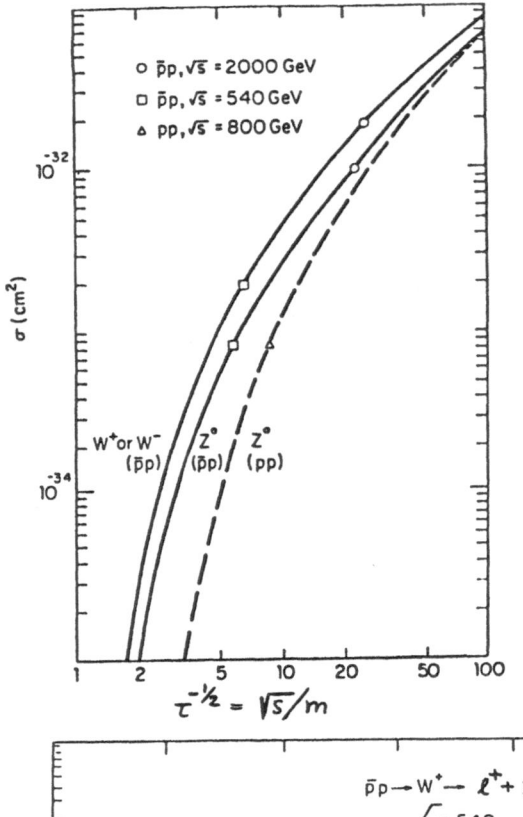

Fig. 14: Predicted production cross sections for W and Z^O in pp and $\bar{p}p$ collisions, as a function of $\tau^{-1/2} = \sqrt{s}/M$. The calculation includes scale-breaking effects, but not the enhancement factor K (from ref. 99).

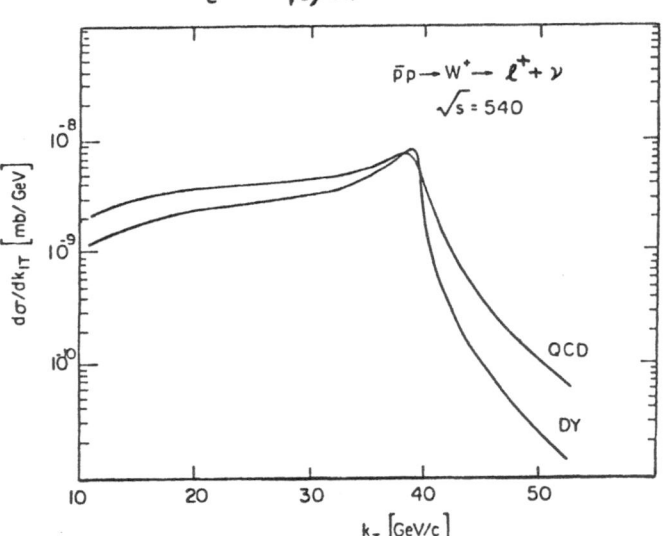

Fig. 15: The Jacobian peak of the single-lepton spectrum from W^+ production and decay ($\bar{p}p \to W^+ \to \ell^+ \nu$) at $\sqrt{s} = 540$ GeV. The Drell-Yan calculation is indicated by "DY", while "QCD" includes the QCD corrections (fig. from ref. 104).

verse momentum. Due to their weak origin, they are expected
to be hidden by the high p_T jets of strong origin("QCD-jets"),
and it will be difficult to disentangle the two classes of
jets.

Leptonic decay modes

Due to the high mass of W and Z the leptons from
their decay will have high momenta in the $\bar{p}p$ c.m.s., and
they are expected to be isolated (i.e. no other high momentum particles nearby in phase space). In the following
we concentrate on these leptonic modes since they are easier
to identify experimentally than the hadronic modes. From
the production cross section and the leptonic branching
ratios of W and Z, one sees that

$$\Gamma(\bar{p}p \to W^+ \text{ or } W^- \to \ell^+\nu \text{ or } \ell^-\bar{\nu})/\sigma(\bar{p}p \to Z^0 \to \ell^+\ell^-) \sim 10 ,$$

i.e. identifiable W events occur much more frequently than
identifiable Z^0 events.

Detection of $Z^0 \to \ell^+\ell^-$

The signature for Z^0 is much clearer than that for the
W: to identify $Z^0 \to \ell^+\ell^-$, it is sufficient to detect and identify two leptons with high p_T, measure their momenta (energies)
and calculate their invariant mass. Evidence for $Z^0 \to \ell^+\ell^-$
will be seen as a peak at the Z^0 mass in the lepton-pair
mass spectrum. The background from other sources of high p_T
leptons ("usual" Drell-Yan dilepton production and leptonic
decay of heavy quarks and quarkonium) are expected to be small
[103].

Detection of $W \to \ell \nu$

For the leptonic decay of the W, the situation is less
favourable than for the Z^0 because of the (anti) neutrino
which is difficult to detect. Luckily enough, kinematics
helps by causing a "Jacobian peak" to appear in the single

lepton transverse momentum spectrum[104,105]. This Jacobian
peak is a distinctive feature of a heavy object decaying
into two light leptons and is due to the fact that the
distribution in transverse momentum k_T of the lepton is
given by the angular distribution in the W rest frame, multi-
plied by a Jacobian factor

$x_T \cdot (1 - x_T^2)^{-1/2}$ (where $x_T = 2k_T/M_W$).

The peak (see fig.15) is smeared out due the finite width of
the W and the transverse momentum Q_T of the W originating
from the intrinsic momentum of the constituents (Fermi motion)
and QCD correction (gluon radiation). This smearing may be
so important that even though the Jacobian peak is still
visible it may become difficult to determine the mass of
the W from the k_T distribution.

If however from observation of missing energy the transverse momentum $\vec{k}_T'$ of the neutrino can be measured in addition
to the transverse momentum $\vec{k}_T$ of the charged lepton, the
distribution in "transverse mass" m_T of the ($\ell\nu$/system can
be studied, where

$$m_T^2 = (k_T + k_T')^2 - (\vec{k}_T + \vec{k}_T')^2.$$

This distribution has a very sharp Jacobian peak at the mass
of the W, whose width and tail at $m_T \rightarrow M_W$ are determined
only by the decay width of the W and the experimental resolution,
and are independent of the transverse momentum of the W.
Therefore, a model independent determination of the W mass
(and width) becomes possible [106,107].

Asymmetry of lepton distribution

Having found peaks in the (transverse) mass distribution,
one can check whether the observed objects are indeed produced
by weak interactions. For IVB's produced in $\bar{p}p$ collisions,
parity violation in production and decay will induce a

forward-backward asymmetry of the lepton spectrum.

For the W this asymmetry is expected to be large [108] because of the pure V-A coupling and can be visualized qualitatively on the basis of helicity arguments. For example, a W^+ will be predominantly produced by collisions of a u_L quark (width helicity opposite to momentum) from the proton and $\bar{d}_L$ (helicity in momentum direction) from the antiproton, which fixes the helicity of the W^+ to point in the same direction as the momentum of the incoming antiproton. Since the outgoing neutrino is lefthanded, its favoured direction of movement is opposite to that direction and therefore the e^+ will tend to be emitted in the direction of the antiproton beam. This may be slightly modified by contributions from sea quarks and gluons, but the essential features remain. Thus, positive leptons from W^+ decay are expected to move preferentially in the antiproton direction, while negative leptons from W^- decay will tend to follow the proton direction.

In the case of the Z^o, this asymmetry is expected to be much smaller because the vector coupling between the Z^o and the leptons is proportional to $(1-4 \sin^2\theta_W) \approx 0.14$. Therefore it will be difficult to measure.

7. SEARCH FOR IVB EVENTS IN THE UA1 AND UA2 EXPERIMENTS

7.1 Detection and Identification of Leptons

In order to detect and identify W's and Z's by their leptonic decay, as discussed in the previous section, it is necessary to detect and identify high p_T charged leptons (e,μ) and neutrinos in a large solid angle and measure their momenta and/or energies with good precision. We will discuss here briefly how this is done in practice, taking the UA1 detector as an example. For more information and previous references, the reader is referred to the experimental publications [32,42] and presentations[43,109].

7.1.1 Charged leptons

A charged lepton is detected as a charged particle track in the central detector ("CD-track"), with supplies a measurement of the momentum vector ($\Delta p/p \sim 20\%$ for p = 40 GeV), the ionization energy loss and the impact point on the calorimeter. The energy deposits in the four samplings of the EM and the two samplings of the hadron calorimeter are used for the identification as a lepton (hadron rejection) and give a measurement of the total energy of electrons.

Electrons

An electron is then defined as a charged particle with "EM signature"; it has to fulfill the following conditions:

(A) Shower-track matching in position:
the position of the centre of gravity of the shower in the calorimeters agrees with the extrapolation of the CD-track to within 3 standard deviations.

(B) Energy-momentum matching:
the momentum of the CD-track agrees with the energy measured in the EM calorimeter to within 3 standard deviations ($|1/p - 1/E| < 3\sigma$).

(C) EM shower containment:
the shower is contained in the EM calorimeter, i.e. $E_H < 600$ MeV, where E_H is the energy deposited in the hadron calorimeter cell which is hit by the extrapolation of the CD-track.

(D) Electron-like shower behaviour:
the shape of the shower (width, longitudinal profile) agrees with that expected for electrons.

(E) CD-track minimum ionizing:
the ionization energy loss in the CD is compatible

with that expected for a single minimum ionizing particle.

Test measurements have shown that nearly all electrons ($\sim$98%) but only a small fraction of hadrons (0.8% for p = 40 GeV) fulfill condition (c).

Muons

While electrons and hadrons are absorbed in the calorimeters ($\sim$70/sinθ radiation lengths, $\sim$7/sinθ absorption lengths), muons traverse them and are detected by the muon-chambers.
The μ-chamber signals are used to reconstruct a "μ-track" (not necessarily a muon), which together with the vertex information yields an additional momentum measurement.

A muon is then defined as a charged particle (min. ionizing in the CD) for which the CD-track parameters agree with those of the μ-track within the deviations expected from multiple scattering, and for which the energy depositions in all calorimeter samplings are compatible with that expected for a single minimum ionizing particle.
For more details, see ref. 109 .

7.1.2 Neutrinos

Neutrinos are detected by measuring the missing energy, exploiting the nearly complete calorimetric coverage (down to θ = 0.2°) of the UA1 detector.

We assign an energy vector $\vec{E}_i$ to every calorimeter cell i, defined by $\vec{E}_i = E_i\vec{u}_i$, where $\vec{u}_i$ is a unit vector pointing from the interaction vertex to the centre of gravity of the energy E_i measured in cell i. For every event, a missing energy vector $\Delta\vec{E}$ is then defined by $\Delta\vec{E} = \sum_i \vec{E}_i$, where the sum is extended over all calorimeter cells.

If we neglect particle masses, momentum conservation implies $\Delta\vec{E} = 0$ for an ideal calorimeter with perfect response and complete coverage (and for events without μ's or neutrinos).

In reality, a certain fraction of the energy remains undetected due to dead zones (e.g. light-guides) and gaps in the calorimeter. In particular, the longitudinal component ΔE_x is affected by energy escaping into the forward direction ($\theta < 0.2°$) and therefore not of much practical use.

The transverse components ΔE_y and ΔE_z however, are "well behaved" for minimum bias and jet-enriched events in which neutrinos and muons are rare: Their distributions are centred around 0, have no prominent tails and their r.m.s. width is found to be

$$\sigma(\Delta E_{y,z}) \sim 0.43 \cdot (\sum_i E_{Ti})^{1/2} \quad ,$$

where $\sum_i E_{Ti} = \sum E_i \sin\theta_i$ is the total transverse energy observed.

The resolution of the missing transverse energy $\Delta E_T = (\Delta E_y^2 + \Delta E_z^2)^{1/2}$ is then

$$\sigma(\Delta E_T) \sim 0.61 \cdot (\sum_i E_{Ti})^{1/2} \quad ,$$

which is typically ~ 4 GeV.

High p_T neutrinos can therefore be detected by the presence of a significant transverse energy, and the neutrino transverse momentum can be measured with a precision of ~ 4 GeV.

7.2 W Search by UA1

7.2.1 Data taking

The data on which this search is based were taken during

the p̄p collider period of Oct. - Dec. 1982. The data sample represents an integrated luminosity of ∿18 nb^{-1}, corresponding to ∿1.2 × 10^9 p̄p collisions.

In addition to a $\int E_T$ and a jet trigger (see[32,35,85] for details), two triggers sensitive to high p_T charged leptons were used:

Electron trigger: at least 10 GeV of transverse energy in two adjacent cells of the central EM calorimeter (gondolas, bouchons).

Muon trigger: a μ-track candidate pointing to the interaction vertex (within ± 150 mrad/sin^{2}θ).

7.2.2 Analysis of electron trigger events

Among the 140 000 electron trigger events, there are ∿2000 which have a localized energy deposition of E_T > 15 GeV in two adjacent gondolas and a well reconstructed high p_T (>7 GeV) track associated with the vertex (sample (I)).

This sample is then subjected to two sequences of further selection criteria: the "electron" and the "missing energy" selection.

Electron selection:

Requirements of isolation of the high p_T track (the sum of transverse momenta of all other tracks entering the two gondolas must be <2 GeV) and position matching between the shower and the track (condition (A)) reduce the sample to 167 events (sample (II)). Imposing the conditions of energy-momentum matching and containment of the EM shower (electron conditions (B) and (C)) leaves 39 events, which are scanned on a high resolution graphic display. After rejecting those events in which the electron candidate is in a jet (i.e. not well isolated) and those where the electron candidate belongs to an e$^+$e$^-$ conversion pair, we are left

464

EVENTS WITHOUT JETS   EVENTS WITH JETS

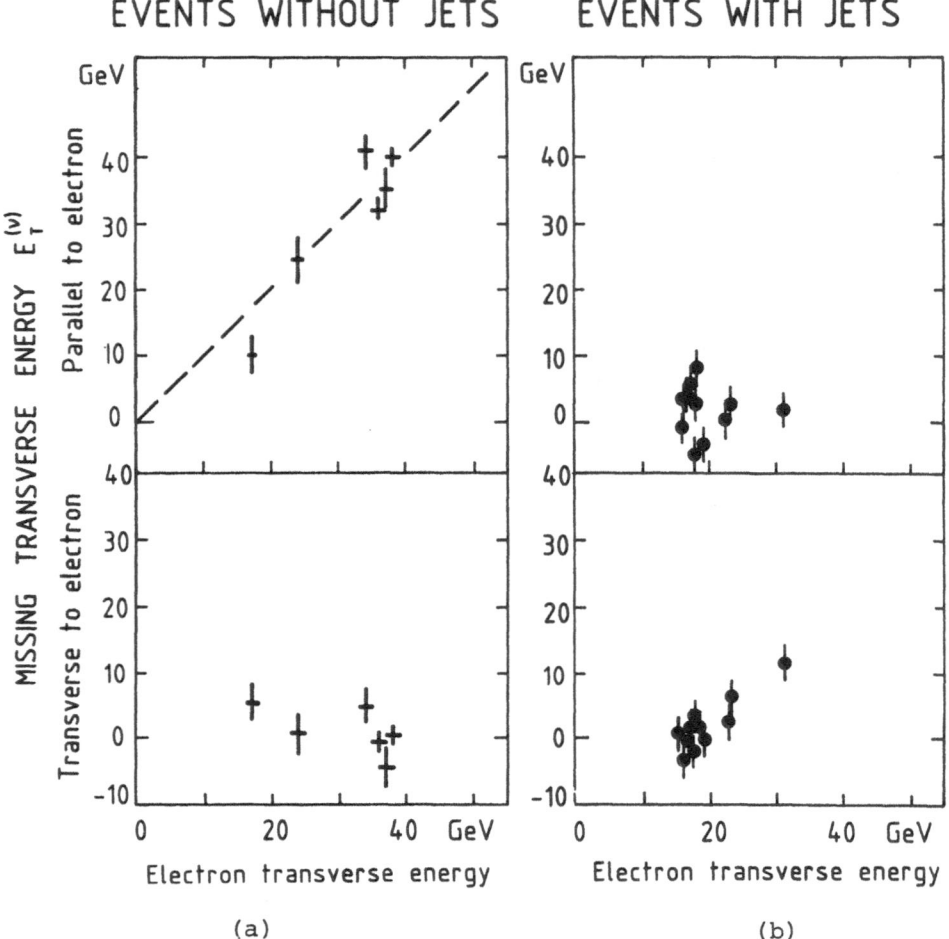

Fig. 16: Events found in the electron selection (UA1).
The components of the missing energy parallel
and perpendicular to the transverse momentum of
the electron candidate are plotted versus the
transverse momentum of the electron candidate;
(a) for events without jets
(b) for events with jets.

with 16 events which can be divided into two distinct classes:

  11 events with one high p_T jet (> 10 GeV)
  opposite (within ± 30°) to the electron candidate,

  and 5 events without visible high p_T jets.
  (A similar selection in the bouchons yields one
  additional event without jet).

The two classes of events show different behaviour
with respect to the components of missing energy parallel
and perpendicular to the transverse momentum of the
electron (see fig.16).

The events with jets have no significant parallel
component. In contrast, the events without jets have a
parallel component of the same magnitude as the electron
momentum, while the perpendicular component is compatible
with zero.

Missing energy (neutrino) selection:

We start again from sample (I) and impose the condition that the high p_T track points to the two gondolas
with the localized energy deposition (E_T > 15 GeV); this
reduces the sample to ∿1000 events. Then instead of
selecting for electron characteristics of the charged track,
we select those events which have missing transverse energy
>15 GeV. From the remaining 70 events those are rejected in
which the high p_T is not isolated, i.e. part of a jet
(checked on the graphic display). This leaves 18 events
which again can be divided into two classes:

  7 events without detectable high p_T jets, and 11
events with a jet opposite to the isolated charged high
p_T track.

Fig. 17 shows that the two classes of events behave
differently with respect to the energy E_H deposited by the
high p_T particle in the hadron calorimeter: For the events

EVENTS WITH JETS

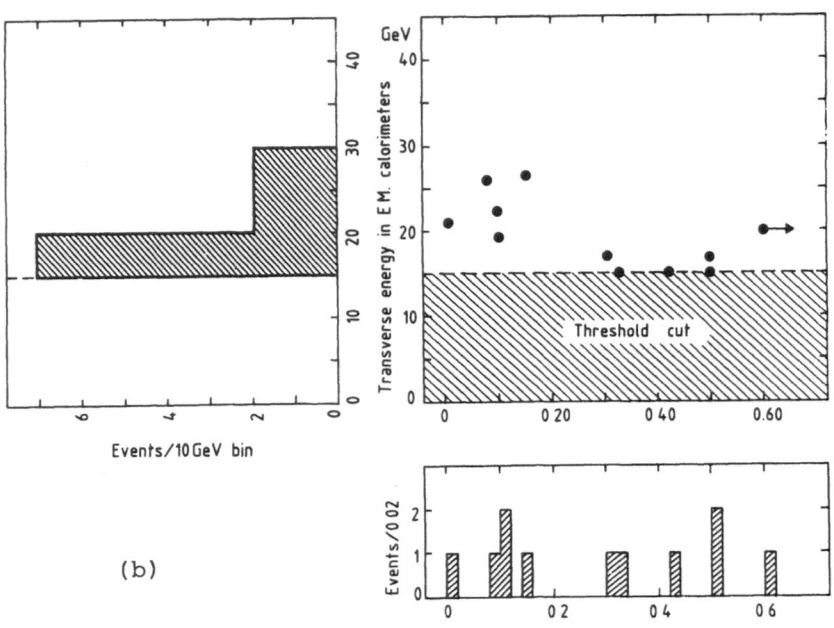

(b)

EVENTS WITHOUT JETS

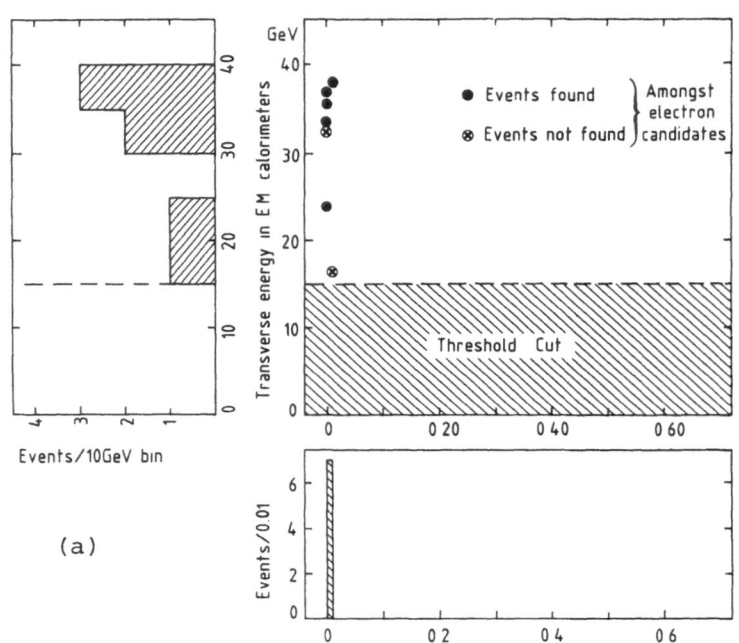

(a)

Fig. 17: Events found in the missing energy selection (UA1)
The transverse energy in the EM calorimeters is
plotted versus the fraction of energy deposited
in the hadron calorimeter $(E_H/(E_H+E_{EM}))$;
(a) for events without jets,
(b) for events with jets.

with jets (fig. 17b), E_H tends to be large, i.e. the high p_T particle is most likely a hadron. On the other hand, for the events without jets (fig. 17a), E_H is negligible, i.e. the high p_T particle could be an electron.

Comparison of "electron" and "neutrino" events without jets:

We have seen that the five events containing a high p_T particle with electron signature show evidence for significant missing energy, i.e. a high p_T neutrino. By looking for high missing transverse energy (i.e. high p_T neutrinos), we find the same five events.

Thus there appears to be a physical process which produces both high p_T charged leptons and high p_T neutrinos, as expected for the leptonic decay of W's. As an example, we show in figs. 18 and 19 two of the events found in the electron and the neutrino selection.

The two additional events found in the neutrino selection, which had been eliminated in the electron selection condition (B) (energy-momentum matching), could be due to, e.g. $\tau^- \to \pi^- \pi^0 \nu_\tau$.

7.2.3 Analysis of muon trigger events

The μ-trigger was only used for part of the data taking (~ 11 nb^{-1}), and the trigger acceptance was smaller than that for electrons. Therefore, if high p_T muons are produced at the same rate as high p_T electrons, we expect to find about three times fewer events with a high p_T muon than with a high p_T electron.

The analysis (described in [109]) yields four events with a μ of $p_T > 15$ GeV, one of which has also significant transverse energy (note that in the calculation of the missing energy, the μ-momentum has to be taken into account in addition to the calorimeter information).

EVENT 2958. 1279.

(a)

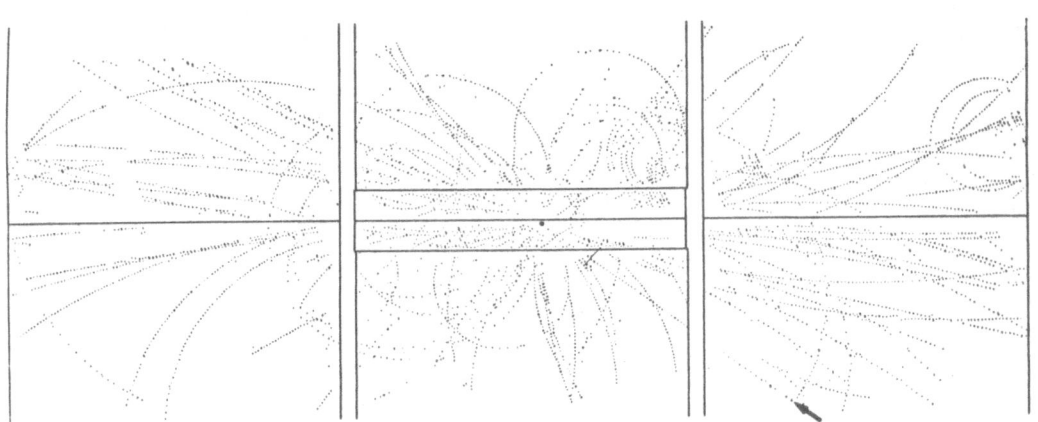

EVENT 2958. 1279.   THR 2.00   PTCD   0.0

(b)

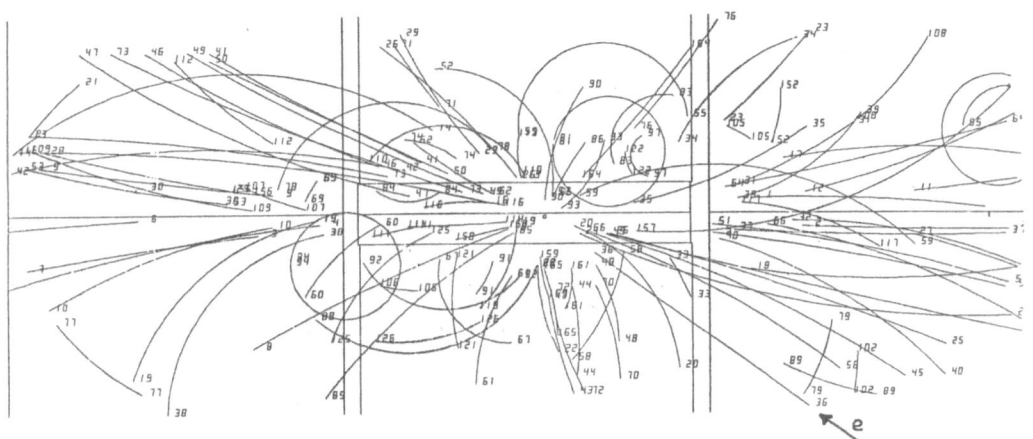

Fig. 18: The central detector information for one of the
events found in both the electron and the missing
energy selection (UA1);
(a) the raw digitizations
(b) the tracks found and reconstructed by the
analysis program.
The electron candidate is indicated.

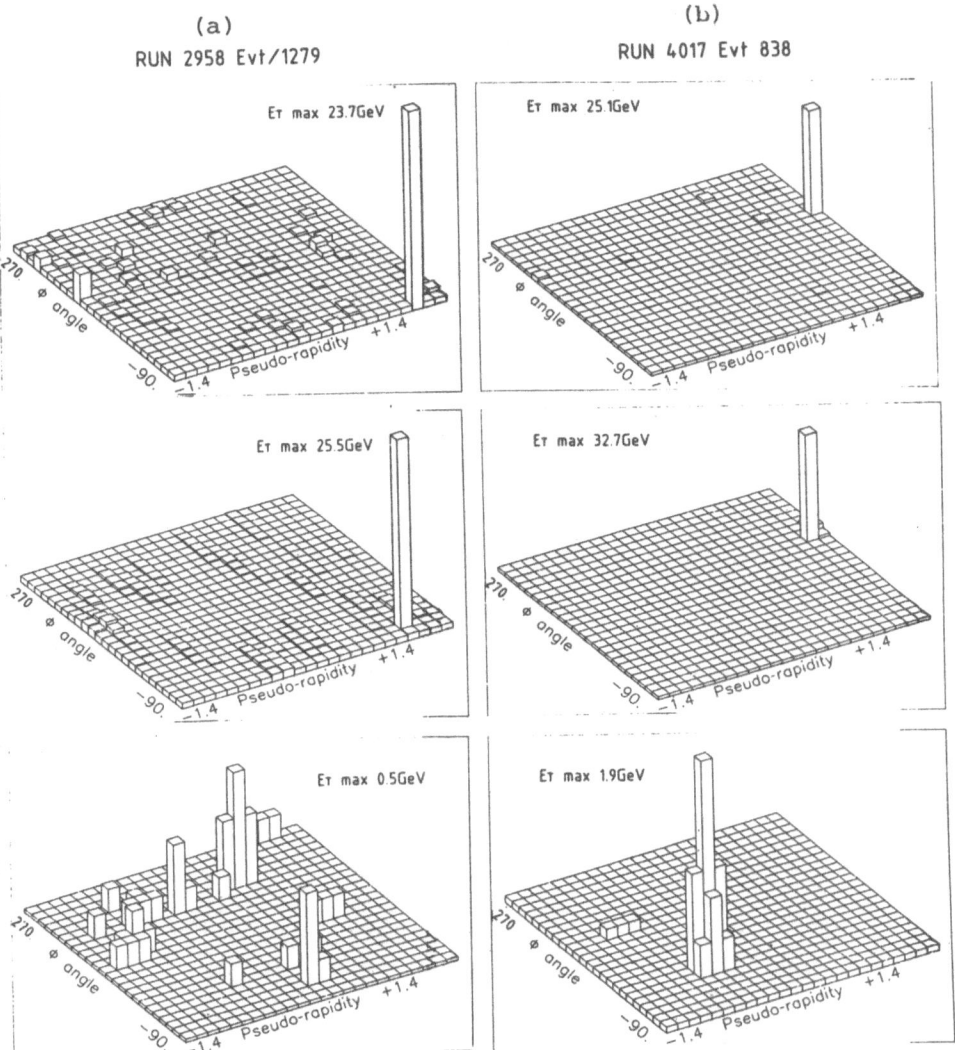

(a)
RUN 2958 Evt/1279

(b)
RUN 4017 Evt 838

Eᴛ max 23.7GeV

Eᴛ max 25.1GeV

Eᴛ max 25.5GeV

Eᴛ max 32.7GeV

Eᴛ max 0.5GeV

Eᴛ max 1.9GeV

Fig. 19: "Lego plot" of two of the events found in both the electron and missing energy selection (UA1). For every bin in pseudorapidity η and azimuth φ, the diagram shows the transverse energy measured in the EM calorimeter (top row), the transverse momentum as measured by the central detector (middle row) and the transverse energy in the hadron calorimeter (bottom row). Note that the vertical scale is different for each diagram;
(a) for the event of fig.18 (high multiplicity)
(b) for a low multiplicity event.

7.3 W Search by UA2

7.3.1 Data taking

The data corresponding to an integrated luminosity of 19 nb^{-1} were taken during the $\bar{p}p$ collider period at the end of 1982, using triggers sensitive to events with large transverse energy in the central and forward calorimeter. Three types of trigger requirements were used:

W-trigger: at least one quartet (2×2) of EM calorimeter cells in which the measured transverse energy exceeds 8 GeV.

Z^0 trigger: presence of two such quartets, each having $E_T > 3.5$ GeV.

$\sum E_T$ trigger: $\sum E_T > 35$ GeV in the central calorimeter.

7.3.2 Analysis

The event selection is similar in spirit to that done by UA1, but differs in details due to the differences in the detector; also there are different procedures for the central and the forward part of the detector. We only give a superficial description and refer to [42,43] for details.

In order to select isolated high p_T electrons, the following conditions are imposed on the data:

· $E_T > 15$ GeV in the cluster of EM calorimeter cells

· containment of the EM shower

· isolation of the EM shower

· presence of a charged particle track, matching in position with the EM shower

· presence of an isolated shower in the preshower counter, matching in position with the charged particle track

· energy-momentum matching (forward calorimeter only)
· shape of the shower compatible with that expected for an electron.

The selection leaves a total of six events, three in the central and three in the forward part of the detector. Finally, only those four events are retained for which $p_T^{miss}/E_T > 0.8$, where p_T^{miss} is the component of the missing transverse energy parallel to the electron transverse momentum (see fig. 20). Fig. 21 shows how one of these events appears in the UA2 detector.

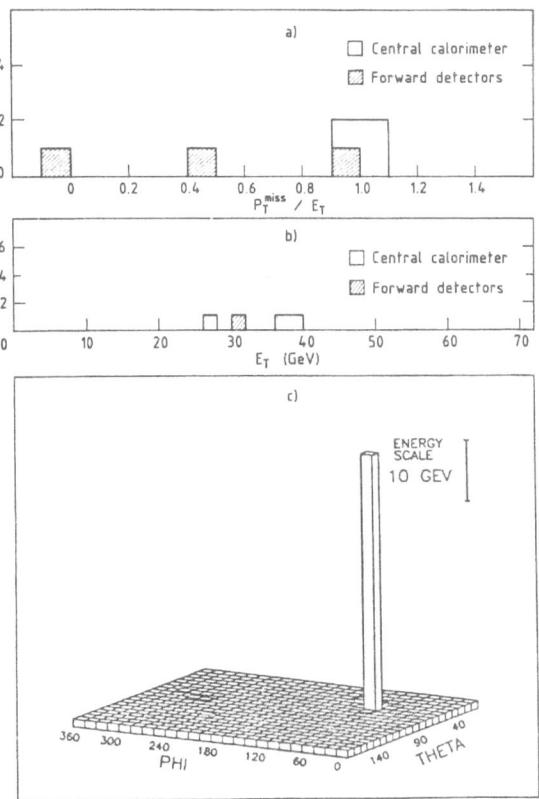

Fig. 20: Events found be electron selection (UA2):
(a) Distribution of p_T^{miss}/E_T, the ratio between the parallel component of missing energy and the transverse energy of the electron candidate.
(b) Transverse energy distribution of the electron. candidates observed in events with $p_T^{miss}/E_T > 0.8$.
(c) The cell energy distribution as a function of polar angle θ and azimuth ϕ for the electron candidate with the highest E_T value.

472

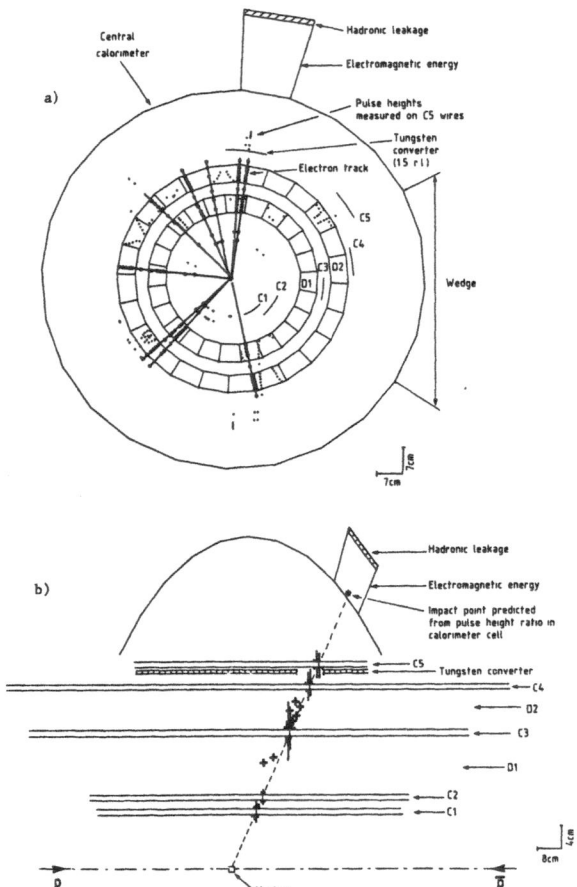

Fig. 21: The event with the electron candidate with the
highest E_T found by UA2.
(a) Transverse view of the event. Shown are hits in
proportional chambers C_1 to C_5 and drift chambers
D_1, D_2, and the tracks reconstructed in the vertex
detector.
(b) longitudinal view of the electron track of the
event shown in (a).

7.4 Interpretation of (Lepton + Neutrino) Events

7.4.1 Origin of (ℓ + ν) events

The use of simple straightforward cuts motivated by the expected signature of leptons in the UA1 and UA2 detectors has allowed to select a sample of events which have an isolated charged lepton of high p_T and a neutrino with approximately equal and opposite transverse momentum. Fig. 22 shows that for most of the events the transverse momenta of the leptons are well above the experimental cut of 15 GeV.

From the known properties of the apparatus (hadron rejection capability) and measured cross section for production of high p_T charged and neutral particles, it can be deduced that these leptons cannot be due to misidentification (hadron punch through), nor can they be explained by known sources of high p_T leptons (decays) (see [32,42,109] for details about background estimation).

The most natural explanation appears to be that they are due to leptonic decays of W bosons.

7.4.2 Comparison with the W → eν hypothesis

Since the interpretation of the two events with the lowest lepton p_T (the electron event found in the bouchon and the μ event) is not completely clear, we restrict ourselves to the five gondola events of UA1 and the four UA2 events (see fig. 22).

Cross section:

Using the predicted W production cross section and branching ratio for W → eν (see sect. 6), and guessing the enhancement factor K to be ∿1.5 and the experimental detection efficiency to be ∿70%, we expect to find about 10 W → eν events in the combined data of UA1 and UA2, which is compatible with the number of events found. Note, however,

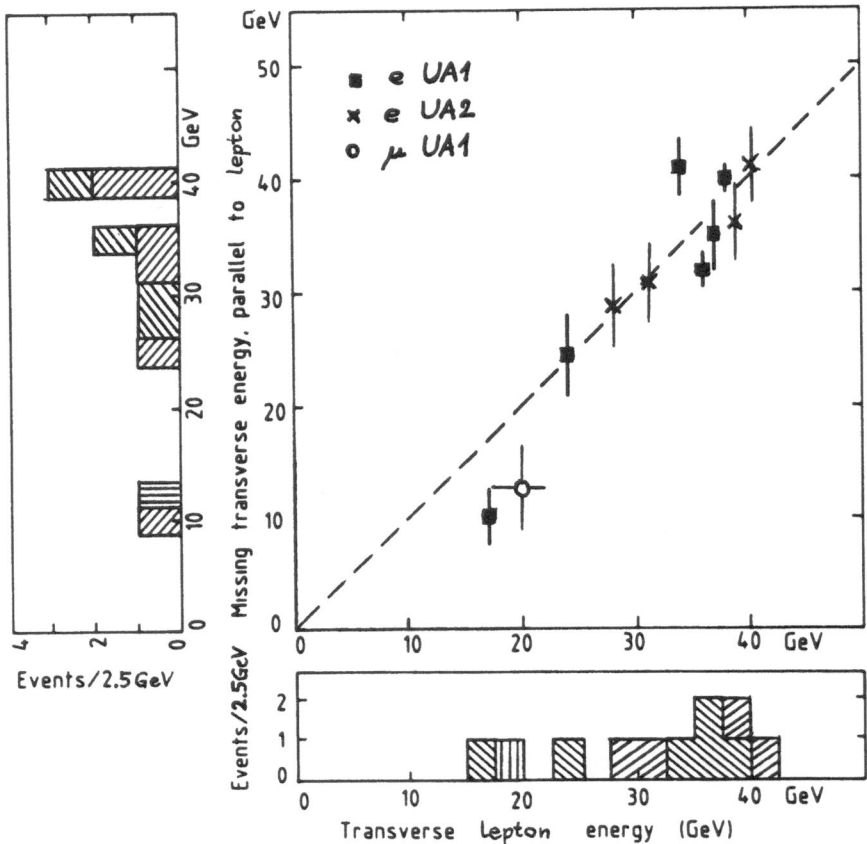

Fig. 22: Events with an isolated charged lepton (e,μ) and significant missing energy (neutrino) found by UA1 and UA2. The missing energy component parallel to the charged lepton transverse momentum is plotted versus the transverse momentum of the charged lepton. The errors on the missing energy for the UA2 points are fixed to ± 4 GeV.

that in view of the uncertainties in both the prediction
and the efficiency of the event selection this agreement
may be fortuitous.

Mass:

From the highest transverse mass of the $(e\nu)$ system
observed by UA1 $(78 \pm 2.2$ GeV$)$, a model-independent lower
limit on the W mass can be determined:

$M_W > 73$ GeV $(90\%$C.L.$)$.

From a fit to the lepton transverse momentum distribution (after correction for the transverse motion of the W),
UA1 obtains

$M_W = 81 \pm 5$ GeV,

while a fit to the transverse mass spectrum (UA1 data) gives
[107]

$M_W = 80 \begin{smallmatrix} +6 \\ -3 \end{smallmatrix}$ GeV.

A fit to be electron transverse momenta of the UA2 events
(taking W motion into account) yields

$M_W = 80 \begin{smallmatrix} +10 \\ -6 \end{smallmatrix}$ GeV .

Finally, an analysis of the combined UA1 and UA2 data [110]
finds

$M_W = 82.3 \pm 4.5$ GeV.

All of these values are compatible with the predictions
from the standard model (see sect. 6).

Asymmetry:

Given the limited statistics, no meaningful comparison

between expectation and observation is possible.

7.4.3 Conclusion about IVB search

Both UA1 and UA2 have observed events whose characteristics agree with those expected for the reaction

$$\bar{p}p \rightarrow W + x$$
$$\quad \hookrightarrow e\nu$$

These events were extracted from the total data sample by straightforward selection criteria; the background to the signal appears to be small. No Z^{o} event has been found, but it seems certain that both experiments have the potential to observe and identify leptonic decays of Z^{o}, provided that enough integrated luminosity can be accumulated.

8. OUTLOOK

8.1 Analysis of Existing Data

Apart from the search for W's and Z's, a concentrated analysis effort is devoted to improving and understanding the systematics of jet finding (see sect. 4), with the aim of extracting jet cross sections and fragmentation functions. This may also make it possible to see a W and Z signal in the dijet mass distribution, if one finds a way to reduce the background of jets from strong hard scattering processes. From studies of events with leptons and jets, one also hopes to learn something about heavy flavour production [111-114].

But of course we will not close our eyes if some other interesting observation pops out of the data.

8.2 Collider Period April - July 1983

As mentioned in sect. 2.3, a number of improvements have been implemented in the PS-AA-SPS complex during the winter shut down 1983 in order to increase luminosity and reliability. The hope is to achieve an integrated luminosity of ~ 100 to 200 nb^{-1}, which should allow the observation of a handful of $Z^0 \rightarrow e^+ e^-$ events by both UA1 and Ua2. Also the experiments have improved their equipment for this data taking period:

UA1 has installed additional iron shielding ("μ-wall") to improve the μ identification.

UA2 has removed the wedge detector and closed the wedge, thus reaching its full design acceptance for leptons from W and Z decay.

UA4 has improved the inelastic rate measurement (by including the UA2 detector) and hopes to measure σ_{tot} to $\pm$ 1 nb.

UA5 has added a calorimeter to identify neutral particles and to provide a trigger on transverse energy. The next data taking is foreseen during a $\bar{p}p$ machine development period in September 1983.

8.3 Future Upgrading of Detector

UA1 plans to implement additions to and improvements of the present detector in order to further increase its versatility:

(a) Magnetization of the iron in the newly installed μ-wall and the addition of track detection in the wall will allow an additional momentum measurement and improve the matching between central detector and μ-chambers.

(b) A vertex detector with $\sim 20 \mu m$ resolution will be useful for the study of new short lived particles by the

478

detection of small displacements of tracks with respect
to the vertex.

(c) New EM calorimeter (replacing the bouchons and the gondolas), non-ageing and with better spatial resolution
should allow the detection of single photons and electrons in jets.

UA2 will probably increase the angular coverage of the
detector to smaller polar angles.

8.4 Collider Upgrading

A number of "small" changes to the collider are being discussed [14] , among them:

- upgrading of the magnet power supply cooling, thus
  permitting to increase the beam energy from 270 GeV
  to ∿400 GeV,

- beam cooling and beam separation in the SPS, which
  will increase the luminosity lifetime,

- a mini β scheme to increase the luminosity.

A major improvement program is also being discussed,
which consists in the construction of a new accumulator ring
(AC) with much larger acceptance than the present AA. In
this scheme the tasks now handled by the AA would be divided
between the two rings: The AA would become a pure stacking
ring and all other operations (precooling and reduction of
momentum spread) would be done in the new AC ring. As a
consequence the movable shutters of the present AA would
not be necessary, thus making the operation more reliable.
This scheme could increase the stacking rate by up to a factor of ∿10.

9. CONCLUSIONS

The CERN $\bar{p}p$ collider is alive and well. After only a few months of effective operation it has become a viable machine and has made it possible to achieve one of its primary aims: the observation of the W boson.
An active physics research program is going on, and we can look forward to many new interesting results in the years (months?) to come.

Note added:

Meanwhile, the collider has reached a new record luminosity of $\sim 2 * 10^{29}$ cm^{-2}s^{-1}. Five Z^{o} events have been observed by UA1[115], and the many W's (their number changing from day to day) confirm the previous observations.

ACKNOWLEDGEMENT

I am grateful to Prof.H.Mitter for the invitation to the Schladming School. I wish to thank the people who designed, built and run the collider, as well as my colleagues in UA1 for making it possible to live this exciting physics adventure.

REFERENCES

1. C. Rubbia, P. McJntyre, D. Cline, Proc. Int. Neutrino Conference Aachen 1976 (Vieweg, Braunschweig 1977) p. 683.
2. S. Weinberg, Phys. Rev. Lett. 19 (1967) 1264; S.L. Glashow, Nucl. Phys. 22 (1961) 579; A. Salam in "Elementary Particle Theory", ed. N. Svartholm (Almquist an Wikrell, Stockholm 1968), p. 367.

For a review, see lectures by G. Ecker at the 21.Int.
Universitätswochen für Kernphysik, Schladming, Acta
Physica Austriaca, Suppl. XXIV, 3-62 (1982).

3. H. Hoffman: The CERN $p\bar{p}$ collider, CERN-EP/81-139, talk at the
   Int.Symposium on lepton and photon interaction,Bonn,August 198
   M. Spiro: The CERN proton-antiproton collider,lecture at the
   XXI. Int. Universitätswochen für Kernphysik, Schladming 1982;
   Acta Physica Austriaca Suppl. XXIV (1982) 125-155.
   P.I.P. Kalmus: The CERN proton-antiproton collider pro-
   gramme, CERN-EP/82-58, presented at the Int. Workshop
   on Very High Energy Interactions in Cosmic Rays, Phila-
   delphia, April 1982. K. Eggert: Results from the SPS
   Antiproton-Proton Collider, presented at the Neutrino
   Conference, Balatonfüred, June 1982.

4. Proceedings of the 3rd Topical Workshop on proton-antiproton collider physics, Roma, January 1983,to be published.

5. For a recent review, see: J. Ellis et al, Ann. Rev. Nucl.
   Part. Sci. 32 (1982) 443.

6. G.I. Budker, Atomnaya Energiy 22 (1967) 346.

7. G.I. Budker et al., Novosibirsk preprint IAF 76-33;
   translation: Experimental study of electron cooling,
   BNL-TR-635; Summary of electron cooling in the USSR, Yellow
   Report CERN 77-08.

8. D. Mohl, G. Petrucci, L. Thorndahl and S. van der Meer,
   Physics Reports 58 (1980) 73.

9. S. van der Meer, Stochastic damping of betatron oscillation in the ISR, CERN-ISR-PO/72-31(1972).

10. P. Bramham et al., Nucl. Instr. Meth. 125 (1975) 201.

11. S. van der Meer: Proton-antiproton colliding beam facility,
    CERN/SPS/423 (1978); Design Study of a proton-antiproton
    colliding beam facility, CERN/PS/AA/78-3.

12. R. Billinge, M.C. Crowley - Milling: The CERN proton-
    antiproton colliding beam facilities, CERN/PS/AA/79-17.

13. The staff of the CERN proton-antiproton project, Phys.
    Lett. 107B (1981) 306.

14. S. van der Meer: Practical and foreseeable limitations
    in luminosity for the collider, presented at the 3rd

topical workshop on proton-antiproton collider physics,
Roma, Jan. 1983.

15. C. Quigg, Rev. Mod. Phys. 94 (1977) 297.

16. R. Horgan and M. Jacob, in Proc. CERN School of Physics,
    Molente (FRG) 1980, Yellow Report CERN 81-04.

17. M. Jacob in: Fundamental Interactions: Cargèse 1981,
    Plenum Publishing Corp., 1982.

18. M. Abud, R. Gatto and C.A. Savoy, Annals of Physics 122
    (1979) 219.

19. G. Goggi: Physics prospects at the SPS antiproton-proton
    collider, CERN-EP/81-08, lectures at the IX. Int. Winter
    Meeting on Fund. Phys.,Siguenza, Spain, Feb. 1981.

20. UA1 proposal: CERN/SPSC/78-6/P92.

21. UA2 proposal: CERN/SPSC/78-8/P93.

22. UA3 proposal: CERN/SPSC/78-15/P96.

23. UA4 proposal: CERN/SPSC/78-105/P114.

24. UA5 proposal: CERN/SPSC/78-70/P108.

25. M. Barranco Luque et al., Nucl. Instr. Meth. 176 (1980)
    175; M. Calvetti et al., Nucl. Instr. Meth. 176 (1980) 255;
    M. Calvetti et al., The UA1 central detector, CERN-EP/82-44;
    M. Calvetti et al., First operation of the UA1 central
    detector, CERN-EP/82-170.

26. A. Astbury et al., Phys. Scripta 23 (1981) 397.

27. K. Eggert et al., Nucl. Instr. Meth. 176 (1980) 217;
    K. Eggert et al., ibid. 233.

28. UA1 Collaboration, G. Arnison et al., Phys. Lett. 107B
    (1981) 320.

29. UA1 Collab., G. Arnison et al., Phys. Lett. 118B (1982) 167.

30. UA1 Collab., G. Arnison et al., Phys. Lett. 118B (1982) 173.

31. UA1 Collab., G. Arnison et al., Phys. Lett. 121B (1982) 77.

32. UA1 Collab., G. Arnison et al., Phys. Lett. 122B (1982) 103.

33. UA1 Collab., G. Arnison et al., Phys. Lett. 122B (1982) 189.

34. UA1 Collab., G. Arnison et al., Phys. Lett. 123B (1982) 108.

35. UA1 Collab., G. Arnison et al., Phys. Lett. 123B (1982) 115.

36. UA1 Collab., G. Arnison et al.: Transverse momentum
    spectrum of neutral electromagnetic particles produced
    at the CERN proton antiproton collider, CERN-EP/82-120;
    presented to the XXI. Int. Conf. on High Energy Physics,
    Paris 1982.

37. A.G. Clark in: Proc. Int. Conf. on Instrumentation for colliding beam physics, SLAC 1982; UA2 Collab., First results from the UA2 experiment, presented at the 2nd Int. Conf. on Physics in Collisions, Stockholm, June 1982.

38. UA2 Collab., M. Banner et al., Phys. Lett. 115B (1982) 59.

39. UA2 Collab., M. Banner et al., Phys. Lett. 118B (1982) 203.

40. UA2 Collab., M. Banner et al., Phys. Lett. 121B (1982) 187.

41. UA2 Collab., M. Banner et al., Phys. Lett. 122B (1982) 322.

42. UA2 Collab., M. Banner et al., Phys. Lett. 123B (1982) 476.

43. UA2 Collab., M. Banner et al.: Preliminary searches for hadron jets and for large transverse momentum electrons at the SPS $\bar{p}p$ collider, CERN-EP/83-23, presented at the 3rd Topical workshop on proton-antiproton collider physics, Roma, Jan. 1983.

44. B. Aubert et al., Phys. Lett. 120B (1983) 465.

45. UA4-Collab., R. Battiston et al., Phys. Lett. 115B (1982)333.

46. UA4-Collab., R. Battiston et al., Phys. Lett. 117B (1982)126.

47. UA4-Collab., presented by G. Matthiae at the 3rd Topical workshop on proton-antiproton collider physics, Roma, Jan. 1983.

48. UA4 Collab., Phys. Scripta 23 (1981) 642.

49. UA5 Collab., K. Alpgård et al., Phys. Lett. 107B (1981)310; UA5 Collab., K. Alpgård et al., Phys. Lett. 107B (1981) 315.

50. UA5 Collab., K. Alpgård et al., Phys. Lett. 115B (1982) 65.

51. UA5 Collab., K. Alpgård et al., Phys. Lett. 115B (1982) 71.

52. UA5 Collab., K. Alpgård et al., Phys. Lett. 121B (1983) 209.

53. UA5 Collab., K. Alpgård et al., Phys. Lett. 123B (1983) 361.

54. UA5/R703 Collab., K. Alpgård et al., Phys. Lett. 112B (1982) 183.

55. UA6 proposal: CERN/SPSC/80-63/P148.

56. UA4-Collab., presented by J. Timmermans and G. Sanguinetti at the Moriond workshop on $\bar{p}p$ collider physics, March 1983.

57. UA1-Collab., presented by F. Ceradini at the 3rd topical workshop on $\bar{p}p$ collider physics, Roma, Jan. 1983; UA1-Collab., presented by C. Hodges at the Moriond workshop on $\bar{p}p$ collider physics, March 1983.

58. UA1-Collab., G. Arnison et al., Elastic and total cross

section measurement at the CERN $\bar{p}p$ collider, CERN-EP/83- submitted to Phys. Lett.

59. M. Block and R. Cahn, Phys. Lett. 120B (1983) 224; M. Block and R. Cahn, Phys. Lett. 120B (1983) 229.

60. J. Burg et al., Phys. Lett. 109B (1982) 124.

61. U. Amaldi et al., Phys. Lett. 66B (1977) 390.

62. L. Baksay et al., Nucl. Phys. B14 (1978); U. Amaldi and K.R. Schubert, Nucl. Phys. B166 (1980) 301.

63. A. Martin, Zeit. f. Physik C15 (1982) 185; A. Martin, in Proc. of the XXI. Int. Conf. on High Energy Physics, Paris 1982.

64. A. Martin: Elastic scattering and total cross sections, CERN-TH 3527, invited talk at the 3rd topical workshop on $\bar{p}p$ collider physics, Roma, Jan. 1983.

65. Reviews of particle production at FNAL and ISR energies are, e.g.: J. Whitmore et al., Phys. Rep. 10C (1974) 273; L. Foà, Phys. Rep. 22 (1975)1; J. Whitmore, Phys. Rep. 27 (1976) 187; G. Giacomelli and M. Jacob, Phys. Rep. 55 (1979) 1.

66. W. Thomé et al., Nucl. Phys. B129 (1977) 365.

67. Z. Koba, H.B. Nielson, P. Olesen, Nucl. Phys. B40 (1972) 317.

68. UA5-Collab., presented by D. Ward at the 3rd topical work- shop on $\bar{p}p$ collider physics, Roma, Jan. 1983.

69. R. Odorico, Phys. Lett. 119B (1982) 151.

70. C.M.G. Lattes et al., Phys. Rep. 65 (1980) 151.

71. L. van Hove, Phys. Lett. 118B (1982) 138.

72. S.M. Berman, J.D. Bjorken, J.B. Kogut, Phys. Rev. 4D (1971) 3388.

73. For review, see e.g. K. Hansen and P. Hoyer, eds., Jets in high energy collisions, Phys. Scr. 19 (1979); R.D. Field, Proc. of Boulder Summer School (1979); P. Darriulat, Ann. Rev. Nucl. Part. Sci. 30 (1980) 159; N.A.Mc. Cubbin, Rep. Prog. Phys. 44 (1981) 1027; G. Wolf, in Proc. XXI. Int. Conf. on High Energy Physics, Paris 1982, J.de Physique 43 Suppl. 12, C3-525(1982); N.A.Mc Cubbin, Hard scatte-

484

ring at ISR energies, in Proc. Conf. Physics in Collision, Stockholm 1982.

74. For a recent review, see P. Söding and G. Wolf, Ann. Rev.
Nucl. Part. Sci. $\underline{31}$ (1981) 231.

75. R. Odorico, Nucl. Phys. $\underline{B199}$ (1982) 189 and ref. 69;
R.D. Field, G.C. Fox, R.L. Kelly, Phys. Lett. $\underline{119B}$ (1982)
439; M. Greco, Phys. Lett. $\underline{121B}$ (1983) 360; A. Nicolaides:
Gluonic radiation and calorimeter physics, Collége de
France preprint LPC 83-02 (1983).

76. H. Bøggild: High p_T jets in hadron collisions, invited talk
at the Europhysics Study Conference on Jet Structure
from Quark and Lepton Interactions, Erice, Sept. 1982,
CERN-EP/82-187.

77. For example: M. Della Negra et al., Nucl. Phys.$\underline{B127}$ (1977)1;
M.G. Albrow et al., Nucl. Phys. $\underline{B145}$ (1978) 305; M.G. Albrow
et al., Nucl. Phys. $\underline{B160}$ (1979) 1; A.L.S. Angelis et
al., Phys. Scr. $\underline{19}$ (1979) 116; A.G. Clark et al., Nucl.
Phys. $\underline{B160}$ (1979) 397; D. Drijard et al., Nucl. Phys. $\underline{B166}$
(1980) 233; D. Drijard et al., Nucl. Phys. $\underline{B208}$ (1982) 1;
D. Drijard et al., Phys. Lett. $\underline{121B}$ (1983) 433.

78. C.de Marzo et al., Phys. Lett. $\underline{112B}$ (1982) 173; C.de Marzo
et al., Nucl.Phys. $\underline{B211}$ (1983) 375.

79. B. Brown et al., FNAL-Conf-82/34-Exp, T.L. Watts et al.,
Proc. Int. Conf. High Energy Physics, Paris 1982, J.de
Physique $\underline{43}$, C3-127.

80. W. Selove et al., Proc. Int. Conf. High Energy Physics,
Paris 1982, J. de Physique $\underline{43}$, C3-131; A. Arenton,
presented at the Europhysics Study Conference on Jet
structure from quark and lepton interactions, Erice,
Sept. 1982.

81. AFS collaboration, T. Åkesson et al., Phys. Lett. $\underline{119B}$
(1982), 185, 193; T. Åkesson et al., Phys. Lett. $\underline{123B}$
(1983) 133.

82. T. Åkesson and H. Bengtsson, Phys. Lett. $\underline{120B}$ (1983) 233.

83. R. Horgan and J. Jacob, Nucl. Phys. $\underline{B179}$ (1981) 441.

84. W. Furmanski and H. Kowalski, CERN-EP/83-21, submitted
to Nucl. Phys.

85. UA1-Collab., presented by J. Sass at the Moriond work-
    shop on p̄p collider physics, March 1983.

86. UA2-Collab., presented by L. Fayard at the Moriond work-
    shop on p̄p collider physics, March 1983.

87. UA1-Collab., presented by V. Vuillemin at the Moriond
    workshop on p̄p collider physics, March 1983.

88. K.H. Mess and B.H. Wiik, DESY 82-011; G. Wolf, ref.73;
    D. Luke, Proc. XXI. Int. Conf. on High Energy Physics,
    Paris 1982, J. de Physique $\underline{43}$ Suppl. 12, C3-67.

89. J. Ellis, M.K. Gaillard, G. Girardi and P. Sorba, Ann.
    Rev. Nucl. Part. Sci. $\underline{32}$ (1982) 443.

90. an introduction to the field can be found in the lectures
    by G. Ecker, ref.2; more details are given in the lectures
    by G. Altarelli at the Schladming School 1982, Acta
    Physica Austriaca, Suppl. XXIV, 1982.

91. J. Kim, P. Langacker, M. Levine, H. Williams, Rev. Mod.
    Phys. $\underline{53}$ (1981).

92. D.Yu. Bardin and V.A. Dokuchaeva, JINR-P2-82-522, submitted
    to Yad. Fiz.; W.J. Marciano and A. Sirlin, Nucl. Phys.
    $\underline{B189}$ (1981) 442; J.F. Wheater and C.H. Llewellyn Smith,
    Phys. Lett. $\underline{105B}$ (1981) 486; Nucl. Phys. $\underline{B208}$ (1982) 27;

93. J.F. Wheater in Proc. XXI. Int. Conf. on HEP, Paris 1982,
    J. de Physique 43 (1982) C3-305;

94. W.J. Marciano and Z. Parsa, in Proc. of Summer Study on
    Elementary Particle Physics and Future Facilities,
    Snowmass, Colorado 1982 (AIP, New York 1983).

95. W.J. Marciano and Parsa, in Proc. Cornell Z° Theory
    Workshop 1981, ed. M. Peskin and S.-H. Tye, CLNS 81-485,
    Cornell Univ. Press, Ithaca, N.Y. 1981.

96. S.D. Drell, T.M. Yan, Ann. Phys. NY $\underline{66}$ (1971) 578; T.M.
    Yan, Ann. Rev. Nucl. Sci. $\underline{26}$ (1976) 199.

97. Recent reviews are R. Stroynowski, Phys. Reports $\underline{71}$ (1981)1;
    I.R. Kenyon, Rep. Prog. in Physics $\underline{45}$ (1982) 1261;
    B. Cox, in Proc. XXI. Int. Conf. on HEP, Paris 1982, J
    de Physique $\underline{43}$ (1982) C3-140; E.L. Berger, ANL HEP CP 82-68
    and 82-72.

98. L.B. Okun and M.B. Voloshin, Nucl. Phys. $\underline{B120}$ (1977) 459;

R.F. Peierls, T.L. Trueman, L.L. Wang, Phys. Rev. D16 (1977) 1397; J. Kogut, J. Shigemisu, Nucl. Phys. B129 (1977) 461.

99. F.E. Paige in Proc. Topical Workshop on the prod. of new particles in super high energy collisions, Madison 1979, ed. V. Barger, F. Halzen, AIP New York, 1979.

100. E. Reya, Phys. Rep. 69 (1981) 195.

101. G. Altarelli, R.K. Ellis, G. Martinelli, Nucl. Phys. B147 (1979) 461.

102. B. Humpert, W.L. van Neerven, Phys. Lett. 93B (1980) 456.

103. S. Pakvasa et al., Phys. Rev. D20 (1979) 2862; M. Chaichian, M. Hayashi, K. Yamagishi, Phys. Rev. D25 (1982) 130.

104. P. Aurenche and J. Lindfors, Nucl. Phys. B185 (1981) 301.

105. F. Halzen et al., Phys. Lett. 106B (1981) 147; F. Halzen et al., Phys. Rev. D25 (1982) 754.

106. W.L. van Neerven, J.A.M. Vermaseren and K.J.F. Gaemers, preprint NIKHEF-H/82-20a, Dec. 1982; J. Smith, W.L. van Neerven and J.A.M. Vermaseren, Phys. Rev. Lett. 50 (1983) 1738.

107. V. Barger, A.D. Martin, and R.J.N. Phillips, RAL preprint RL-83-011, Jan. 1983.

108. M. Perottet, Ann. Phys. NY 115 (1978) 107; J. Finjord et al., Nucl. Phys. B182 (1981) 427.

109. UA1 collaboration, presented by T. Hansl-Kozanecka at the Moriond workshop on p̄p collider physics, March 1983.

110. P. Minkowski, CERN-TH 3519, February 1983.

111. A recent review is: F. Muller, Hadroproduction of heavy flavours, lectures at the Erice and Kupari summer schools, 1982.

112. D. Dibitonto, presentation at the Moriond workshop on p̄p collider physics, March 1983.

113. R.M. Godbole, S. Pakvasa and D.P. Roy, Tata preprint TIFR/TH/83-8, submitted to Phys. Rev. Letters; D.P. Roy, Tata preprint TIFR/TH/83-15.

114. A. Martin, presentation at the Moriond workshop on p̄p collider physics, March 1983, Durham preprint DTP/83/12, and references quoted therein.

115. UA1-Collab., G. Arnison et al., CERN-EP/83-73, submitted to Phys. Letters B.

Acta Physica Austriaca, Suppl. XXV, 489–547 (1983)
© by Springer-Verlag 1983

IMPLICATIONS OF QUANTUM CHROMODYNAMICS FOR
LOW AND MEDIUM ENERGY NUCLEAR PHYSICS[+]

by

E. WERNER
Institut für Theoretische Physik
Universität Regensburg
D-8400 Regensburg, W.-Germany

1. LECTURE

Introduction

During the last ten years QCD has developed into a serious
candidate on which a theory of strong interactions could
be built. This is based on two salient features of QCD:
(i) The asymptotic freedom predicted by the theory for short
distance dynamics allows to make quantitative calculations
for processes where high four-momentum transfers are encountered (hard processes). A wealth of experimental data
had been analyzed in this way leading to a very impressive
agreement between experiment and QCD [1].
(ii) QCD yields a long distance behaviour, being characterised
by strong coupling phenomena which prevent the liberation
of colour and leads to the prediction that only colour singlet objects can exist as free particles in nature. This explains in a natural way the great success of the constituent quark models based on unitary symmetry. It explains
qualitatively why hadrons are built from particles which do

─────────
[+]Lectures given at the XXII. Internationale Universitätswochen für
Kernphysik,Schladming,Austria,February 23-March 5,1983.

not exist as free particles in nature and it tells us - also
qualitatively - how these particles interact amongst each
other and what is the nature of the forces which keep hadrons
together. Thus it delivers the basis for the construction
of models of hadronic structure where baryons and mesons
are treated as objects consisting of three quarks or quark-
antiquark pairs respectively interacting via coloured non-
abelian gauge fields.

Since in the strong coupling regime QCD constitutes an
extremely complicated relativistic many body problem, there
exists up to now no comprehensive theoretical treatment which
would allow to build a really microscopic theory of hadronic
structure and of hadronic interactions based on the original
QCD Lagrangian. There are however various attempts to extend
QCD into the strong coupling region. The most important of
these attempts are:

(a) The 1/N expansion which gives astonishingly good results
    already for $N \simeq 3$ [2].
(b) Operator expansion methods, where the calculation of
    correlation functions is split up into a perturbative
    determination of certain coefficients and an empirical
    determination of nonperturbative vacuum expectation values
    [3]. QCD sum rules which can be derived from this approach
    are a very useful tool in applying QCD to hadronic structure
    problems.
(c) Analytical solutions of the classical Yang-Mills-equations
    (instantons) have been used to investigate qualitatively
    possible confinement mechanism and the nature of the
    effective quark-quark interaction induced by large scale
    gauge field fluctuations [4].
(d) Lattice gauge calculations in which starting from the QCD
    Lagrangian one calculates quantities like Wilson-loop inte-
    grals or quark propagators in a lattice space via Monte
    Carlo techniques in order to investigate the confining pro-
    perties of QCD [5].

Though none of these approaches is yet so far developed that a genuine microscopic theory of long distance phenomena in hadronic structure and interactions could evolve from it at the moment, they all have influenced the development of phenomenological models.

These phenomenological models are the product of our present inability to treat QCD in the strong coupling domain properly which leaves us only with two options: either we abstain at all from the attempt to investigate hadronic structure beyond the limits set by the present state of the art of long distance QCD dynamics, or we try to extract as much information as possible from the above mentioned approaches and use it as an input in hadronic models. The idea behind phenomenological models is not only to apply them to problems of hadronic structure but also to the question of interactions between hadrons. The latter field is particularly challenging from the point of view of nuclear physics. In nuclear physics we have amazingly successful models for the interaction between nucleons and between nucleons and Δ-resonances which only use mesonic degrees of freedom for the exchanged particles and which do not make explicit use of the $q\bar{q}$-structures of the exchanged mesons [6]. On the other hand, the $q\bar{q}$-structure of mesons implies that - except for the pion - the size of the exchanged meson is comparable to or shorter than its range. How can this be reconciled with the apparent success of a model which exchanges structureless particles coupled to baryons with simple formfactors?

Quite generally in the problem of interactions between hadrons there is a direct impact of the quark and gluon structure of hadrons on nuclear physics since the question how hadrons behave when they begin to touch each other is intimately related to the question of hadronic structure itself. Also the behaviour of a system consisting of many nucleons is decisively determined by the size and structure of the constituent particles and their mutual polarizibilities.

Therefore, before one begins to study subtle effects of QCD on nuclear structure properties one has to come to a better understanding of nucleonic properties in the small q domain. It seems that presently this aim can be obtained only with phenomenological models.

Bag Models

An extreme realization of ultraviolet asymptotic freedom on one hand and infrared strong coupling with colour confinement on the other hand is in the various types of bag models, where quarks and gluons move freely in the interiour of a sphere of Radius R (R ∿ 1 fm) and are confined by a boundary condition being equivalent to an infinite jump of a scalar potential at r = R.

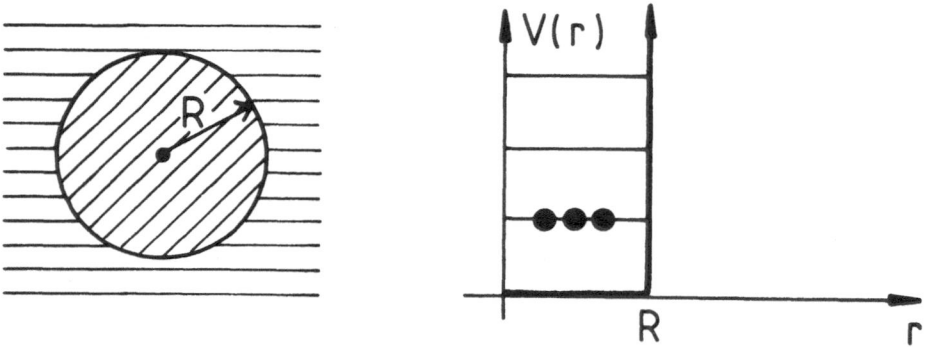

Fig. I.1: Bag models as an extreme realization of asymptotic freedom and infrared confinement.

The groundstate of a hadron is defined by three quarks (baryons) or a quark and an antiquark (mesons) occupying the lowest single particle states compatible with the quantum numbers of the hadron. The single particle states are obtained through the solution of the Dirac-equation in the potential well. The contribution of the gluonic field to the hadronic energy is taken into account phenomenologically through a volume energy density B, which is assumed to be universal for all hadrons. The total energy of the hadron is then

$$E(R) = N \frac{X}{R} + (\frac{4\pi}{3} R^3) B .$$

Here N is the number of quarks and antiquarks and X is a dimensionless quantity which is related to the eigenvalues of the single particle state; for all quarks and antiquarks in the 1 s1/2 level one has X = 2.04.

The condition $\frac{\partial E}{\partial R} = 0$ fixes R in terms of the parameter B. This stability condition implies that the gluonic energy is 1/4 of the total energy of the hadron irrespective of the type of hadron. For comparison with potential models which will be discussed later on we note the form of the Lagrangian density for the MIT-model [7]:

$$L_{MIT}(x) = \frac{i}{2}(\bar{\psi}(x) \overset{\leftrightarrow}{\partial} \psi(x)) \theta(R-r) - B\theta(R-r) - \frac{1}{2} \bar{\psi}(x)\psi(x)\delta(r-R) ,$$

where for brevity the θ- and δ-function are written in non-covariant form.

Chiral Bag Model

The original bag models fail completely in one important property: they do not respect chiral symmetry which - after isospin invariance - is the best respected symmetry of hadronic interactions. This is unavoidable in a model consisting only of quarks confined by a scalar potential: upon reflection at the confining potential quarks change chirality, so that chirality is no more a good quantum number. This leads to nonconservation of the axial current

494

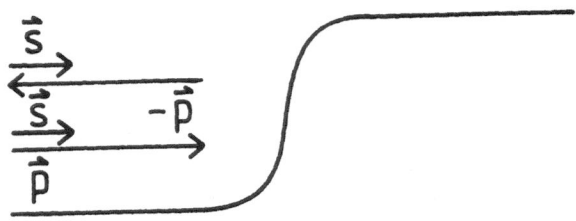

<u>Fig. I.2:</u> Nonconservation of chirality in a scalar potential.

carried by the quarks, as is seen from

$$\partial_\mu A_\lambda^\mu(x) = \partial_\mu [\bar\psi(x) \gamma^\mu \gamma_5 \frac{\tau_\lambda}{2} \psi(x)]$$

$$= M(\vec{r}) \bar\psi(x) i\gamma_5 \frac{\tau_\lambda}{2} \psi(x) \quad ; \tag{I.1}$$

by virtue of the Dirac equation

$$(i\not{\partial} - M(\vec{r}))\psi(x) = 0 ; \quad x = \vec{r}, t \quad .$$

In the MIT-model eq. (I.1) leads to a surface term for the divergence of the axial current

$$\partial_\mu A_\lambda^\mu(x) = \delta(r-R)\frac{1}{i}\bar\psi(x)\gamma_5 \frac{\tau_\lambda}{2}\psi(x) \quad . \tag{I.2}$$

The nonconservation of the axial current on the hadron surface has been repaired in essentially three different ways within the framework of hybrid models, consisting of quarks and elementary classical pion and σ-fields:

(A) <u>Chiral MIT-bag [8].</u>

This is a generalization of the σ-model in which the r.h.s. of eq. (I.2) is interpreted as a source term for the $(\sigma, \vec{\pi})$ field. The Lagrangian density of this model

$$L(x) = (\tfrac{i}{2}\bar{\psi}(x)\overset{\leftrightarrow}{\not{\partial}}\psi(x)-B)\theta(R-r) - \tfrac{\lambda}{2}\bar{\psi}(x)[\sigma(x)+i\vec{\tau}.\vec{\pi}\gamma_5]\psi(x)\delta(r-R)$$

$$+ \tfrac{1}{2}(\partial_\mu\sigma)^2 + \tfrac{1}{2}(\partial_\mu\pi)^2 , \qquad \lambda = \frac{1}{\sqrt{\sigma^2+\pi^2}} , \qquad (I.3)$$

is chirally invariant and leads to a conserved axial current

$$A^\mu_\lambda = (\bar{\psi}(x)\gamma^\mu\gamma_5\tfrac{\tau_\lambda}{2}\psi(x))\theta(R-r) - \pi_\lambda(x)\partial^\mu\sigma(x) + \sigma(x)\partial^\mu\pi_\lambda(x) .$$

The classical solution that was obtained for this model leads to $\vec{\pi}$-and σ-fields which are discontinous at the hadron sur-face. The physical meaning is hard to interpret, because the solution conserves neither isospin nor angular momentum. The σ- and $\vec{\pi}$-fields exist everywhere in space and lead to a local nonzero mass for the quarks also inside the bag.

(B) <u>Bag models with excluded pions</u>

Many of the difficulties in the solution of model (A) have been avoided by taken seriously the two-phase picture suggested by QCD according to which in the interior of a hadron there is asymptotic freedom for the quarks and chiral symmetry in the sector of u- and d-quarks, whereas in the exterior region free quarks and gluons can not exist and chiral symmetry is spontaneously broken in the quark sector leading to a non-zero order parameter $\langle\bar{q}q\rangle$ and to the appearance of a pseudoscalar Goldstone boson which is interpreted as the pion. Since in such a picture quarks on one hand and σ- and $\vec{\pi}$-mesons on the other hand can only communicate via their interaction at the hadron's surface it is suggested that a perturbative calcu-lation should be possible in which the pion field on the ha-dron surface is a measure of the expansion parameter. Such a program has been carried through in the perturbative treatment of the chiral MIT-model [9] and in the Little bag model of the Stony Brook group [10]. Both approaches use a non-linear sigma-model in which the sigma field is eliminated by the introduction of a new pion field. The interaction of the pion field with the quark core is via a surface source term. In the spirit of the above mentioned two-phase model pions are allowed to exist only in the outside phase; therefore the pion field is necessarily discontinous at the bag surface. This disconti-niuty has an important effect on the prediction of the axial-vector coupling g_A of the model: it leads to a pion contri-bution $g_A^{(\pi)}$ which is 1/2 of the quark-core contribution $g_A^{(c)}$; since $g_A^{(c)}=1.09$ is already rather close to the experimental value, one obtains notoriously too large values for $g_A^{(c)}+g_A^{(\pi)}$ $= \frac{3}{2}g_A^{(c)}$. This is a mere artifact of the extreme two-phase picture; if the sharp surface separating the chiral phase and the broken symmetry phase is replaced by a smooth one, $g_A^{(\pi)}$ vanishes identically for $m_\pi \neq 0$. This is seen as follows: In the case where the quarks act as a fixed source for the pion field this field is determined by the equation

$$(\Delta-m_\pi^2)\pi_\lambda(\vec{r}) = F(r)\sum_i \bar\psi_i(\vec{r})i\gamma_5\tau_\lambda\psi_i(\vec{r}) \tag{I.4}$$

where $F(r)$ follows from PCAC: $F(r) = -\frac{M(r)}{f_\pi}$. The solution of (I.4) satisfying the appropriate boundary condition can be written as

$$\pi_\lambda(\vec{r}) = \tau_\lambda(\vec\sigma.\hat{r})h(r) . \tag{I.5}$$

Asymptotically $h(r)$ goes as

$$h(r) \underset{r\to\infty}{\to} const.\frac{e^{-m_\pi r}}{r^2}(1 + m_\pi r) . \tag{I.6}$$

The contribution of the pion field to the axial charge g_A of the nucleon N becomes

$$g_A^{(\pi)}<\vec\sigma_N \frac{\vec\tau_N}{2}> = f_\pi\int_V d^3r\nabla[(\vec\sigma.\hat{r})h(r)\vec\tau]_N$$

$$= f_\pi\int_S d\hat{S}[(\vec\sigma.\hat{r})h(r)\vec\tau]_N .$$

The last equality is true only, if the function $(\vec{\sigma}\cdot\hat{r})h(r)$ and its gradient is continous inside the integration volume. For a continous source distribution the pion field and its gradient are continous everywhere: the only contribution is then from the surface at infinity; from (I.6) one obtains $g_A^{(\pi)} = 0$ for $m_\pi \neq 0$; for the chiral bag model one obtains a contribution from the bag surface which is $g_A^{(\pi)} = \frac{1}{2}g_A^{(c)}$.

(C) The_cloudy_bag_model [11].

There are essentially two modifications as compared to the previous models:

(a) pions are not excluded from the interior of the bag for two reasons: (1) it is assumed that creation of $q\bar{q}$-pairs carrying the quantum numbers of the pion can take place everywhere in the bag via an residual interaction though preferentially in the surface region. (2) The concept of a rigid, well defined surface is no more realistic as soon as pions can be created on it which carry away momenta of typically several hundred MeV/c.

(b) the pion field is quantized

In all chiral models hadrons are made up of a quark core surrounded by a cloud of pions (and eventually) σ-mesons. Therefore all calculated properties receive in general quark-core and mesonic contributions. The importance of the pion cloud depends sensitively on the size of the quark core. This is shown for the case of the pion contribution to the nucleon mass in fig. I.3 for different models. It is also seen that for core sizes which appear physically acceptable ($R \leq 0.8$ fm) the selfenergy corrections due to the pion become so large that perturbative calculations appear at least doubtful. This is not astonishing since the scale dependence of QCD is not properly accounted for by these models and because the structure of the pion is neglected.

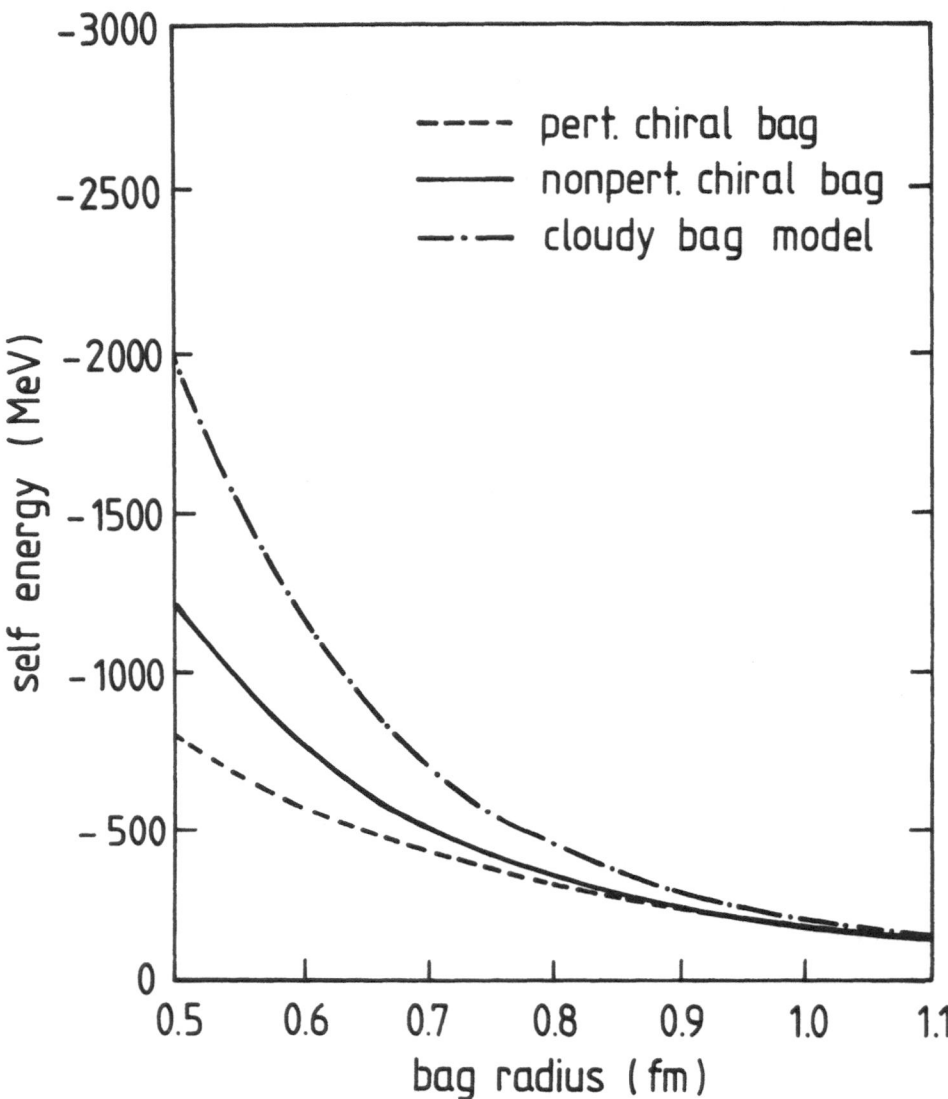

Fig. I.3: Pion induced selfenergy corrections of the nucleon as a function of the bag radius for 3 different bag models.

In all models the pion is put into the theory "by hand" as an almost Goldstone boson in order to compensate the breaking of chiral symmetry which happens on the quark level as a consequence of confinement. This pion has no internal structure and there is no possibility to calculate empirical properties of the pion, as its mass m_π, its weak decay constant f_π and its r.m.s. radius r_π. On the other hand the bag models predict mesons as $q\bar{q}$-states. In lowest order the meson mass is just given by the sum $\varepsilon_q + \varepsilon_{\bar{q}}$ of the eigenvalues of the Dirac equation for the states occupied by the quark and antiquark respectively. The lightest mesons are obtained with masses between 600 and 700 MeV depending on the parameters of the model. This applies also to the pion mass; the meson masses can still be changed by residual interactions and by center of mass corrections. But these changes are not sufficient to bring the pion mass down to the experimental value and there is no possibility to make any connection with the Goldstone pion.

2. LECTURE

In this lecture we want to discuss attempts to go beyond bag models and to make a closer connection of quark models of hadrons with the underlying QCD. I shall present two different approaches:

(1) the work of Goldman and Haymaker [12]
(2) the model that has been developed by the Regensburg group [13]

(1) The work of Goldman and Haymaker

The Goldman Haymaker model (GHM) does not give up the bag concept where nonperturbative effects are treated through a boundary condition, but is breaks chiral symmetry on the quark level prior to imposing confinement. To this end an effective Lagrangian density is introduced

$$L = L_{kin} + L_{QCD;pert.} + L_B + L_{Inst.}$$

where L_{kin} is the free quark and gluon part, $L_{QCD;pert.}$ takes into account the perturbative quark-gluon and gluon-gluon interactions, L_B describes the nonperturbative confining effects and L_{Inst} contains small instanton effects which are included in order to solve the U(1) problem (π-η mass splitting).

Conceptually, the order of grouping the above terms in the MIT approach is

$$L_{MIT} = L_{MIT}^o + L_{MIT}^1 \quad ,$$

$$L_{MIT}^o = L_{kin} + L_B \ , \quad L_{MIT}^1 = L_{QCD;pert.} + L_{Inst.} \quad ,$$

whereas in the GHM the grouping is

$$L_{GHM} = L_{GHM}^o + L_{GHM}^1 \quad ,$$

$$L_{GHM}^o = L_{kin} + L_{QCD;pert.} + L_{Inst.} \ ,$$

$$L_{GHM}^1 = L_B \quad .$$

Of course, this concept makes sense only, if L_{GHM}^o is able to yield already bound states on its own which limits the applicability to the meson sector.

The interaction part $L_{QCD,pert.} + L_{Inst.}$ is approximated by a separable interaction. The relevant parameters of the interaction are a strength parameter α and a cutoff Λ which is necessary to regularize ultraviolet divergent integrals. Λ^{-1} is a measure of the range of the separable $q\bar{q}$ interaction. Starting from zero mass quarks one can generate finite quark masses and this breaks chiral symmetry on the quark level. On the other hand one can solve the Bethe-Salpeter equation in order to obtain meson propagators and wavefunctions. As it must be the pion is obtained with $m_\pi = 0$; the pion decay constant f_π is obtained from the wavefunction in relative co-ordinates at $\vec{r}_q - \vec{r}_{\bar{q}} = 0$. It depends on Λ and α.

Λ and α are then fixed in such a way that one obtains a quark mass $m_q \sim 300$ MeV and that the experimental value for $f_\pi = 93$ MeV is reproduced. The determination of α and Λ is not unique since the quark propagator contains a dynamical mass $m_D(q)$ which describes the approach to the asymptotic freedom domain; unfortunately $m_d(q)$ is not well known.

Typically one obtains $\alpha \sim 0.75$ for $\Lambda = 1$ GeV. The value of α which is comparable to the parameter α_s of the quark-gluon vertex indicates that strong interaction effects are needed to obtain the above results. Therefore the meaning of the separation of the effective Lagrangian into L_{GHM}^o and L_{GHM} is not quite clear.

With this reservation the finite pion mass can be explained as an effect of the confining part L_B which compresses (in r-space) the exponentially decaying wavefunction to $|\vec{r}_q - \vec{r}_{\bar{q}}| \leq R$ where R is the bag radius. One thus obtains the pion energy as a function of R. The final result of the numerical calculations in shown in fig. II.1.

(2) The Regensburg model

We begin our discussion with the Lagrangian density of QCD for quark fields ψ_f interacting with gluon fields G^a

$$L_{QCD}(x) = \sum_f \bar{\psi}_f(x)[i\not{D} - m_f]\psi_f(x) - \frac{1}{4}F_{\mu\nu}^a(x)F_a^{\mu\nu}(x) ,$$

$$D_\mu = \partial_\mu - ig\frac{\lambda_a}{2}G_\mu^a(x) ,$$

$$F_{\mu\nu}^a(x) = \partial_\mu G_\nu^a(x) - \partial_\nu G_\mu^a(x) + gf^{abc}G_\mu^b(x)G_\nu^c(x) , \qquad (II.1)$$

f and a are flavour and colour indices, λ_a designates the standard SU(3) colour matrices. We restrict the discussion to up and down quarks (flavour SU(2)). With $m_u = m_d = 0$ the Lagrangian is invariant under the chiral transformation

$$\psi(x) \rightarrow e^{i\gamma_5\vec{\tau}\cdot\frac{\vec{\theta}}{2}}\psi(x) ,$$

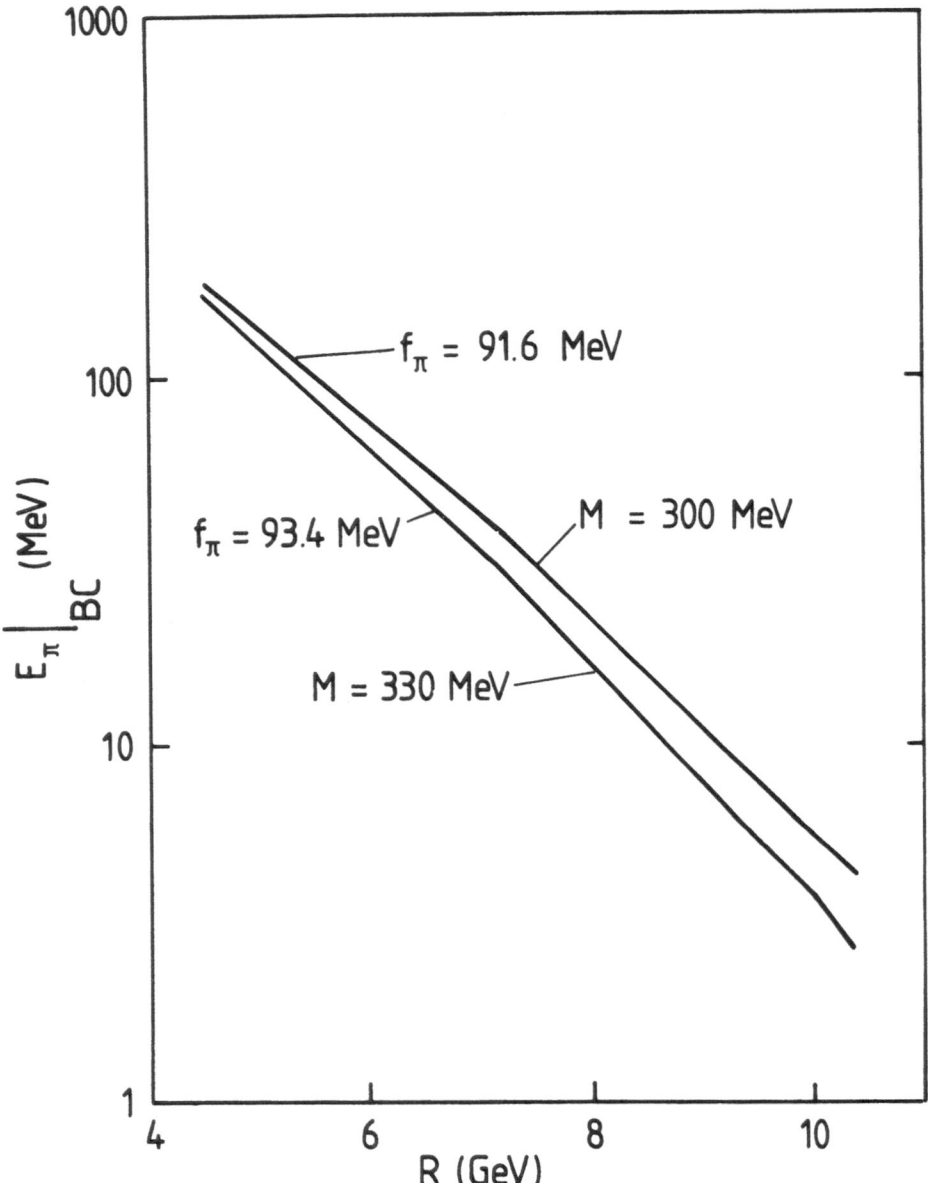

Fig. II.1: Pion decay constant f_π and pion mass m_π in the Goldman-Haymaker model.

$$\bar{\psi}(x) \rightarrow \bar{\psi}(x)e^{i\gamma_5 \vec{\tau} \cdot \frac{\vec{\theta}}{2}} \quad , \tag{II.2}$$

acting in the space (u,d). The effect of small masses m_u and m_d will be discussed later on. The chiral symmetry of L_{QCD} is dynamically broken by the mechanism which leads to confinement. Certainly, we do not know the details of this mechanism, but we know that it leads to some complicated, nonlocal selfenergy operator, $M(x,x')$ acting on the quarks; this selfenergy prevents quark propagation over space-times distances $\Delta x > 1/\Lambda$, Λ being the infrared QCD-cutoff. The equation of motion for the quark field operator ψ_f

$$i \slashed{\partial} \psi_f(x) - \int d^4 y M(x,y) \psi_f(y) = 0 \tag{II.3}$$

leads to the following expression for the divergence of the axial current

$$\partial_\mu [\bar{\psi}(x) \gamma^\mu \gamma_5 \frac{\tau_\lambda}{2} \psi(x)]$$

$$= \int [M(x,y) \bar{\psi}(x) i \gamma_5 \frac{\tau_\lambda}{2} \psi(y) + M^+(x,y) \bar{\psi}(y) i \gamma_5 \frac{\tau_\lambda}{2} \psi(x)] d^4 y \neq 0 \quad . \tag{II.4}$$

This spontaneous breaking of chiral symmetry on the quark level must be compensated by the appearance of a mass zero Goldstone boson which is not yet the physical pion, but hopefully becomes the pion after the inclusion of some minor effects.

Our incapability to treat properly the dynamics following from the Lagrangian (II.1) in the strong coupling domain leads us to try something much simpler which is very familiar to nuclear physicists, namely the elimination of the boson fields via which the fermions interact. If such an operation is possible, it leads to a considerable reduction of the degrees of freedom that have to be treated explicitely, at the cost - of course - of introducing a very complicated retarded interaction between the fermions.

In a linear theory like QED such an elimination is formally always possible, since the sourceterm for the boson
fields contains only the fermion field operators. In QCD the
formal elimination of gluon degrees of freedom from the
equations of motion is in general not possible - due to the
inherent nonlinearity of the theory. There are, however, two
limiting cases where the gluon fields can be eliminated:

(i) The elimination is quite trivial and analogous to the
QED case in the weak coupling limit; it leads to the
colour electric and magnetic interactions mediated by
one gluon exchange.

(ii) The elimination is also possible - though not at all
trivial - in the instanton model, as has been shown by
't Hooft [4]. The effective quark-quark interaction which
is obtained in this way has the property of being invariant under the chiral transformation (II.2) - a feature
which is not unexpected, given the chiral symmetry of
L_{QCD}. For an instanton located at the point x_o one obtains
a quark-quark interaction (for flavour SU(2)) which is
of the form

$$(\delta_{ij}\delta_{kl}-\delta_{ik}\delta_{jl}) \int (\bar{\psi}_i(x_i)(1\pm\gamma_5)\psi_j(x_j))(\bar{\psi}_k(x_k)(1\mp\gamma_5)\psi_l(x_l))$$

$$\times K(x_i,x_j,x_k,x_l;x_o)d^4x_id^4x_jd^4x_kd^4x_l \qquad (II.5)$$

where i,j,k,l are flavour indices.

Here K is the kernel of the four-fermion interaction; the
range of the interaction (in the variables x_i-x_o) is given
by the size of the instanton.

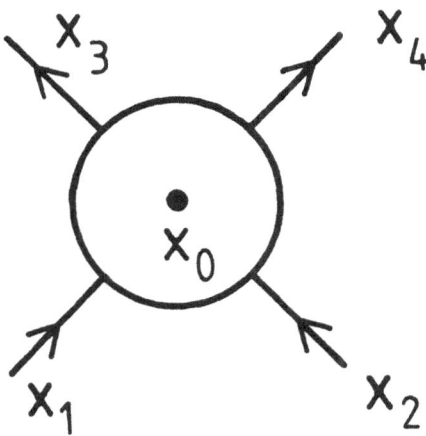

<u>Fig. II.2:</u> Four quark vertex of the effective interaction.

With the identity $\delta_{ij}\delta_{kl}-\delta_{ik}\delta_{jl}=\frac{1}{2}[\delta_{ij}\delta_{kl}-\vec{\tau}_{ij}\cdot\vec{\tau}_{kl}]$ the expression (II.5) can be rewritten as

$$\int\{[(\bar{\psi}_i(x_i)\psi_i(x_j))(\bar{\psi}_k(x_k)\psi_k(x_l))-(\bar{\psi}_i(x_i)\gamma_5\vec{\tau}\psi_i(x_j))(\bar{\psi}_k(x_k)\gamma_5\vec{\tau}\psi_k(x_l))]$$

$$-[(\bar{\psi}_i(x_i)\vec{\tau}\psi_j(x_j))(\bar{\psi}_k(x_k)\vec{\tau}\psi_l(x_l))-(\bar{\psi}_i(x_i)\gamma_5\psi_j(x_j))(\bar{\psi}_k(x_k)\gamma_5\psi_l(x_l))]\}$$

$$\times K(x_i,x_j,x_k,x_l;x_o)d^4x_i\ldots\ldots d^4x_l \quad . \qquad (II.6)$$

It can be easily checked that (II.6) is invariant under the chiral transformation (II.2); it is important to note that the two pieces

$$(\bar{\psi}\psi)(\bar{\psi}\psi)-(\bar{\psi}\gamma_5\vec{\tau}\psi)(\bar{\psi}\gamma_5\vec{\tau}\psi) \quad \text{and} \quad (\bar{\psi}\vec{\tau}\psi)(\bar{\psi}\vec{\tau}\psi)-(\bar{\psi}\gamma_5\psi)(\bar{\psi}\gamma_5\psi) \qquad (II.7)$$

are separately invariant under (II.2). In the limit of vanishing size of the instanton the interaction (II.6) reduces to a four-point interaction.

In order to obtain the Lagrangian density of the interaction one has to integrate over all instanton locations
x_0 and sizes d weighted with the density $\rho_I(x_0,d)$ of instan-
tons of size d at x_0. ρ_I can be calculated reliably only in

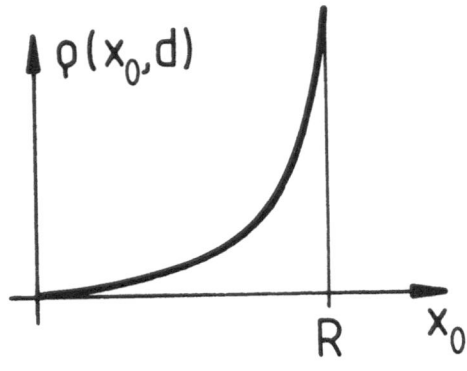

Fig.II.3: Qualitative be-
haviour of instan-
ton density as a
function of the in-
stanton location.

the - rather uninteresting - perturbative domain; however,
qualitatively it is clear, that ρ as a function of x_0 rises
very rapidly when x_0 approaches the surface region of the
baryon from inside. This leads to the picture of an effective
quark-quark interaction which is not just a nonlocal two-
body interaction but an interaction the strength of which
depends strongly on the distance of the interacting quarks from
the hadrons center. This is of course due to the fact that the
eliminated gluon field has its own dynamics and creates a
kind of medium in which the quarks move.
Certainly, the whole proceeding is very qualitative and it is
based on the instanton model which is not applicable in the
strong coupling limit of QCD, but is gives at least an idea
how things could go.

A picture which is qualitatively very similiar emerges
if one assumes that quarks inside a hadron interact via the
exchange of multigluon objects to which one can ascribe an
effective mass M_G; the strength which couples quarks to these
multigluon configuration and the effective mass M_G should both
increase strongly towards the surface region leading to a

shortrange interaction with exploding strength. If the principle of chiral invariance of the effective quark-quark interaction is used as an additional constraint for the construction,
one arrives at the following form for the interaction kernel

$$K = \sum_{\rho} \int d^4x_1 \ldots d^4x_4 K_{\rho}(x_1, \ldots, x_4)$$

$$\times [\bar{\psi}(x_1) \Gamma_{\rho} \psi(x_2)][\bar{\psi}(x_3) \Gamma_{\rho} \psi(x_4)] \tag{II.8}$$

where the sum is over all chiral invariant combination of
isospin and Dirac matrices. Possible colour factors are
omitted in (II.8), because nothing can be said about them.
The simplest assumption which one can make and which has also
the advantage of the least number of free parameters is to
assume that in the mean only colour singlet multigluon configuration contribute to the interactions. In this case
the effective interaction is colour independent. As far as
the spacetime and isospin-dependence is concerned we choose
also the most simple form which requires the introduction of
only one kernel function $K_1(x_1 \ldots x_4)$, leading to the following interaction-kernel

$$K = \int d^4x_1 \ldots d^4x_4 K(x_1 \ldots x_4)[\bar{\psi}(x_1) \psi(x_2) \bar{\psi}(x_3) \psi(x_4)$$

$$+ \bar{\psi}(x_1) i\gamma_5 \vec{\tau} \psi(x_2) \bar{\psi}(x_3) i\gamma_5 \vec{\tau} \psi(x_4)] \quad . \tag{II.9}$$

It should be kept in mind that our derivation is based
on the principle of invariance of the effective interaction
under the transformation (II.2) and is therefore restricted
to flavour SU(2). The inclusion of strange quarks would change
the form of the interaction; this change is most drastically
seen for the instanton induced interaction which for flavour
SU(3) becomes a six-point interaction. The spectroscopic
application of (II.8) is therefore limited.

The property of the effective interaction which is typical for its origin in QCD is its variable "strength" which increases, if one goes from the interior of the hadron towards the surface. This reflects the basic aspects of ultraviolet asymptotic freedom and infrared strong coupling which, in a description based on an effective Lagrangian, must be contained in a corresponding property of the effective interaction.

An interaction to the form (II.9) but with a zero range kernel and with a constant coupling strength g instead of the function G(r) has been used long before the invention of QCD by Nambu and Jona-Lasinio [14] for the discussion of dynamical breaking of chiral symmetry starting from massless fermions.

The results of the Nambu-Jona-Lasinio (NJL) model which are relevant in our discussion can be summarized as follows: (i) Starting from the Lagrangian density

$$L_{NJL}(x) = \bar{\psi}(x) i\gamma_\mu \partial^\mu \psi(x) + g[(\bar{\psi}(x)\psi(x))^2 + (\bar{\psi}(x) i\gamma_5 \vec{\tau}\psi(x))^2] \ ,$$

$$(II.10)$$

the fermion selfenergy $\sum(p)$ is calculated in the meanfield approximation, i.e.

$$\sum = \quad \bigcirc \quad + \quad \bigcirc \qquad\qquad (II.11)$$

where the intermediate propagator is treated selfconsistently:

$$S_F^{int.}(p) = \frac{1}{\not{p} - \sum(p)} \ . \qquad\qquad (II.11a)$$

The resulting selfconsistency condition for $\sum(p)$ leads to a fourmomentum independent fermion mass m (consequence of the meanfield approximation which neglects retardation effects) which depends on the coupling strength g in the manner shown in fig. II.4

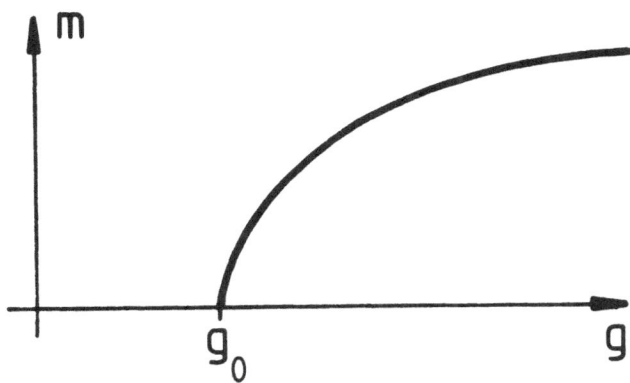

<u>Fig. II.4:</u> Selfconsistent fermion mass of the NJL-model.

For $g < g_0 = \frac{4}{5} \frac{\pi^2}{\Lambda^2}$ the fermions remain massless, whereas for $g > g_0$ they acquire a finite mass. Λ is a cutoff parameter which must be introduced in order to remove the divergence of the loop integral resulting from the four-point interaction. For $g > g_0$ chiral symmetry is spontaneously broken in the fermion sector.

(ii) The required appearance of a massless pseudoscalar Goldstone boson is furnished by the pseudoscalar part of the interaction

$$g(\bar{\psi}(x) i \gamma_5 \vec{\tau} \psi(x))^2 .$$

Looking at the homogenous Bethe-Salpeter equation in the fermion-antifermion sector

$$(II.12)$$

one finds that the selfconsistency condition for a finite
fermion mass is identical to the condition for a zero mass
of the lowest pseudoscalar meson state satisfying eq. (II.12).
At the same time the interaction in other fermion-antifermion channels - there are nonzero exchange matrix elements
in the vector and pseudovector sector - lead to small mass
shifts compared to the unperturbed mass 2 m for scalar,
vector and pseudovector particles.

In our model the point-interaction with coupling
strength g is replaced by the nonlocal interaction (II.9);
the kernel of this interaction is not translational invariant,
because it contains the effects of the running coupling strength
which increases with increasing distance from the center x_o
of the hadron. With this effective interaction we can
construct in analogy to the NJL model a quark selfenergy $\sum_q$
which is position dependent and which leads in the selfcon-
sistant mean field approximation to a relation connecting
the "effective running coupling strength" of the interaction
kernel - what this means will be explained below - to the
position dependence of $\sum_q(\vec{r})$ i.e. to the confinement potential.
Furthermore we investigate the pseudoscalar $q\bar{q}$-channel; there
the pseudoscalar part of the interaction leads to a dynamical
description of the pion which appears as a superposition of
$q\bar{q}$-configurations in

$(1s1/2)_q \otimes (1s1/2)_{\bar{q}}$, $(1p1/2)_q \otimes (1p1/2)_{\bar{q}}$, $(1p3/2)_q \otimes (1p3/2)_{\bar{q}}$ ...

states; this superposition makes the resulting state highly collective and leads to a pion mass $m_\pi \ll 2m_q$.

In order to carry through this program we simplify first the interaction kernel by making the following ansatz:

$$K = \int dx^4 d^4 y K(x,y) [\bar{\psi}(x) \psi(x) \bar{\psi}(y) \psi(y)$$

$$+ \bar{\psi}(x) i\gamma_5 \vec{\tau} \psi(x) \bar{\psi}(y) i\gamma_5 \vec{\tau} \psi(y)] \qquad (II.13)$$

with $K(x,y) = \sqrt{G(|x-x_o|)} \; d_\Lambda(x,y) \sqrt{G(|y-x_o|)}$.

Here x_o is the position of the center of the hadron; $G(|x-x_o|)$ is a position dependent coupling strength; the notation is such that for a zero range kernel the running coupling strength becomes $G(|x - x_o|)$. (The interaction (II.9) has the form of a local two-body interaction). The dimension of $G(|x - x_o|)$ is (length)^2. $d_\Lambda(x,y)$ is a function, the range of which in the variable $|x-y|$ is determined by the extension Λ^{-1} of the multigluon configurations which mediate the interaction. The range Λ^{-1} of the interaction leads to an effective cut-off Λ in fourmomentum space. The NJL four-point interaction of eq. (II.9) is obtained with the replacement $d_\Lambda(x,y) \to \delta^4(x-y)$.

The range parameter Λ should be of the order of ~ 1 GeV. The behaviour of the coupling strength $G(|x-x_o|)$ can of course not be calculated from QCD; we only know that its qualitative properties should be determined by asymptotic freedom for $|x-x_o| \to 0$ and by a rapid increase as $|x-x_o| \to R$ where R is the hadronic size.

Confining Potential as a Selfconsistent Mean Field

In order to obtain the equivalent of the NJL-selfcon-sistency equation (II.11) we assume that the motion of the quarks inside the hadron can be described by a Dirac equation (stationary solution)

$$[\gamma_o E_\alpha + i\vec{\gamma}\cdot\vec{\nabla} - M(\vec{r})]\psi_\alpha(\vec{r}) = 0; \quad \alpha = \{n,j,l,m\}. \tag{II.14}$$

The potential (or local mass) M(r) is considered to be a functional of the effective interaction from which it is dynamically generated. In the mean field approximation the quark selfenergy $\Sigma(x,x')$ is determined by the selfconsistency equation

which contains as essential ingredients the interaction kernel $K(x_1,\ldots,x_4)$ and the quark propagator S_F in the effective field Σ. The latter depends on Σ through the energy eigenvalues E_α and the eigenfunctions; to obtain its explicit form we expand the quark field operator $\psi(x)$ in terms of the particle and antiparticle solutions of eq. (II.14)

$$\psi(x) = \sum_\alpha [u_\alpha(\vec{r})e^{-iE_\alpha t}b_\alpha + v_\alpha(\vec{r})e^{iE_\alpha t}d_\alpha^+] \tag{II.15}$$

where b_α and d_α^+ are quark annihilation and antiquark creation operators, respectively. Since all our considerations are restricted to u- and d-quarks the wavefunctions and energies do not depend on flavour. The propagator S_F becomes

$$S_F(x,x') = \sum_\alpha [u_\alpha(\vec{x})\bar{u}_\alpha(\vec{x}')e^{-iE_\alpha(t-t')}\theta(t-t')$$

$$- v_\alpha(\vec{x})\bar{v}_\alpha(\vec{x}')e^{iE_\alpha(t-t')}\theta(t'-t)] \quad . \qquad (II.16)$$

The positive and negative energy solutions of eq. (II.14) are

$$u_{njlm}(\vec{r}) = \begin{pmatrix} ig_{njl}(r)/r \\ \\ (\vec{\sigma}\cdot\hat{r})f_{njl}(r)/r \end{pmatrix} \phi_{jlm}(\hat{r}) \quad ;$$

$$v_{njlm}(\vec{r}) = (-1)^{j+m-1} \begin{pmatrix} \vec{\sigma}\cdot\hat{r}if_{njl}(r)/r \\ \\ g_{njl}(r)/r \end{pmatrix} \phi_{jl-m}(\hat{r}) \quad , \quad (II.17)$$

where $\phi_{jlm}(\hat{r}) = \sum_{m_l m_s} (lm_l\frac{1}{2}m_s|jm)Y_{lm_l}(r)\chi_{\frac{1}{2}m_s}$.

The selfconsistency equation becomes particularly simple in the zero range limit for the interaction kernel (i.e. $d_\Lambda(x,y) = \delta^4(x-y)$ in eq. (II.13)). One then obtains a local potential $M(r)$ satisfying the equation

$$M(r) = 2dG(r) \sum_{n,l,j} \frac{2j+1}{4\pi r^2} [g^2_{njl}(r) - f^2_{njl}(r)] \quad . \qquad (II.18)$$

The cutoff procedure for a d_Λ of finite range Λ^{-1} would effectively lead to a termination of the sum over all (quark and antiquark) intermediates states $\alpha = (njl)$ in eq. (II.18) at energies or momenta of order Λ. This is equivalent to choosing a finite model space, the procedure which we shall

adopt here. Furthermore d is a degeneracy factor[+]. The confining potential M(r) replaces the mass in the NJL scheme
and receives its primary (direct) contribution from the scalar
part of the interaction, eq. (II.13). Exchange terms from
the Fierz transformed interactions would simply change the
factor d.

Since the coupling strength G(r) is quantitatively not
well known we prefer to choose first a reasonable form for
the potential M(r) and then calculate G(r) from eq. (II.18).
Unfortunately, none of the nonperturbative methods mentioned
in the introduction has allowed a reliable calculation of
the potential M(r) for light quarks. Therefore we made the
ansatz $M(r) = cr^n$ for the potential and determined c and a
best choice for the exponent n from the requirement to obtain
a good fit of:

(i) the quark core contribution to the charge form factor of
    the proton;
(ii) low energy baryon spectra (see Lect. 3 for details).

The best results were obtained with n = 2 and n = 3. Once
n is fixed the r.m.s. radius of the quark core determines
c [13] for n = 2 (3) one obtains c = 0.83 GeV fm^{-2}
(1.25 GeV fm^{-3}). To take account of the cut-off Λ we restricted the sum in eq. (II.18) to the states 1 s 1/2, 1 p 1/2
and 1 p 3/2 which corresponds roughly to $\Lambda \sim 0.5$ GeV. In
fig. II.5 we show the results for M(r) and G(r) together with
the quark density (for better comparison). One sees that
there is an interior region where the potential (local mass)
is small compared to the eigenvalue and where the effective
interaction is weak. One the other hand in the surface region

---

[+] d=2 for two flavors from direct terms of the interaction,
eq. (II.13) and d=5/2 including exchange terms. An additional factor of 3 would come from color. We propose to absorb
it in the coupling strength G.

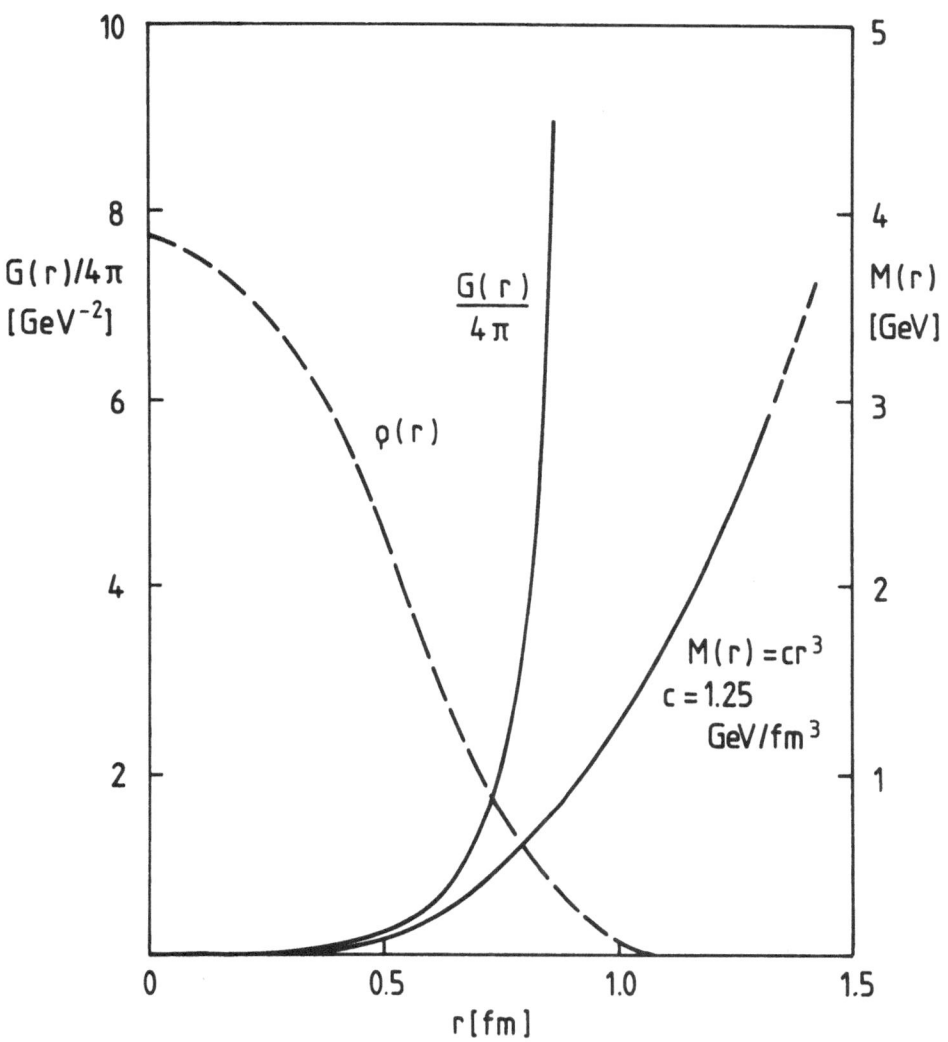

<u>Fig. II.5:</u> Relationship between confining potential M(r)
and running coupling strength.

G(r) increases very rapidly and gives rise to strong interaction effects. Note that one obtains a dimensionsless coupling strength, $\alpha(r) = g^2(r)/4\pi = \Lambda^2 G(r)/4\pi$ of order unity
in the surface. This is the region where the confining potential breaks chiral symmetry as can be seen by examining the
divergence of the quark axial current,

$$\partial_\mu A_\lambda^\mu(x) = M(r)\bar{\psi}(x) i\gamma_5 \tau_\lambda \psi(x)$$

the non-conservation of $A_\mu(x)$ being proportional to $M(r)$.

We want to emphasize that the previous treatment is based
on the mean field approximation; it is therefore valid only
as long as for a given single particle state $|\alpha\rangle$ with energy
E_α there are no nearly competing $qq\bar{q}$ - or even more complicated - configurations with comparable energy. This restricts
the applicability to low lying states.

Remembering the PCAC result (I.4) we observe that the
pion quark coupling is localized in the surface region. This
is exemplified for the case of the nucleon in fig. (II.6)
which shows the πNN form factor in coordinate space. The
interaction (II.13) can readily be used to calculate first
order shifts of baryon masses due to the residual qq-inter-
action. These shifts are typically of the order of 20 - 40 MeV;
this can be interpreted in the sense that the independent
particle nature of the baryonic wave function is only slightly
affected by the residual interaction.

The q$\bar{q}$-Structure of the Goldstone Pion

In lowest order all mesons states are obtained by putting the
quark and the antiquark each in one particular single particle
state $|\alpha_q\rangle$ and $|\alpha_{\bar{q}}\rangle$ respectively, so that the meson state
and energy is

$$|M\rangle = |\alpha_q \otimes \alpha_{\bar{q}}\rangle_M = |(q\bar{q})_a\rangle M \quad ,$$

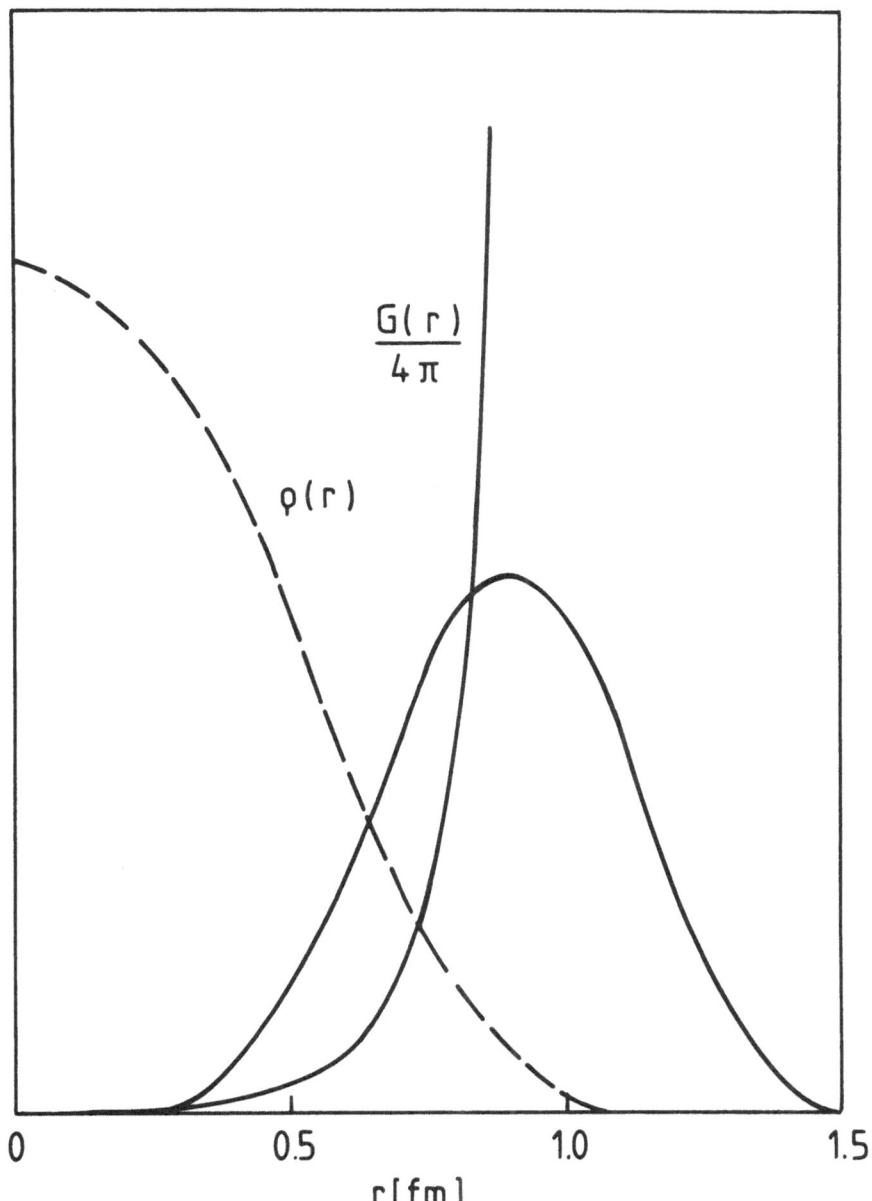

Fig. II.6: The pion source function $\rho_\pi(r)$ for the
nucleonic groundstate.

$$E_M = E_{\alpha_q} + E_{\alpha_{\bar{q}}} = E_a \quad . \tag{II.19a}$$

where the subscript a is an abbreviation for the set of quantum numbers α_q and $\alpha_{\bar{q}}$; the subscript M stands for the quantum numbers to which α_q and $\alpha_{\bar{q}}$ must be coupled. With the full Lagrangian taken into account, the $q\bar{q}$-pair moves in the confining potential under the additional influence of the effective interaction. The meson state then becomes

$$|M\rangle = \sum_a A_a |(q\bar{q})_a\rangle \quad . \tag{II.19b}$$

The r-space amplitude $\chi_a(\vec{x},\vec{y}) = \langle\vec{x},\vec{y}|(q\bar{q})_a\rangle$ becomes

$$\chi_a(\vec{x},\vec{y}) = N_a[u_{njl}(\vec{x})\bar{v}_{n'j'l'}(\vec{y})]_M \quad , \tag{II.20}$$

N_a is a normalization factor.

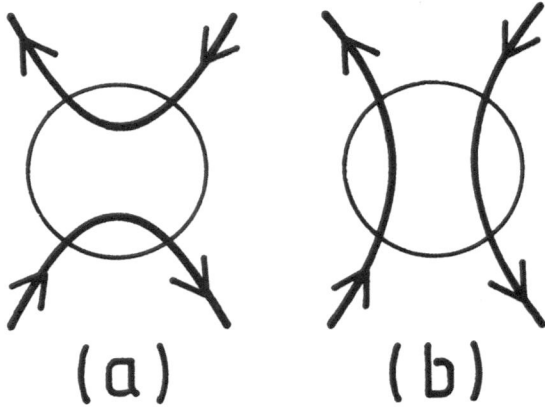

Fig. II.7a,b: Direct and exchange part of the $q\bar{q}$-interaction.

The pair density is $\rho_a = \text{Tr } \chi_a^+\chi_a$. We introduce the $q\bar{q}$ center-of-mass $\vec{R} = (\vec{x}+\vec{y})/2$ and relative coordinate $\vec{z} = \vec{x} - \vec{y}$. If $\vec{R}$ is fixed at the origin, $u(\vec{z}/2)$ and $v(-\vec{z}/2)$ satisfy the Dirac equation, eq. (2.1), with $M(r)$ replaced by $M(z/2)$ which implies confinement in the relative coordinate. A similar (though not identical) procedure has been described in ref. 12.

Consider now the matrix elements derived from the action, eq. (II.10). With a coupling strength $G(r)$ vanishing at the origin, direct matrix elements from $<q\bar{q}|\bar{\psi}i\gamma_5\tau\psi|0><0|\bar{\psi}i\gamma_5\tau\psi|q\bar{q}>$ of the type, Fig. (II.7a) will not contribute (except for fluctuations of the $q\bar{q}$ center of mass), but exchange pieces (Fig. (II.7b)) obtained from the Fierz-transformed interaction will do so. The relevant matrix elements reduced to space coordinates and written for the π^0 channel become

$$W_{ab} = -\frac{1}{2}\int d^3R\,d^3z\,K(\vec{R} + \frac{\vec{z}}{2}, \vec{R} - \frac{\vec{z}}{2}) \times$$

$$<(q\bar{q})_a|\bar{\psi}(\vec{R}+\frac{\vec{z}}{2})i\gamma_5\tau_3\psi(\vec{R}-\frac{\vec{r}}{2})|0><0|\bar{\psi}(\vec{R}-\frac{\vec{z}}{2})i\gamma_5\tau_3\psi(\vec{R}+\frac{\vec{z}}{2})|(q\bar{q})_b> \ .$$

$$(II.21)$$

The factor in front includes a standard factor of 2 times 1/4 from exchange. The integral over the center-of-mass coordinate $\vec{R}$ of the pair involves the kernel K. Its evaluation would require a detailed specification of the distribution d_Λ, but we employ the simplifying assumption that the c.m. integral leaves us with $G(z/2)$ times a slowly varying function of z which we replace by a constant λ, expected to be of order unity. Finally

$$W_{ab} = -4\lambda\int d^3r\,D_a(r)G(r)D_b(r) \qquad\qquad (II.22)$$

where

$$D_a(r) = <(q\bar{q})_a|\bar{\psi}(\vec{r})i\gamma_5\tau_3\psi(-\vec{r})|0> \ .$$

The eigenvalue equation for the pion energy E_π and the amplitudes A_a is wellknown in many-particle physics where it is used in connection with collective particle-hole excitations. It is known as the eigenvalue equation of the random phase approximation (RPA). For our purposes it can be obtained from the Bethe-Salpeter-equation for a pair of particles which move under the influence of a confining potential and interact via a nonretarded two-body kernel; under the mentioned assumptions the BS-equation can be reduced to a Salpeter equation which can be brought into following form

$$(E_\pi^2 - E_a^2)A_a = 2E_a \sum_b W_{ab} A_b \quad . \tag{II.23}$$

Before we present the results of the numerical solution of equation (II.23) we want to show how a pionic soft mode developes in the $|(q\bar{q})_a>$ - space (schematic model).

Consider a model space of $q\bar{q}$-pairs in $(1s\frac{1}{2})^2$, $(1p\frac{1}{2})^2$ and $(1p\frac{3}{2})^2$ configurations. The strong peaking of the interaction in the surface suggests that a separable approximation. $W_{ab} \approx -\gamma\, D_a(R)D_b(R)$ should be appropriate, with $D_a(R)$ taken at a point R in the surface, and γ proportional to $G(R)$. Then the RPA eigenvalue equation simply becomes

$$1 = \sum_a \frac{2W_{aa}E_a}{E_\pi^2 - E_a^2} \tag{II.23a}$$

with the diagonal matrix elements

$$W_{aa} = -\, 4\lambda(2j+1)\int_0^\infty dr\, \frac{G(r)}{4\pi r^2}\, [g_{njl}^2(r) - f_{njl}^2(r)]^2 \quad . \tag{II.24}$$

We note the strong correspondence between W_{aa} and the confining potential $M(r)$, eq. (II.18), a feature which is characteristic to the Nambu and Jona-Lasinio scheme.

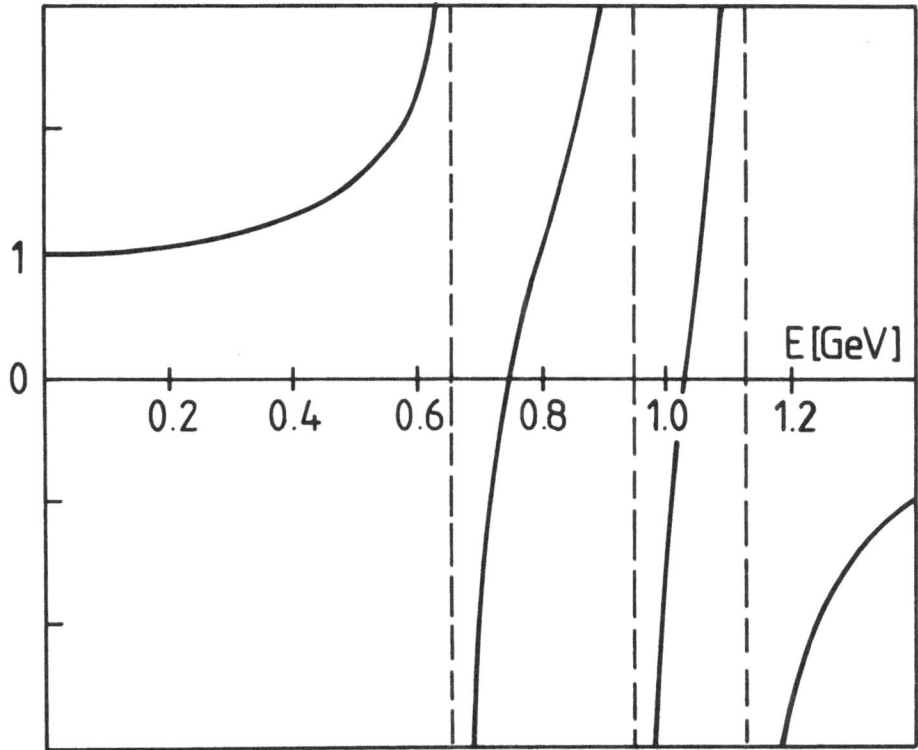

Fig. II.8: Graphical solution of the eigenvalue equation
(II.23a).

The graphical solution of eq. (II.23a) within the model
space discussed above, is displayed in fig.(II.8). The unperturbed energies E_a are corrected for c.m. motion. The pion
mode in this picture becomes a strongly "collective" superposition of $q\bar{q}$ basis states. The Goldstone theorem requires
that this mode ends up at zero energy, which is the case for
$\lambda = 0.9$ (see eq. (II.24).

The finite pion mass is supposed to be generated by finite, but small current quark masses m_u and m_d, with $m_u + m_d$ about 10-20 MeV. Therefore, after having fixed λ by the requirement to obtain the Goldstone pion with $m_\pi = 0$ from eq. (II.23) with $m_u = m_d = 0$, we solve this equation once more with m_u, $m_d \neq 0$. The empirical pion mass is reproduced for $m_u + m_d \approx 23$ MeV.

QCD sumrules [3] predict a relation

$$(m_u + m_d)\langle\bar{u}u + \bar{d}d\rangle = -m_\pi^2 f_\pi^2 \ . \qquad (II.25)$$

Here $f_\pi = 95$ MeV is the decay constant for the weak decay $\pi \to \mu\nu$ and $\langle\bar{q}q\rangle$ is the expectation value of the scalar field in the nonperturbative QCD vacuum. The value predicted for $\langle\bar{d}d\rangle = \langle\bar{u}u\rangle$ ranges between $-(160\ \text{MeV})^3$ and $-(250\ \text{MeV})^3$. In fig. (II.9) we show the behaviour of the product $f_\pi^2 m_\pi^2$ calculated from our solution as a function of $m_u + m_d$ and compare it with the result of the QCD sumrule. The functional relationship (II.25) between $m_u + m_d$ and m_π is reproduced amazingly well.

Table II. 1

Pion properties following from the solution of eq. (II.23) with $m_u + m_d = 24$ MeV

	Calc.	Exp.
$m_\pi+$	147 MeV	139,6 MeV
f_π	88 MeV	93 MeV
$\sqrt{\langle r_\pi^2\rangle}$	0.54 fm	0.66 ± 0.02 fm

The strong collectivity of the pionic $q\bar{q}$ mode has important consequences for the intrinsic wave function $\phi_\pi(r)$ of the pion. For example, the pion in our separable model has 47 % $(1s_{1/2})^2$, 32 % $(1p_{3/2})^2$ and 21 % $(1p_{1/2})^2$ $q\bar{q}$-components. The rms radius is about 0.4 fm, after c.m. corrections.

Consequently, $\phi_\pi(r = 0)$ is reduced as compared to standard quark models which interpret the pion as a 100 % $(1s_{1/2})^2$ $q\bar{q}$-pair. We find [15] that this has considerable impact on the $\pi \to \mu\nu$ decay rate which is proportional to $|\phi_\pi(r = 0)|^2$. Part of the van Royen-Weisskopf paradox [16] (which states that the pion decay constant f_π and the pion rms radius cannot be matched consistently in standard models) may in fact be removed by the collectivity of the pion without necessarily requiring that a major fraction of the pion wave function must be components other than $q\bar{q}$, as proposed in ref. 17.

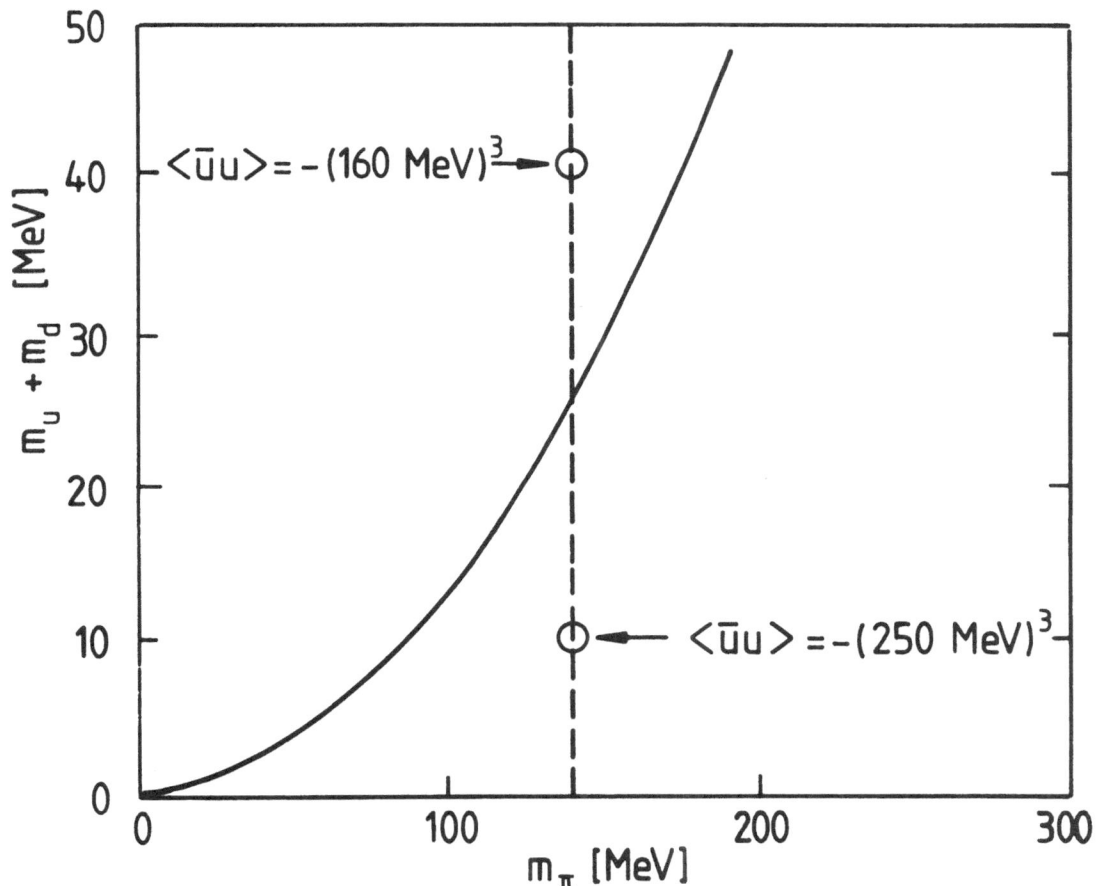

Fig. II.9

3. LECTURE

Results: Hadronic Properties and Implications for the Pionic
Part of the Nucleon-Nucleon Force

The chiral, relativistic potential model discussed in the
second lecture surrounds hadrons automatically with a pion
cloud. Therefore, in general hadronic properties receive
contributions from the quark core and from the pion cloud. In
the following we concentrate mainly on nucleonic properties
because we possess much more experimental data on the nucleon
than on any other elementary particle.

The nucleon is the lowest energy state of three u and/or
d quarks in the confinement potential, i.e. the state
$1s\frac{1}{2} \otimes 1s\frac{1}{2} \otimes 1s\frac{1}{2}$. This wavefunction contains an appreciable
amount of spurious center-of-mass motion. In any applications
it is important to remove this center-of-mass spuriority
and to construct translationally invariant wave functions. This
can be done by applying a projection operator,

$$P(\vec{Q}) = \frac{N(\vec{Q})}{(2\pi)^6} \int dp_1^3 dp_2^3 dp_3^3 \delta^3 (\sum_{i=1}^{3} \vec{P}_i - \vec{Q}) |\vec{P}_1\vec{P}_2\vec{P}_3><\vec{P}_1\vec{P}_2\vec{P}_3| \qquad (III.1)$$

to the three-quark wave function. The projected wave function,

$$\psi_{\vec{Q}}(\vec{r}_1\vec{r}_2\vec{r}_3) = <\vec{r}_1\vec{r}_2\vec{r}_3|P(\vec{Q})|\alpha_1\alpha_2\alpha_3> \qquad (III.1a)$$

is translation invariant in the sense that the center-of-mass
moves as a plane wave with momentum $\vec{Q}$, the normalization $N(\vec{Q})$
being determined by

$$\int dr_1^3 dr_2^3 dr_3^3 \psi_{\vec{Q}'}^+, \psi_{\vec{Q}} = (2\pi)^3 \delta^3 (\vec{Q} - \vec{Q}') \quad .$$

However, $\psi_{\vec{Q}}$ is not Lorentz invariant, because the small component associated with the center-of-mass motion is not
treated properly by eq. (III.1). Therefore this procedure

can only be applied to situations where the nucleons as a whole moves non-relativistically, i.e. $|\vec{Q}|^2/4m_N^2 << 1$, where m_N is the nucleon mass. The numerical treatment shows that depending on the details of the potential the subtraction of the center-of-mass energy reduces the nucleon mass from $3 \cdot E_{1s\frac{1}{2}}$ to approximately $2 \cdot E_{1s\frac{1}{2}}$.

In order to fix the potential parameters and to see if our model is capable of yielding satisfactory electromagnetic properties of the nucleon, we investigate electromagnetic formfactors.

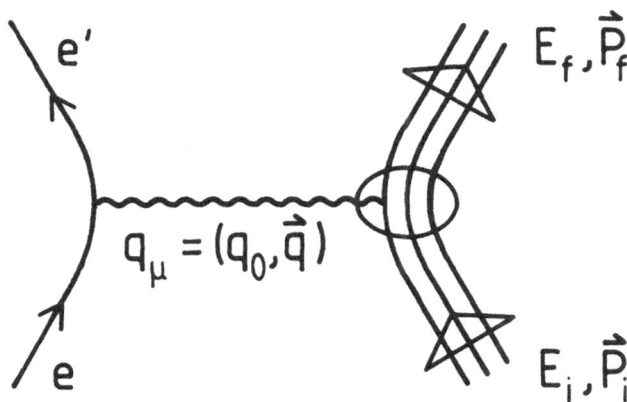

<u>Fig. III.1:</u> Kinematic variables in the Breit frame.

The calculation of electromagnetic form factors is done in the Breit frame (see fig. III.1) where $|\vec{P}_i| = |\vec{P}_f|$ and $q_\mu = (0, \vec{q})$ with $\vec{q} = \vec{P}_f - \vec{P}_i$, i.e. the energy transfer is zero. The quark current is

$$J_\mu(x) = e_q \bar{\psi}(x) \gamma_\mu \psi(x) .$$

The nucleon form factor $F_\mu(q)$ is defined by

$$(2\pi)^3 \delta^3(\vec{P}_f - \vec{P}_i - \vec{q}) F_\mu(q) = \langle \psi_{\vec{P}_f} | \sum_{i=1}^{3} e_i e^{i\vec{q}\cdot\vec{r}_i} \gamma_0 \gamma_\mu | \psi_{\vec{P}_i} \rangle \qquad \text{(III.2)}$$

where the $\psi_{\vec{P}}$ are momentum projected wave functions of the type, eq. (III.1a). To leading order in $q^2/4m_N^2$,

$$F_0(q) = e\, G_E(q^2) \quad ,$$

$$\vec{F}(q) = \frac{e}{2m_N} [\vec{q} \delta_{fi} G_E(q^2) + i\langle \vec{\sigma}\times\vec{q}\rangle_{fi} G_M(q^2)] \qquad . \qquad \text{(III.3)}$$

Without center-of-mass corrections, the proton charge and magnetic form factors become

$$G_E(q^2) = \int_0^\infty dr\, j_0(qr) [g^2(r) + f^2(r)] \quad , \qquad \text{(III.4a)}$$

$$G_M(q^2) = -\frac{4m_N}{|\vec{q}|} \int_0^\infty j_1(qr) f(r) g(r) dr \quad . \qquad \text{(III.4b)}$$

The modification of these simple expressions due to c.m. corrections is presented in the Appendix.

For comparison with the measured proton charge form factor, one has to correct for pion cloud effects. In order to isolate these effects, it is useful to discuss the iso-vector form factor, isoscalar contributions to $G_E(q^2)$ being small at $\vec{q}^2 \lesssim 0.4$ GeV^2. Pion cloud contributions to the isovector form factor are dominated by ρ meson intermediate states. To a good approximation [18] the ρ-dominance can be taken into by the ansatz:

$$G_E(q^2) = G_E^{core}(q^2) \frac{1}{1 + \vec{q}^2/m_V^2} \qquad \text{(III.5)}$$

where $[G_E]_{core}$ is now the quark core contribution and $m_V \approx m_\rho$

(see also ref. 19).

Results for $G_E(q^2)$ are shown in fig. (III.2). The importance of proper center-of-mass corrections, becomes obvious from this figure.

Similar results are obtained for a confining potential $M=cr^2$ with $c = 0.83$ GeV fm^{-2}. A linear potential $M = cr$ does however not give a reasonable fit; also higher powers $n \geq 4$ yield less satisfactory results.

The rms radius of the quark core is given by

$$<r^2> = - 6\frac{d}{dq^2} G_E^{core}(q^2)/_{q^2=0} \quad . \tag{III.6}$$

Results with and without proper projection on total c.m. momentum are shown in table 1. It is seen that c.m. corrections generally reduce the rms radius.

$\dfrac{M(r)}{c}$	$\begin{array}{c} cr^2 \\ 0.83 \text{ GeV } fm^{-2} \end{array}$	$\begin{array}{c} cr^3 \\ 1.25 \text{ GeV } fm^{-3} \end{array}$
$<r^2>_o^{1/2}$[fm]	0.64	0.60
$<r^2>_{cm}^{1/2}$[fm]	0.55	0.51
$<r^2>_p^{1/2}$[fm]	0.86	0.84

Table III.1: Quark core rms radii without (subscript "o") and with (subscript "cm") center-of-mass corrections for two types of confining potentials. The last line shows the proton charge radius including cm corrections after applying meson cloud corrections.

We conclude from this discussion that nucleon quark core rms radii $\lesssim 0.6$ fm are consistent with both the charge

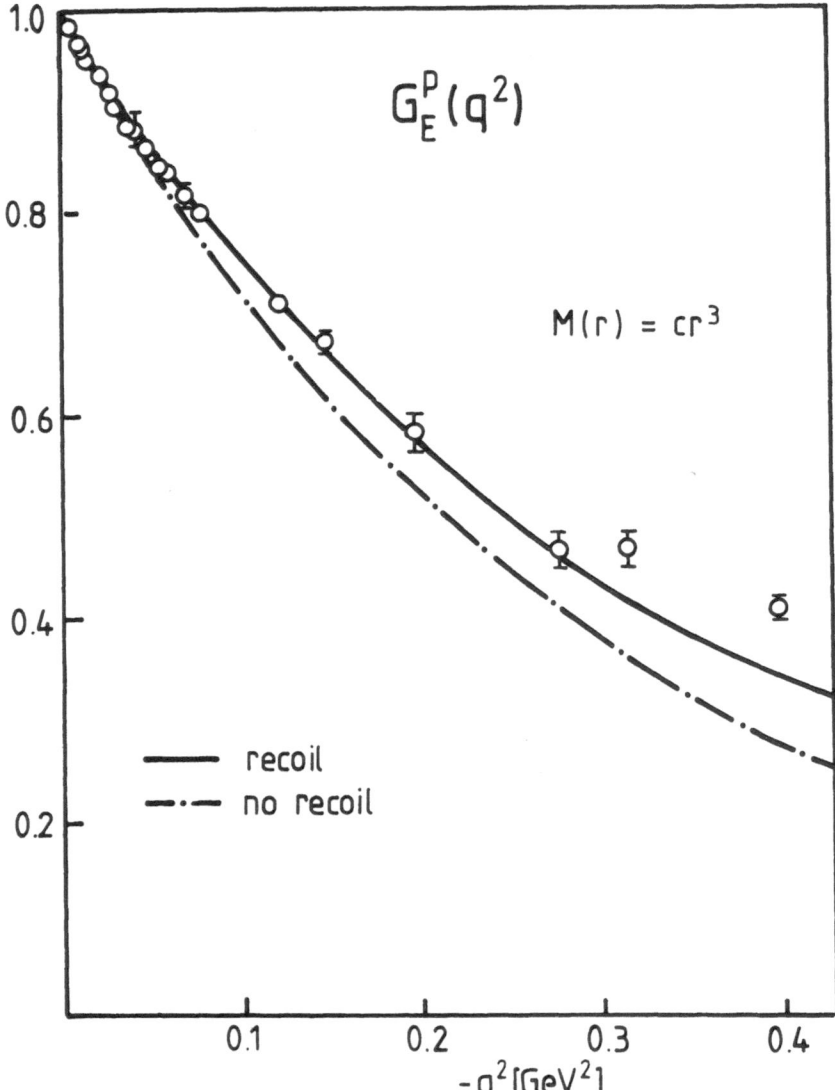

Fig. III.2: Proton charge form factor $G_E(q^2)$ obtained with
a confining potential $M(r)=cr^3$ with c=1.25 GeV fm^{-3};
the lowest eigenvalue in this potential is
E_o = 540 MeV. Solid and dashdotted curves show results with and without c.m. corrections. Pion
cloud corrections are included with m_v=730 MeV.
(For more details see R. Tegen, R. Brockmann
and W. Weise, ref. 13).

form factor and mass. We consider this as an upper limit since a more refined treatment of (attractive) pion cloud contributions to the nucleon mass will require a further slight reduction of the core rms radius.

Magnetic moments of the quark core are given by

$$\mu = \frac{e}{2m} \, G_M^{core}(0) \quad .$$ (III.7)

Without center-of-mass corrections, one obtains from eq.(III.46)

$$\mu_p = -\frac{3}{2}\mu_N = -\frac{2}{3}e \int_0^\infty rf(r)g(r)dr \quad .$$ (III.7a)

The c.m. corrected expression can be found in Appendix A. Table III.2 shows that about half of the empirical proton magnetic moment comes from the quark core. Most of the other half is expected to be contributed by the pion cloud.

M(r)	μ_p	
	without c.m. corr.	incl. c.m. corr.
cr^3	1.48	1.41
cr^2	1.51	1.47

Table III.2: Quark core contribution to proton magnetic moment (in units of $(e/2m_N)$ for different types of confining potentials, with and without inclusion of center-of-mass corrections. Parameters c as in table III.1

Axial current and axial charge of the quark core

The axial current of the quark core part is given by

$$A_\lambda^\mu(x) = \bar{\psi}(x)\gamma^\mu\gamma_5 \frac{\tau_\lambda}{2} \psi(x) \qquad\qquad (III.8)$$

where the ψ are quark fields. The nucleon axial form factor $F_A(q^2)$ is defined in the Breit frame by

$$(2\pi)^3\delta^3(\vec{P}_f-\vec{P}_i-\vec{q})(F_A(q^2))_\lambda^\mu =$$

$$= \langle\psi_{\vec{P}_f}| \sum_{i=1}^{3} (\gamma_0\gamma^\mu\gamma_5 \frac{\tau_\lambda}{2})_i e^{i\vec{q}\cdot\vec{r}_i}|\psi_{\vec{P}_i}\rangle \ . \qquad\qquad (III.9)$$

The axial charge or axial vector coupling constant g_A can be obtained from the relation:

$$g_A\langle\sigma_z^N \frac{\tau_3^N}{2}\rangle = \lim_{\vec{q}\to 0}(F_A(q^2))_{\lambda=3}^{\mu=3} \qquad\qquad (III.9a)$$

where $\vec{\sigma}^N$ and $\vec{\tau}^N$ on the left hand side refer to nucleon (rather than quark) spin and isospin. Without center-of-mass corrections, one obtains

$$g_A = \frac{5}{3}[1 - \frac{4}{3} \int_0^\infty dr\, f^2(r)] \qquad\qquad (III.10)$$

where $f(r)$ is the lower component of the $1s_{1/2}$ quark wave function. The more elaborate result including c.m. correction is given in Appendix A.

Results for g_A for massless quarks are shown in table III.3. These numbers should be compared with the MIT bag model result, $g_A = 1.09$. Non-relativistic quark models yield $g_A = 5/3$, as can be seen from eq. (III.10). The experimental result is $g_A = 1.26$. We observe that confining potential models approach this value rather closely. Corrections to the values given in table III.3 for the relativistic potential model result from finite quark masses (between +0.03 and +0.05) and from changes of the wave function of the quark core due to coupling to pions (+0.02). As was already discussed in Lecture 1

M(r)	g_A	
	without c.m. corr.	incl. c.m. corr.
cr^3	1.21	1.15
cr^2	1.26	1.21
MIT Bag	1.09	
Exp.	1.26	

Table III.3: Quark core contributions to g_A obtained for mass-less quarks confined in different types of potentials M(r), with and without c.m. corrections. Parameters c as in table III.1. (For more details see: R. Tegen, M. Schedl and W. Weise, ref. 12).

there is no direct contribution from the pion cloud for potential models with a continous surface.

For scalar confining potentials, a simple relation between μ_p and g_A can be derived. From eqs. (III.7a) and (III.10) one finds

$$\mu_p = - \frac{3}{2} \mu_N = \frac{m}{2E_o} (1 + \frac{3}{5} g_A) \quad . \qquad (III.11)$$

Refinements due to c.m. corrections are given in Appendix A. The constraints implied by eq. (III.11) tell us that, unless the quark core contribution to g_A is substantially larger than the empirical $g_A = 1.26$, μ_p will always be smaller than the empirical value $\mu_p = 2.79$ (e/2m_N). With $g_A \approx 5/4$ one obtains $\mu_p \sim 1.5$. This that substantial contributions to μ_p must come from the pion cloud and/or from other sources.

Next I want to discuss in more detail properties of the pion
as they follow from the collective $q\bar{q}$ wavefunction discussed
in the second lecture.

The pion wavefunction belonging to the eigenvalue
equation (II.23) is

$$|\pi\rangle = \sum_{i,j} (X_{ij} b_i^+ d_j^+ + Y_{ij} b_i d_j) |0\rangle \qquad (III.12)$$

where the operators $b_i^+ (b_i)$ and $d_j^+ (d_j)$ create (annihilate)
quarks and antiquarks in single particle states i and j
respectively.
For the configuration space $(1s\frac{1}{2})^2$, $(1p\frac{1}{2})^2$ and $(1p\frac{3}{2})^2$ the
probabilities are: 0.35, 0.20 and 0.45 respectively.

The electromagnetic formfactor of the pion is defined
by the relation (Breitframe)

$$(2\pi)^3 \delta^3(\vec{P}_f - \vec{P}_i - \vec{q}) F_\mu(q) = \langle\psi_{\vec{P}_f}| \sum_{i=1}^{2} e_i e^{i\vec{q}\cdot\vec{r}_i} \gamma_0 \gamma_\mu |\psi_{\vec{P}_i}\rangle , \qquad (III.13)$$

from which one obtains the r.m.s. (core) radius for $\pi^+ = |u\bar{d}\rangle$

$$\langle r_\pi^2\rangle = \sum_{i,j} (X_{ij}^2 - Y_{ij}^2)(\frac{1}{3}\langle r^2\rangle_i + \frac{2}{3}\langle r^2\rangle_j) . \qquad (III.14)$$

The numerical result is $r_\pi^{core} = 0.54$ fm. For comparison the
result for the $1s\frac{1}{2} \otimes 1s\frac{1}{2}$ "pion" is $r_\pi = 0.524$ fm. The latest
experimental values are

$$\langle r_\pi^2\rangle = 0.46 \pm 0.011 \text{ fm}^2 \text{ (ref. 20)} ,$$

$$\langle r_\pi^2\rangle = 0.44 \pm 0.03 \text{ fm}^2 \text{ (ref. 21)} ,$$

which gives $\sqrt{\langle r_\pi^2\rangle} = 0.66 + 0.02$ fm.
Just as for the case of baryons one has to separate the contribution of the $q\bar{q}$ core from the mesonic cloud which intervene
via intermediate vector mesons. The assumption of ρ-dominance

leads to

$$F_\pi(q^2) = F_\pi^{core}(q^2) \frac{1}{1+q^2/m_\rho^2} \qquad (III.15)$$

yielding

$$\langle r^2 \rangle \overset{\sim}{=} \langle r_\pi^{2core} \rangle + \frac{6}{m_\rho^2} = \langle r_\pi^{2core} \rangle + 0.42 \text{ fm}^2$$

i.e. the pion core should be very small with

$$r_\pi^{core} \simeq 0.2 \text{ fm}.$$

For the pion decay constant one needs the pion wavefunction in relative coordinates at the origin $\vec{r}_q - \vec{r}_{\bar{q}} = \vec{r} = 0$.

With $\chi(\vec{r}_q, \vec{r}_{\bar{q}}) = \bar{\chi}(\vec{R})\chi_{rel}(\vec{r})$ one obtains

$$\chi_{rel}(\vec{r}=0) = N \sum_{i,j} (X_{ij} - Y_{ij}) u_i(0)\bar{v}_j(0) \quad ,$$

and from

$$f_\pi = \frac{1}{\sqrt{m_\pi}}\cos\theta_{Cabbibo} \sqrt{\frac{3}{2}} \frac{1}{4\pi} \frac{\sum_{ij}(X_{ij}-Y_{ij})\sqrt{2j_i+1}[g_i(0)g_j(0)+f_i(0)f_j(0)]}{\sqrt{\sum_{i,j}(X_{ij}^2 - Y_{ij}^2)N_{ij}^2}}$$

$$N_{ij}^2 = \frac{2j_i+1}{4\pi} 8 \int_0^\infty r^2[g_i(r)g_j(r)+f_i(r)f_j(r)]dr \quad . \qquad (III.16)$$

Numerically one finds $f_\pi = 88$ MeV ($f_\pi^{exp.} = 93$ MeV).
Next we turn to the pion-nucleon coupling. As has been already discussed chiral symmetry on the level of quarks and pions dictates the following equation for the (elementary) pion field

$$(\square + m_\pi^2)\pi_\lambda(x) = \frac{M(\vec{r})}{f_\pi} \bar{\psi}(x) i\gamma_5\tau_\lambda\psi(x) \quad . \qquad (III.17)$$

The pion source function

$$J_{5,\lambda}(x) = \frac{M(\vec{r})}{f_\pi} \sum_{i=1}^{3} \bar{\psi}_i(x) i\gamma_5\tau_\lambda\psi_i(x) \qquad \text{(III.17a)}$$

peaks at the nucleon surface (see fig. II.6). In the static limit one defines the pion-nucleon form factor $G_{\pi NN}(q^2)$

$$iG_{\pi NN}(q^2)<\vec{\sigma}^{(N)}\cdot\vec{q}\ \tau_\lambda^{(N)}> = 2m_N<N|\int dr^3 e^{i\vec{q}\cdot\vec{r}}J_{5,\lambda}(\vec{r})|N> \ .$$

$\vec{\sigma}^{(N)}$ and $\tau_\lambda^{(N)}$ are nucleon spin- and isospin operators. For the pion-nucleon coupling constant $G_{\pi NN} = G_{\pi NN}(q^2{=}0)$ one obtains

$$G_{\pi NN} = -\frac{10}{q}\frac{2m_N}{f_\pi}\int_0^\infty dr\ r\ M(r)g(r)f(r)\ . \qquad \text{(III.18)}$$

Using the Dirac equation the integral on the r.h.s. can be rewritten to give

$$\int_0^\infty dr\ r\ M(r)g(r)f(r) = E_0 \int_0^\infty dr\ r\ g(r)f(r)$$

$$+ \frac{1}{2}\int_0^\infty dr\ f^2(r)\ .$$

Comparing this with eqs. (III.10, 11) for g_A and μ_p one obtains a Goldberger-Treiman relation

$$G_{\pi NN} = \frac{m_N}{f_\pi} g_A^{(core)} \qquad \text{(III.19)}$$

where $g_A^{(c)}$ is the axial charge for the quark core.

With the empirical value for $f_\pi = 93$ MeV and our calculated value for $g_A \sim 1.2$ one obtains $G_{\pi NN} \equiv 12$; this has to be compared to the empirical value $G_{\pi NN}^{(emp.)} = 13.3$ (following from $G_{\pi NN}^2/4\pi = 14$).

Given the πNN-coupling we can investigate the effect of

the pion cloud on nucleonic properties and on the pion induced
part of the nucleon-nucleon interaction.

The pion cloud leads to selfenergy corrections of the
nucleon (and Δ) mass of the type shown in fig. III.3.

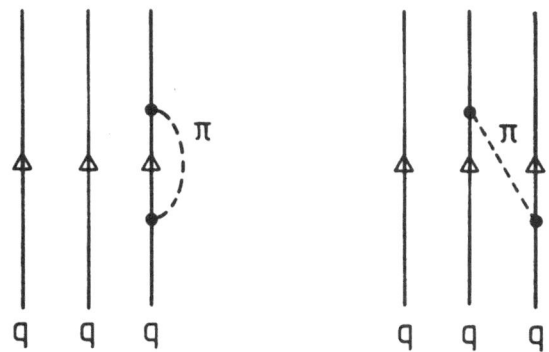

Fig. III.3: Diagrams contributing to the pionic selfenergy.

With the πNN-coupling of eq. (III.17a) one obtains the results shown in fig. (III.4a). The nucleon and Δ-resonance case
differs only by the spin and isospinfactors of the matrix
elements connecting the intermediate state to the initial
and final state respectively. The different magnitude of the
selfenergy for nucleons and Δ's leads to a splitting of
m_Δ and m_N. In the course of the calculation one needs the
pion propagator not only in the exterior phase but also in
the interior of the baryon. Of course in the interior the
"almost Goldstone pion" cannot exist. It will rather dis-
solve into a noncollective $q\bar{q}$ pair moving in the confining
potential of the baryon. Qualitatively we took this into
account by ascribing a mass $2E_o$ to the "pion" in the interior region. The mass difference $2E_o - m_\pi$ which reflects the
loss of attraction which binds the pion with a small mass

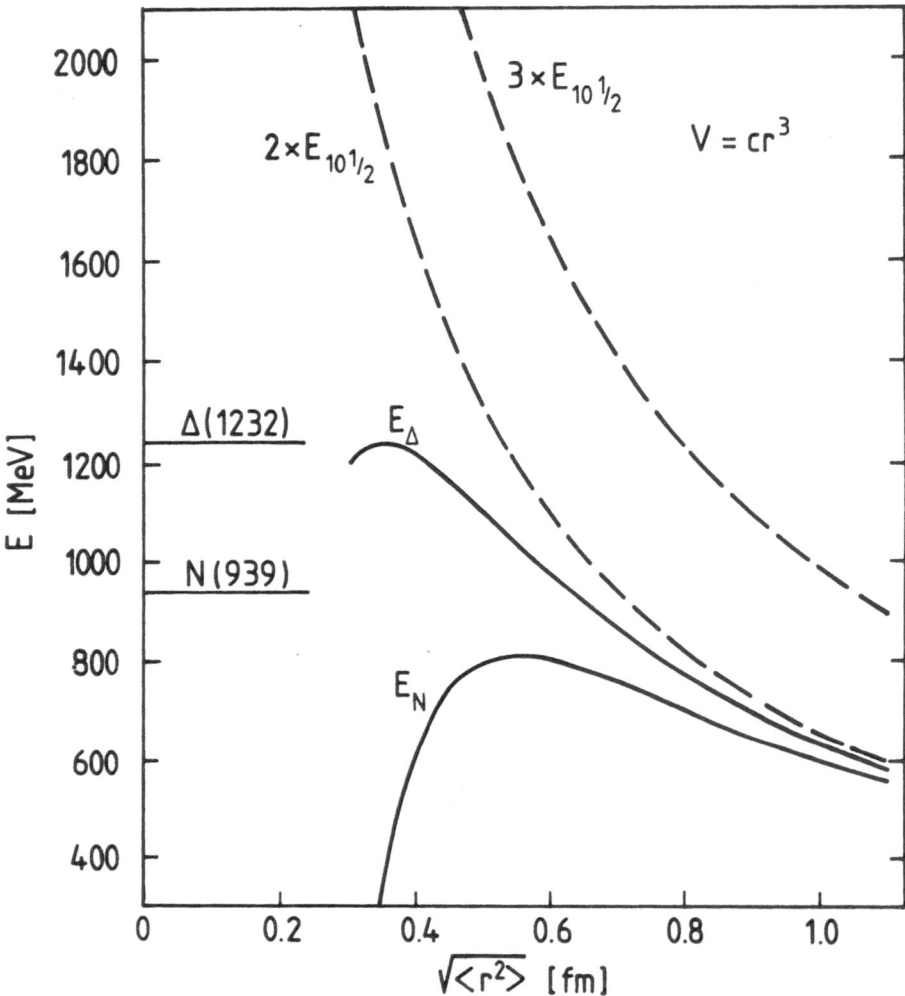

Fig. III.4: Pionic selfenergy corrections for the nucleon and the Δ-resonance as a function of the core size.

acts like a repulsive potential and leads to a quenching of the pion amplitude in the interiour. The results depend rather sensitively on the r.m.s. radius of the quark core. However,

it is seen that for reasonable values of the r.m.s. radius
between $\cdot 5$ and $\cdot 6$ fm one obtains a splitting $E_\Delta - E_N \sim 200 -$
250 MeV. This number is further increased by $20 - 30$ MeV,
if one makes a first order calculation of the energy shift
of E_Δ and E_N due to the residual quark-quark interaction of
eq. (II.10) and b ~ 30 MeV through one gluon exchange with
$\alpha_s = 0.2$.

In conventional bag models the Δ-N mass difference is
usually explained in terms of a colour-magnetic fine structure which is obtained via one gluon exchange with an arti-
ficially large value for $\alpha_s \gtrsim 2$ [22] for which there is no
theoretical foundation.

It is interesting to see how much the quark core
wavefunction is changed by the presence of the pion cloud. A
qualitative measure for this is the magnitude of the residue
Z_0 of the pole of the three-quark propagator corresponding
to all 3 quarks in the groundstate. From the energy dependence of the selfenergy one finds $Z_0 \sim 0.7$. Thus the unperturbed groundstate is found with an amplitude of ~ 0.84 in
the baryonic wavefunction.

The pion cloud correction to the magnetic moments of
neutrons and protons can be obtained from the diagrams shown
in fig. III.5.

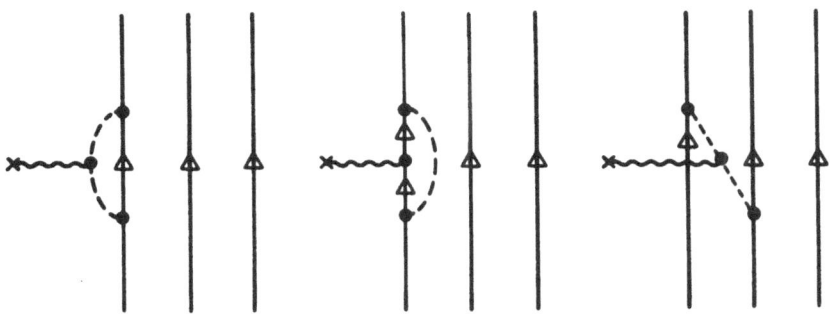

Fig. III.5: Diagrams for the pion cloud vertex corrections.

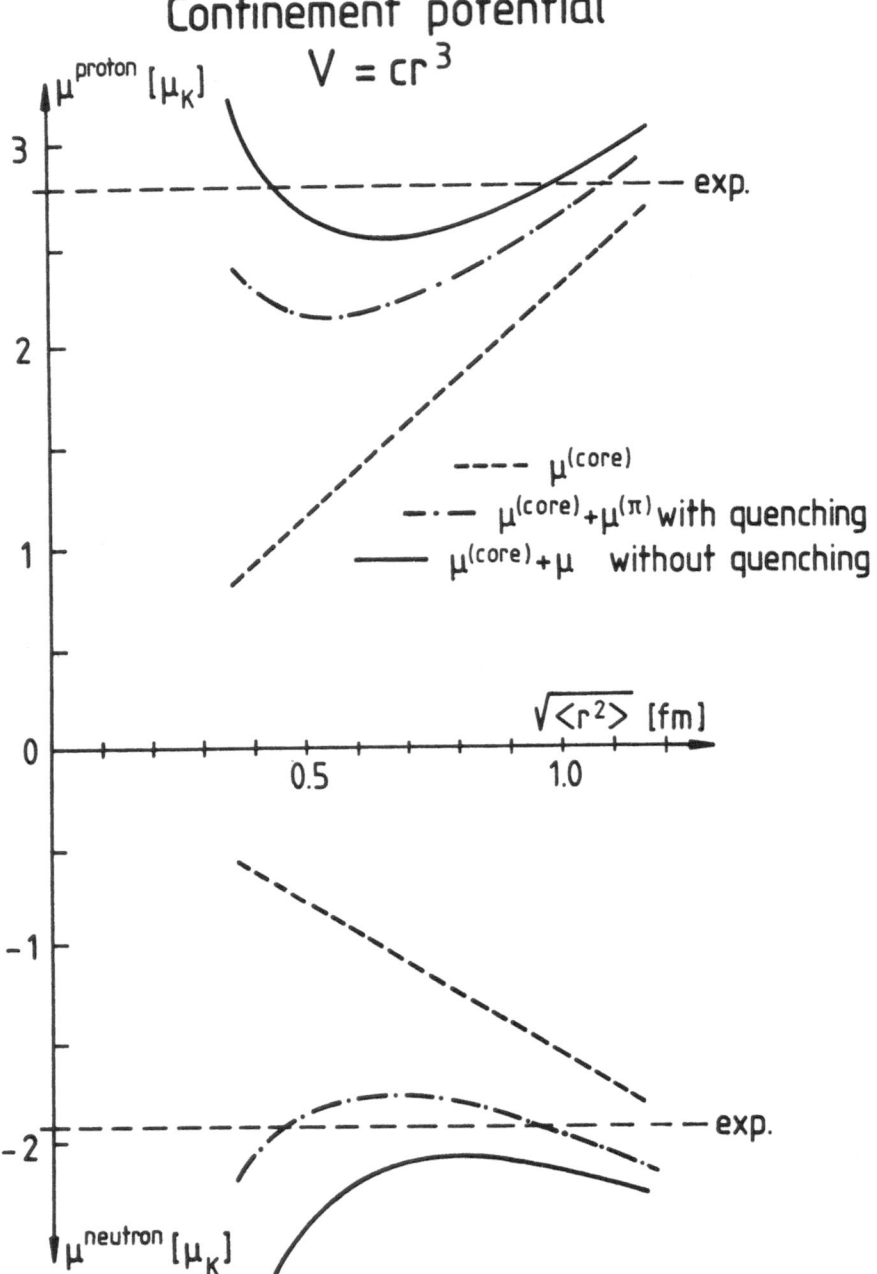

Fig.III.6: Magnetic moments of proton and neutron as a function
of the core radius. The result with quenching is
obtained with a repulsive pion potential $V_\pi(r) = 0$;
$r > 0.6$ fm, $V_\pi(r) = 2E_0$; $r \leq 0.6$ fm.

The results for the magnetic moments are shown in fig. II.6
as a function of the r.m.s. radius of the quark core. One
sees - as has been already infered from eq. (III.11) - that
there is a considerable contribution from the pion cloud,
bringing the total magnetic moments $\mu= \mu^{(core)} + \mu^{(\pi)}$ rather
close to the experimental values.

The πNN-formfactor (III.17a) and the pion quenching in
the interior of the nucleons also has remarkable consequences
for the one pion exchange potential (OPEP) between two nucleons:
We write the OPEP for two pointlike nucleons in the form

$$V_{OPE}(\vec{r}) = (\vec{\tau}_1 \cdot \vec{\tau}_2)(\vec{\sigma}_1 \cdot \vec{\sigma}_2)[V_{c1}(r) + V_{c2}(r)]$$

$$+ (\vec{\tau}_1 \cdot \vec{\tau}_2) S_{12}(\hat{r}) V_{tensor}(r)$$

with

$$V_{c1}(r) = - f^2 m_\pi \frac{\delta(\vec{r})}{3m_\pi^2} \; ; \; V_{c2}(r) = \frac{f^2 m_\pi}{12\pi} \frac{e^{-m_\pi r}}{m_\pi r} \; ;$$

$$V_{tensor}(r) = \frac{f^2 m_\pi}{12\pi}[1 + \frac{3}{m_\pi r} + \frac{3}{m_\pi^2 r^2}] \frac{e^{-m_\pi r}}{m_\pi r} \; .$$

in figs. III.7-9 we show the change of the three potential
types V_{c1}, V_{c2} and V_{tensor} due to the formfactor and due to
quenching effects. For comparison we also show the tensor
part of the Paris potential; this curve represents an upper
limit for the OPE tensor potential since it includes the repulsive effect of two-pion contributions.

In conclusion one can say that chiral, relativistic
potential models with underlying QCD phenomenology seem to
be good candidates for the description of large distance properties of hadrons (lowlying baryon and meson states). They
provide automatically a self-consistent baryon-pion coupling
which is necessary to build up a pion cloud around the quark
core; this pion cloud is required by the chiral symmetry of
the underlying QCD Lagrangian in the sector of u- and d-quarks

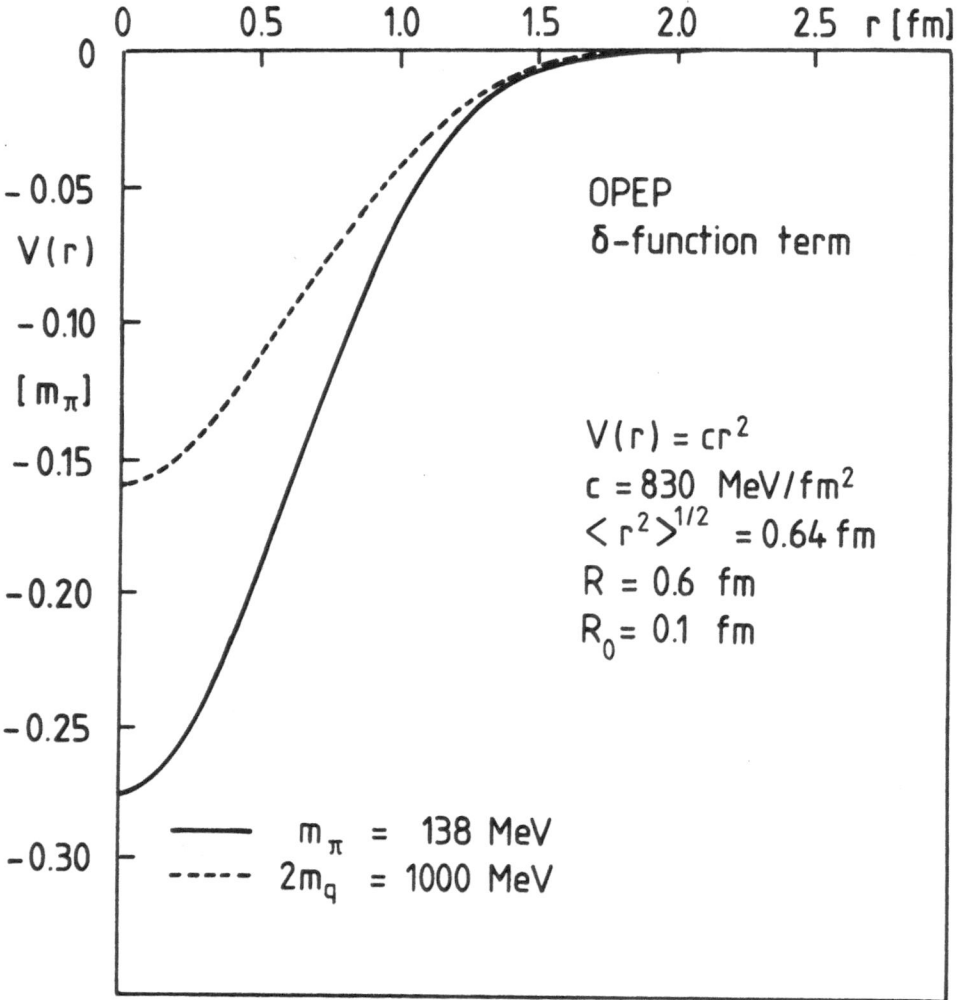

Fig. III.7: "δ-function" part $V_{c,1}(r)$ of the OPEP.
In Figs. III.7 - III.9 the results with quenching
are obtained with a repulsive pion potential which
rises linearly between r = 0.65 and r = 0.55 fm from
V_π = 0 to V'_π = 1 GeV and is V_π = 1 GeV for 0 ≤ r ≤ 0.55 fm

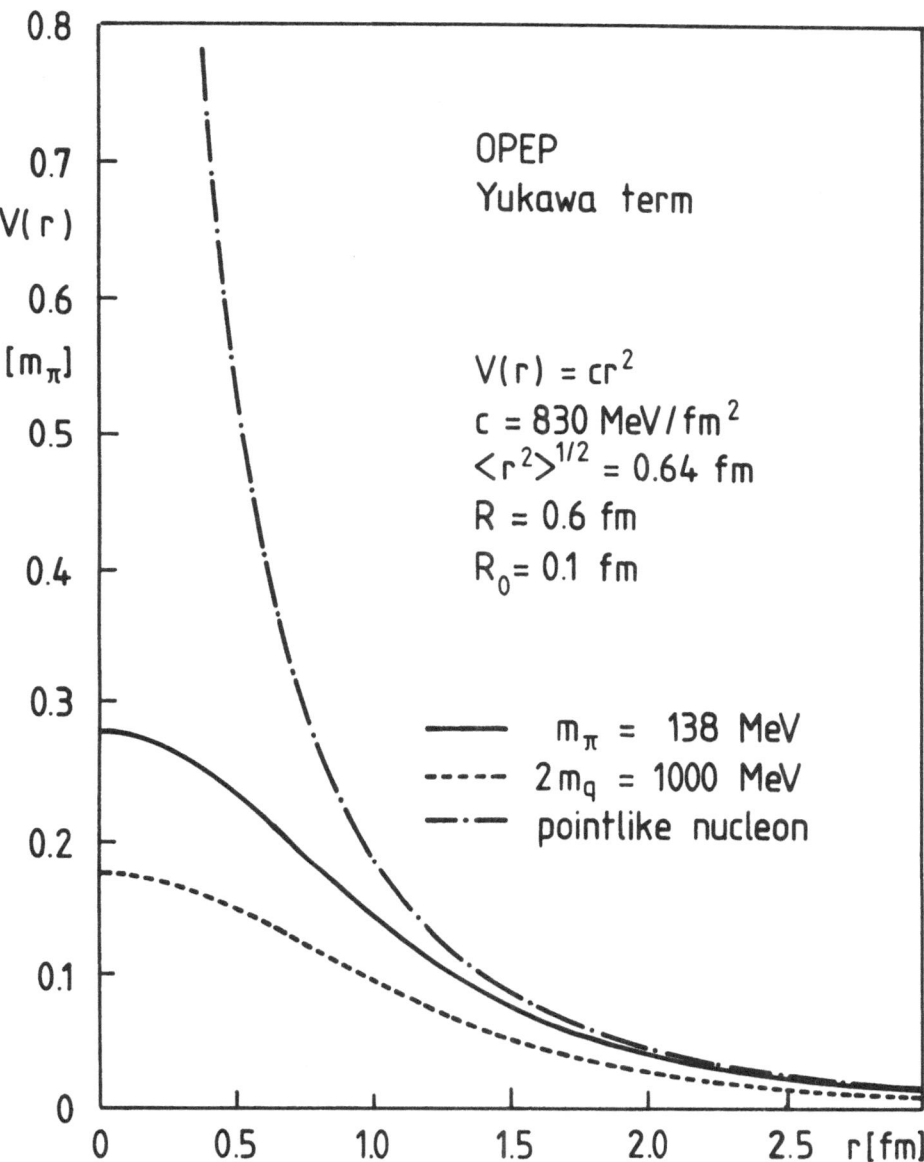

Fig. III.8: "Yukawa part" $V_{c,2}(r)$ of the OPEP.

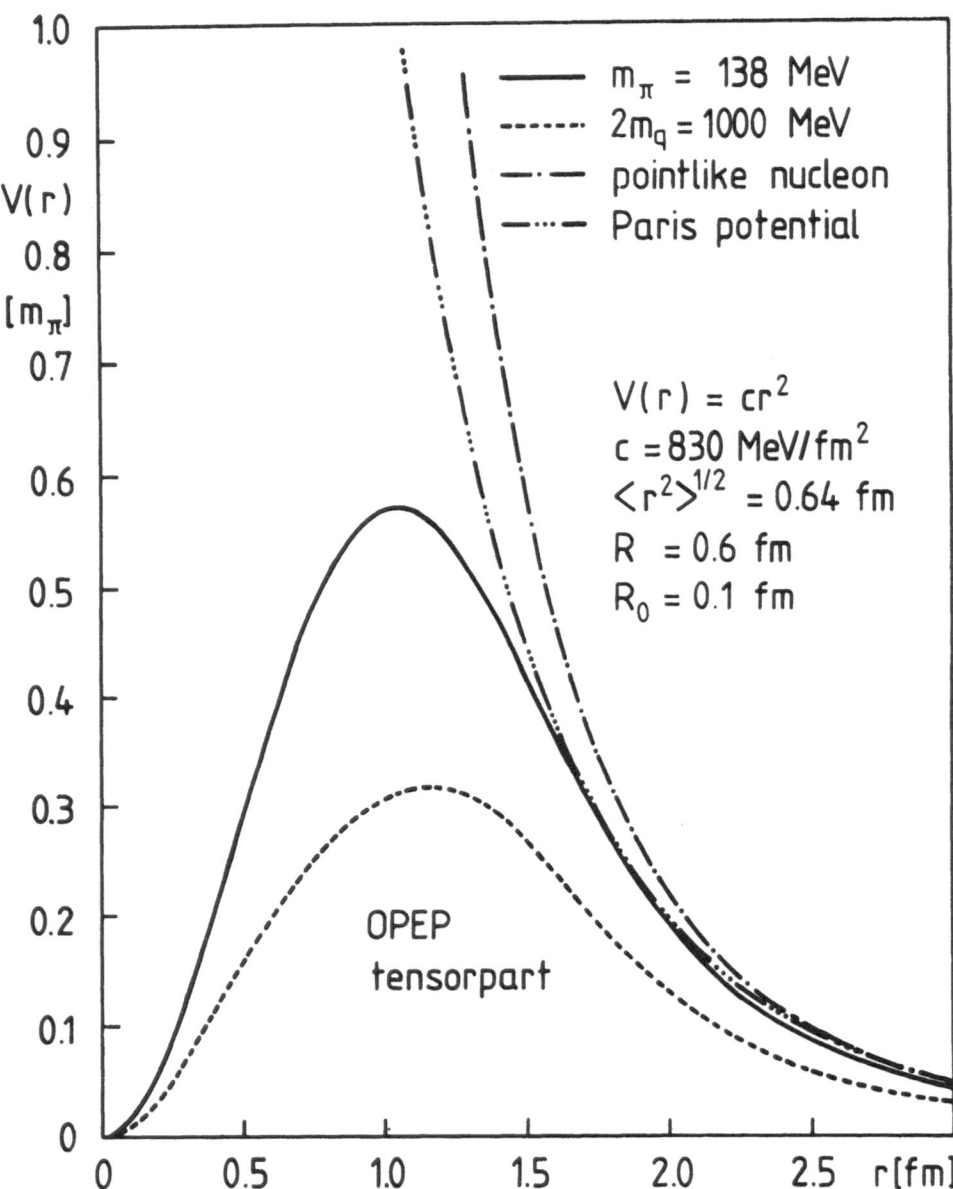

Fig. III.9: Tensor part of the OPEP.

and also by a number of experimental data on low mass hadrons.

Whereas the pion-baryon coupling is dealt with, essentially, a structureless, elementary pion, we have also made a step towards a microscopic description of the pion in terms of a highly collective $q\bar{q}$-state which shows up the connection with the "almost Goldstone boson" nature of the pion.

The essential message for nuclear physics is the relative smallness of the quark core of baryons with $\sqrt{<r^2>} \sim 0.5$ fm, which is very assuring given the overwhelming amount of data advertising a picture of relatively free motion of nucleons in nuclei. Also, the one-pion exchange can be established microscopically as the source of the long range part of the nucleon-nucleon interaction. On the other hand there seems to be no room for the exchange of vector mesons in the sense of asymptotic states, since such an exchange could take place only at a distance where the two quark cores already overlap.

The model also yields a microscopic prescription for the calculation of selfenergy- and vertex corrections due to the pion cloud and provides a very satisfactory explanation of the Δ-N mass difference.

The most urgent improvements and extensions that have to be made are:

(i)   Pion-baryon coupling and pion cloud properties have to be calculated with the collective $q\bar{q}$-pion instead of the elementary, structurelss pion.

(ii)   The model has to be extended to the flavour SU(3) case which requires a generalisation of the chiral invariant interaction.

(iii) The problem of the short-range NN-interaction has to be investigated and the mechanism which simulates vector meson exchange has to be understood.

(iv)   The problem of momentum exchange with the gluonic background must be solved.

APPENDIX: Center-of-Mass Corrections to Nucleon Electromagnetic
and Axial Form Factors

Following ref. 13, we present details of c.m. corrections for electromagnetic and axial form factors. The translation invariant three-quark wave functions eqs. III.1, III.1a are

$$\psi_{\vec{Q}}(\vec{r}_1\vec{r}_2\vec{r}_3) = \frac{N(\vec{Q})}{(2\pi)^6} \int d^3p_1 d^3p_2 d^3p_3 \delta^3 (\sum_j \vec{p}_j - \vec{Q}) \exp[i\sum_j \vec{p}_j \vec{r}_j]$$

$$\times \phi(\vec{p}_1)\phi(\vec{p}_2)\phi(\vec{p}_3) \ , \tag{A.1a}$$

$$\phi(\vec{p}) = \int d^3r \ e^{-i\vec{p}\cdot\vec{r}} u(\vec{r}) \ , \tag{A.1b}$$

$$N(\vec{Q}) = [\int d^3r \ e^{-i\vec{Q}\cdot\vec{r}} \rho^3(r)]^{-1/2} \ , \tag{A.1c}$$

$$\rho(\vec{r}) = \int \frac{d^3k}{(2\pi)^3} e^{i\vec{k}\cdot\vec{r}} \phi^+(\vec{k})\phi(\vec{k}) \ , \tag{A.1d}$$

where eq. (A.1c) is true for three quarks in the same orbits. For a current $\bar{\psi}(x)\Gamma\psi(x)$ where Γ is either γ_μ or $\gamma_\mu\gamma_5 \frac{\tau\lambda}{2}$, the corresponding form factor (up to a momentum conserving δ-function) becomes in the Breit frame:

$$F(\Gamma,q^2) = \sum_{j=1}^{3} N^2(\frac{q}{2}) \int d^3r \ e^{i\vec{q}\cdot\vec{r}_j} \ \bar{u}_j(\vec{r})\Gamma\tilde{u}_j(\vec{r}) \ , \tag{A.2}$$

where $\tilde{u}_j$ now contains the effective of the two recoiling quarks if the j-th quark couples to the current in question. More precisely,

$$\tilde{u}_j(\vec{r}) = \int \frac{d^3r}{(2\pi)^3} e^{i\vec{p}\cdot\vec{r}} \phi_j(\vec{p}) W_{kl}(\vec{p}) \tag{A.3}$$

where $W_{kl}(\vec{p}) = \int d^3r \ e^{-i\vec{p}\cdot\vec{r}} \hat{\rho}_k(\vec{r})\hat{\rho}_l(\vec{r}) \ ,$

$$\hat{\rho}_k(\vec{r}) = \int \frac{d^3p'}{(2\pi)^3} e^{i\vec{p}'\cdot\vec{r}} \bar{\phi}_k(\vec{p}')\phi_k(\vec{p}') \quad \text{etc.}$$

We introduce upper and lower components of the c.m. corrected
wave functions by

$$\tilde{u}(\vec{r}) = \begin{pmatrix} i \; \tilde{g}(r)/r \\ \vec{\sigma} \cdot \hat{r} \; \tilde{f}(r)/r \end{pmatrix} \quad \phi(\hat{r}) . \tag{A.4}$$

In what follows, we shall always restrict ourselves to the
lowest 1s 1/2 - orbit. Now, gauge invariance of the electromagnetic current can be shown to imply

$$\frac{\tilde{f}(r)}{\tilde{g}(r)} = \frac{f(r)}{g(r)} . \tag{A.5}$$

We introduce the quantity

$$\kappa(r) \equiv \tilde{f}(r)/f(r) = \tilde{g}(r)/g(r) ,$$

$$<\kappa> = \int_0^\infty dr \; \kappa(r) \; [g^2(r) + f^2(r)] . \tag{A.6}$$

The following expressions are then obtained for the quark core
charge form factor, magnetic moment and axial charge:

$$G_E(q^2) = \left(\frac{N(a/2)}{N(o)} \right)^2 \frac{1}{<\kappa>} \int_0^\infty dr \; j_0(qr)\kappa(r)[\; g^2(r) + f^2(r)], \tag{A.7}$$

$$\mu_p = -\frac{3}{2} \mu_n = \frac{2e}{3<\kappa>} \int_0^\infty dr \; r \; \kappa(r)g(r)f(r) , \tag{A.8}$$

$$g_A = \frac{5}{3} [1- \frac{4}{3<\kappa>} \int_0^\infty dr \; \kappa(r) \; f^2(r)] . \tag{A.9}$$

For scalar confining potentials, using the Dirac equation,
a relationship of the following type can be established between
μ_p and g_A:

$$\mu_p = \frac{e}{4E_0} \{1 + \frac{3}{5} g_A + \frac{2}{3<\kappa>} \int_0^\infty dr \; r \; \kappa'(r) \; [g^2(r)-f^2(r)]\}, \tag{A.10}$$

where E_O is the 1s1/2 quark energy. In the preceeding equations, the standard expressions without center-of-mass corrections are obtained in the limit $\kappa \equiv 1$.

REFERENCES

1. E. Reya, Phys. Rep. 69 (1981) 195; G. Altarelli, Phys. Rep. 81 (1982) 1.

2. G. 't Hooft, Nucl. Phys. B72 (1974) 461; E. Witten, Nucl. Phys. B160 (1979) 57; S. Coleman, 1979 Erice Lecture.

3. M. A. Shifman, A.I. Vainshtein and V.I. Zakharov, Nucl. Phys. B147 (1979) 385, Nucl. Phys. B147 (1979) 448, Nucl. Phys. B163 (1980) 43; E.V. Shuryak, CERN preprint TH 3351 (1982); R.A. Bertlmann, CERN preprint TH 3440 (1982).

4. A.A. Belavin et al., Phys. Lett. 59B (1975) 85; G. 't Hooft, Phys. Rev. D14 (1976) 3432; C.G. Callan, R. Dashen and D.J. Gross, Phys. Rev. D17 (1978), 2717, Phys. Rev. D19 (1979) 1826; R.D. Carlitz, Phys. Rev. D17 (1978) 3225.

5. K.G. Wilson, Phys. Rev. D10 (1974) 2445; J.M. Drouffe and C. Itzykson, Phys. Rev. 38C (1978) 133; J. Kogut, Rev. Mod. Phys. 51 (1979) 659; M. Creutz, Phys. Rev. D21 (1980) 2308, Phys. Rev. Lett. 45 (1980) 313; A. Hasenfratz and P. Hasenfratz, Phys. Lett. 93B (1980) 165; P. Hasenfratz et al., Phys. Lett. 95B (1980) 299; J. Kogut et al., Phys. Rev. Lett. 48 (1982) 1140.

6. see for example R. Vinh Mau in "Mesons in Nuclei" Vol. I P.148 North Holland 1979, ed. by M. Rho and Denys Wilkinson.

7. A. Chodos et al., Phys. Rev. D9 (1974) 3471, Phys. Rev. D10 (1974) 2599; T. De Grand et al., Phys. Rev. D12 (1975) 2060; K. Johnson, Acta Phys. Pol. B6 (1975) 865; R.L. Jaffe and K. Johnson, Com. Nucl. Part. Phys. 7 (1977) 107; P. Hasenfratz and J. Kuti, Phys. Rev. 40 (1978) 76; J. Kuti, Proc. of the 1977 CERN-JINR School of Physics, Nafplion, Greece 1977.

8. A. Chodos and C.B. Thorn, Phys. Rev. D12 (1975) 2733;

T. Inoue and T. Maskawa, Progr. Theoret., Phys. $\underline{54}$ (1975) 1833.

9. R.L. Jaffe, 1979 Erice Lectures; M.M. Musakhanov, Chiral Bag Model ITEP I 79 (1980).

10. G.E. Brown and M. Rho, Phys. Lett. $\underline{82B}$ (1979) 177; G.E. Brown, M. Rho and V. Vento, Phys. Lett. $\underline{84B}$ (1979) 383; V. Vento et al., Nucl. Phys. $\underline{A345}$ (1980) 413.

11. S. Théberge, A.W. Thomas and C.A. Miller, Phys. Rev. $\underline{D22}$ (1980) 2838, Phys. Rev. $\underline{D23}$ (1981) 2106 (E); A.W. Thomas, S. Théberge and G.A. Miller, Phys. Rev. $\underline{D24}$ (1981) 216; G.A. Miller, S. Théberge and A.W. Thomas, Com. Nucl. Part. Phys. $\underline{10}$ (1981) 101; A.W. Thomas, CERN preprint TH 3368 TRI-PP-82-29.

12. T.J. Goldman and R.W. Haymaker, Phys. Lett. $\underline{100B}$ (1981) 276, Phys. Rev. $\underline{D24}$ (1981) 724.

13. W. Weise and E. Werner, Phys. Lett. $\underline{101B}$ (1981) 223; R. Tegen, R. Brockmann and W. Weise, Z. Phys. $\underline{A307}$ (1982) 339; R. Brockmann, W. Weise and E. Werner, Phys. Lett. $\underline{122B}$ (1983) 201; R. Tegen, M. Schedl and W. Weise, Phys. Lett. $\underline{B}$, to be published.

14. Y. Nambu and G. Jona-Lasinio, Phys. Rev. $\underline{122}$ (1961) 345, Phys. Rev. $\underline{124}$ (1962) 246.

15. V. Bernard, R. Brockmann, W. Weise and E. Werner, to be published.

16. R. van Royen and V. Weißkopf, Nuovo Cim. $\underline{50}$ (1967) 617, Nuovo Cim. $\underline{51}$ (1967) 583.

17. S. Brodsky and G.P. Lepage, Physica Scripta $\underline{23}$ (1981) 945.

18. G. Höhler et al., Nucl. Phys. $\underline{B114}$ (1976) 505.

19. G.E. Brown, in: Progr. in Part. and Nucl. Phys., D. Wilkinson, ed. Vol. 8, p. 147.

20. E.P. Dally et al., Phys. Rev. Lett. $\underline{48}$ (1982) 375

21. A. Quenzer et al., Phys. Lett. $\underline{76B}$ (1978) 512.

22. T. de Grand et al., Phys. Rev. $\underline{D12}$ (1975) 2060; K. Johnson, Acta Physica Polonica $\underline{B6}$ (1975) 865; K. Johnson in: Fundamentals of Quark Models" (Eds. I. Barbour and A. Davies) Scott. Univ. Summer School in Physics.

Acta Physica Austriaca

The Boltzmann Equation
Theory and Applications

Edited by **E. G. D. Cohen** and **W. Thirring**
(Supplementum 10)

1973. 85 figures and 1 portrait. XII, 642 pages.
Cloth DM 148,—, S 1020,—. ISBN 3-211-81137-0

The Schrödinger Equation

Edited by **W. Thirring** and P. Urban
(Supplementum 17)

1977. 13 figures and 1 portrait. VII, 224 pages.
Cloth DM 66,—, S 472,—. ISBN 3-211-81437-X

Quantum Dynamics: Models and Mathematics

Edited by **L. Streit**
(Supplementum 16)

1976. 13 figures. X, 239 pages.
Cloth DM 66,—, S 472,—. ISBN 3-211-81414-0

Current Problems in Elementary Particle and Mathematical Physics

Edited by **P. Urban**
(Supplementum 15)

1976. 59 figures. VI, 638 pages.
Cloth DM 148,—, S 1020,—. ISBN 3-211-81401-9

Contacts Between High Energy Physics and Other Fields of Physics

Edited by **P. Urban**
(Supplementum 18)

1977. 170 figures. VI, 897 pages.
Cloth DM 208,—, S 1488,—. ISBN 3-211-81454-X

Facts and Prospects of Gauge Theories

Edited by **P. Urban**
(Supplementum 19)

1978. 181 figures. VI, 889 pages.
Cloth DM 208,—, S 1488,—. ISBN 3-211-81514-7

Quarks and Leptons
as Fundamental Particles

Edited by **P. Urban**
(Supplementum 21)

1979. 184 figures. V, 716 pages.
Cloth DM 149,—, S 1070,—. ISBN 3-211-81564-3

Field Theory and Strong Interactions

Edited by **P. Urban**
(Supplementum 22)

1980. 245 figures. V, 815 pages.
Cloth DM 166,—, S 1190,—. ISBN 3-211-81615-1

New Developments
in Mathematical Physics

Edited by **H. Mitter** and **L. Pittner**
(Supplementum 23)

1981. 54 figures. VII, 701 pages.
Cloth DM 158,—, S 1130,—. ISBN 3-211-81676-3

Electroweak Interactions

Edited by **H. Mitter**
(Supplementum 24)

1982. 88 figures. V, 474 pages.
Cloth DM 98,—, S 690,—. ISBN 3-211-81729-8

10 % reduction for subscribers to "Acta Physica Austriaca"

Prices are subject to change without notice